INTERNATIONAL SERIES OF MONOGRAPHS IN

ELECTROMAGNETIC WAVES

EDITORS: A. L. CULLEN, V. A. FOCK AND J. R. WAIT

VOLUME 15

ANTENNAS IN INHOMOGENEOUS MEDIA

Antennas in Inhomogeneous Media

BY

JANIS GALEJS

Senior Scientist, Applied Research Laboratory,
Sylvania Electronic Systems, Waltham, Massachusetts, U.S.A.

PERGAMON PRESS

OXFORD · LONDON · EDINBURGH · NEW YORK

TORONTO · SYDNEY · PARIS · BRAUNSCHWEIG

Pergamon Press Ltd., Headington Hill Hall, Oxford
4 & 5 Fitzroy Square, London W.1

Pergamon Press (Scotland) Ltd., 2 & 3 Teviot Place, Edinburgh 1

Pergamon Press Inc., 44–01 21st Street, Long Island City, New York 11101

Pergamon of Canada Ltd., 207 Queens Quay West, Toronto 1

Pergamon Press (Aust.) Pty. Ltd., 19a Boundary Street,
Rushcutters Bay, N.S.W. 2011, Australia

Pergamon Press S.A.R.L., 24 rue des Écoles, Paris 5^{e}

Vieweg & Sohn GmbH, Burgplatz 1, Braunschweig

First edition 1969

Library of Congress Catalog Card No. 68-21384

PRINTED IN GREAT BRITAIN BY A. WHEATON & CO. EXETER

08 013276 6

CONTENTS

LIST OF TABLES

ACKNOWLEDGMENTS

THIS monograph summarizes the results of the author's recent work at the Applied Research Laboratory of Sylvania Electronic Systems. The author is particularly indebted for support to Dr. J. E. Storer, Director of the Laboratory. Among his colleagues, Drs. R. V. Row, S. R. Seshadri, and R. M. Wundt have contributed with a number of discussions. Appreciation is expressed to Mr. W. Rotman of the Air Force Cambridge Research Laboratory, and to Dr. A. Shostak of the Office of Naval Research, for their encouragement in the investigation of a number of topics.

The author acknowledges several helpful discussions with Prof. R. W. P. King of Harvard University, and valuable comments by Dr. J. R. Wait, Editor of the Series on Electromagnetic Waves.

LIST OF PRINCIPAL SYMBOLS

A, B amplitude of trial functions or a constant
a, b, c radii
$a_{nm}, A_n, A^m_{nq}, A(u, v)$ amplitude coefficients of TE modes
$b_{nm}, B_n, B^m_{nq}, B(u, v)$ amplitude coefficients of TM modes
$B = \mathrm{Im}(Y)$ susceptance
B magnetic induction
B_0 static magnetic induction
C capacitance; a constant; Euler's constant (0.5772 . . .)
$Ci(z)$ cosine integral
$Cin(z) = C + \log z - Ci(z)$
$C(z)$ combination of sine and cosine integrals defined by (2.47)
Cs, Ct defined following (11.77)
c free space velocity of electromagnetic waves
d thickness
da area element
dv volume element
D denominator or abbreviations defined by (8.17) or (12.18)
$\mathscr{D}$ differential operator defined by (12.21)
$\mathbf{E}$ electric field
$E(z)$ combinations of sine and cosine integrals defined by (2.45)
$e = 2.71828 \ldots$ base of the natural logarithm
e particle charge; denotes "electric" as a subscript
F_A, F_B functions in impedance formulations
$F(R), G(R), F(u, v), F(w)$ functions in impedance formulations
$F_n(z), G_n(z)$ functions used in Section 2.2
$f_A(z), f_B(z)$ trial functions for antenna current or aperture fields
f frequency in c/s
$\mathfrak{G}$ a dyadic function
G scalar function; conductance
G_r, G_L radiation conductance, loss conductance
g_A, g_B functions in impedance formulations
$\mathbf{H}$ magnetic field
$H_n^{(m)}(x)$ Hankel function of kind m and order n
$H(\lambda), H(\lambda_n), H(u, v), H(n, m)$ functions in impedance formulations
h_j metric coefficients
I electric current

Idl electric dipole moment (ampere meters)
I_n, I_A, I_B abbreviations of integral expressions
i_x unit vector in x direction
i_e electric current density (amperes/meter^2)
$i_n = I_n/I_0$ relative current amplitudes in Section 2.4
J_{se}, J_ϕ, J_x, J_z electric surface current density (amperes/meter)
J_{sm} magnetic surface current density (volts/meter)
$J_n(x)$ Bessel function of first kind and order n
K_γ constants in Sections 2.3 and 7.12
Kdl magnetic dipole moment (volts/meter)
K_n coefficients in Section 2.4
$K_n(x)$ modified Bessel functions of second kind and order n
k wave number
k_0 wave number of free space
k_e, k_p wave numbers of electromagnetic and plasma waves in Chapter 11
k_A, k_B wave numbers of trial functions
$k_z^2 = k_0^2\epsilon_3 - \rho^2$ in Section 12.2
k_a shape factor defined by (4.48)
l length
log logarithm of base e
m particle mass; denotes "magnetic" as a subscript
N number of turns in Chapter 4; particle density in Chapters 11 and 12; numerator
N_n normalizing factor for radial functions
n unit vector in a normal direction
P power
P_r radiated power
P_L power loss
P_s power carried by surface wave
p radiation power factor in Chapter 4; scalar pressure in Chapter 11; denotes "plasma" as a subscript in Chapter 11
P functions in Sections 11.3 and 12.2
Q functions in Chapters 8 and 11
R resistance
R_r, R_L radiation resistance and loss resistance
R_s resistance associated with a sinusoidal current distribution
R reflection coefficient
R_a, R_b reflection coefficients of TE and TM modes
R_c reflection coefficient of scalar pressure in Chapter 11
$R_n(\rho)$ radial function comprised of a combination of Bessel functions
$r = R_r/R_L$ in Section 4.2
S Poynting's vector
$Si(z)$ sine integral

$S(z)$ combination of sine and cosine integrals defined by (2.46)
S_n parameters for computing loop impedance in Chapter 10
$\tan \delta = \sigma/(\omega\epsilon_r)$ loss tangent
U, W functions in Chapter 11
u acoustic velocity in Chapter 11
$u(x)$ unit step function
V voltage
$\mathbf{V}$ velocity in Chapter 11
v_{ph} phase velocity
w width
$X = \text{Im}(Z)$ reactance
X_s reactance associated with sinusoidal current distribution
X amplitude ratio of $\bar{E}_{zj}$ in the two modes in Section 12.2
$Y = G + iB$ admittance
Y_0 characteristic admittance
$y_s = Y/Y_0$ normalized admittance
y_s shunt admittance per unit length
$Z = R + iX$ impedance
Z_s impedance computed with sinusoidal current distribution; impedance with symmetrical antenna excitation
Z_a impedance with antisymmetrical antenna excitation
Z_0 characteristic impedance
z_s series impedance per unit length
x, y, z cartesian coordinates
u, v, w transform variables
ρ, ϕ, z circular cylindrical coordinates
r, θ, ϕ spherical coordinates
α attenuation constant
α_n coefficients in Section 2.4
β phase constant; radial wave number in Section 12.1
β_c defined following (7.2)
β_x, β_y parameters characterizing the cross-sectional field variation in rectangular waveguides.
$\gamma = -\alpha + i\beta = ik$ propagation constant
γ_{NM} parameters of impedance formulations
Δ function in a variational impedance formulation defined by (2.40) or (7.18); perturbation of a real wave number in Section 11.1.2; abbreviation defined by (11.52) in Section 11.2
δ skindepth; function defined by (2.16)
δZ variation of Z
$\delta(x)$ delta function
δ_{ij} Kronecker delta ($\delta_{ij} = 1$ if $i = j$, $\delta_{ij} = 0$ if $i \neq j$)
ϵ complex permittivity

$\epsilon_0 = (1/36\pi) \times 10^{-9}$ Farad/meter permittivity of free space
ϵ_r real part of permittivity
$\epsilon_1, \epsilon_2, \epsilon_3$ elements of the dyadic permittivity defined by (12.3)
$\epsilon_m = 1$ if $m = 0$, $\epsilon_m = 2$ if $m \neq 0$
$\epsilon = w/2$ half-width
ξ transform variable
η efficiency
$\Lambda(x)$ remainder term defined by (6.15) or (7.33)
λ_n eigenvalues
λ_D Debye length in Chapter 11
$\lambda = 2\pi/k$ wavelength
μ permeability
$\mu_0 = 4\pi \times 10^{-7}$ Henry/meter—permeability of free space
ν collision frequency
$\mathbf{\Pi}_e, \mathbf{\Pi}_m$ Hertz potential (vector) of the electric and magnetic type
$\rho = \sqrt{(u^2+v^2)}$, where u and v are transform variables in Section 12.2
σ conductivity in mho/meter
Φ scalar function used for deriving TM modes and defined by (1.20)
Ψ scalar function used for deriving TE modes and defined by (1.19)
$\Omega = 2 \log (2l/a)$ or $2 \log (4l/\epsilon)$ expansion parameter
$\Omega_c = \omega_c/\omega_p$ normalized cyclotron frequency
$\Omega_u = \omega_u/\omega_p$ normalized upper hybrid resonance frequency
$\Omega_n(x)$ Lommel–Weber function in Section 2.4
$\omega = 2\pi f$ angular frequency in radians per second
ω_c cyclotron frequency, defined following (12.6)
ω_p plasma frequency, defined following (12.6)
$\omega_u = \sqrt{(\omega_c^2+\omega_p^2)}$ upper hybrid resonance frequency

CHAPTER 1

INTRODUCTION

1.1. Scope and Limitations of the Treatment

Most antenna texts emphasize antenna performance in the classical free space environment, but a number of important topics involving dielectric loading, buried antennas, antennas in a reentry environment or in the ionosphere require a more specialized treatment. This monograph presents an introduction to methods on analyzing antennas in such inhomogeneous media. This involves complex geometrical configurations in which strictly assumed antenna current or field distributions are not justified and which generally require formulations in terms of integral equations. Variational formulations are selected for the treatment of most configurations. Such formulations appear to provide a straightforward mechanism for determining approximations to antenna current or field distributions, even when the current distributions can be expected to vary between relatively wide limits as shown for cavity-backed slot antennas or for linear antennas in the presence of dielectric antenna loading. The zero or first order iterative solutions of integral equations for the antenna current distributions are shown to be inadequate in several examples, whereas the variational solution appears to offer an accuracy comparable to higher order iterative solutions in geometries calculated in the past. However, very long antennas are not treated in this monograph.

The developments are devoted almost exclusively to calculation of antenna impedances; antenna radiation patterns not being considered. Most of the antenna geometries can be described in rectangular or circularly cylindrical coordinates; spherical geometries being mentioned only briefly, and conical antennas omitted altogether. Emphasis is on dielectric media but also included are discussions of antennas in cold (incompressible) and warm (compressible) isotropic plasma and in cold anisotropic plasma. The equations describing wave propagation in ferrite media are quite similar to the equations characterizing magnetoionic media [Epstein, 1956]. However, ferrite media are not specifically treated in this monograph.

It is assumed that the reader is familiar with basic electromagnetic field theory and antenna fundamentals. No attempt is made to duplicate the fundamental theoretical developments available in most modern texts on electromagnetic theory. This applies in particular to the introductory material of this chapter, where the usual field equations and boundary conditions, the Hertz vector and the properties of elementary electric and magnetic sources are defined by summarizing the results found in more detailed developments of Stratton [1941]. The decomposition of fields in transverse electric (TE) and transverse magnetic (TM) components is discussed following Collin [1960]; and antenna impedances and admittances are defined following Harrington [1961] and King [1956]. However, similar fundamental information is also available in several other references listed at the end of this chapter [Jordan, 1950; King, 1953; Schelkunoff, 1943; Smythe, 1950; Wait, 1959].

In Chapter 2, past work on thin-wire antennas in free space is summarized. The topics discussed are the EMF method for linear antennas of Carter [1932], the integral equation formulation and the iteration method of Hallen [1938]. The variational formulation of Storer [1952] is used extensively in subsequent developments for antennas in more complex environments. For loops in free space the Fourier series solution [Storer, 1956] is shown to give the same impedance as a variational formulation if identical trial functions of the antenna current are used.

In Chapter 3 Booker's concept of complementary antennas [1946] is derived from the properties of elementary electric and magnetic sources. Examples are given that consider radiation from rectangular waveguides [Lewin, 1951] and coaxial lines [Levine and Papas, 1951] terminated in an infinite flange.

Dielectric loading of small antennas is discussed in Chapter 4. It is shown that small loops are less sensitive to the presence of lossy material in the vicinity of the antenna than are dipole-type antennas.

The transmission line theory of insulated antennas in a lossy medium is reviewed in Chapter 5, and the self and mutual impedance of wires are calculated. The transmission line theory has also been used to calculate the surface excitation from a buried veritcal wire.

The essential characteristics of thin insulated wires in the interface of two media are briefly explained in Chapter 6. Slots in a dielectric interface are discussed using the Hallen's iterative method for narrow slots, but the integral equation has been solved using Fourier transforms for wider slots.

Cavity-backed slot antennas are discussed in Chapter 7, and the variational solution is shown to be superior to simple iterative solutions for rectangular slots. Annular slots backed by cylindrical and coaxial cavities are also discussed.

Slot antennas covered with a stratified dielectric or an isotropic plasma are treated in Chapter 8. More general solutions show that the impedance of the rectangular waveguide can be satisfactorily computed by assuming the principal waveguide mode for the aperture field distribution and by neglecting the presence of any longitudinal electric field components. This conclusion is also substantiated by impedance measurements. In the cases considered, the annular slot antenna either is covered by plasma layers or represents a buried, cavity-backed slot. In this geometry, the antenna impedance is found to be more sensitive to aperture field variations, particularly for wider apertures.

Flat strip antennas in a stratified dielectric are discussed in Chapter 9. The formulations of self and mutual impedance can be adapted for treatments of homogeneous media, dielectric slabs, and insulated antennas in a lossy half space. For homogeneous media, the numerical results are correlated with other work. For antennas in a dielectric slab, the observed changes of the antenna radiation efficiency are explained principally by varying amounts of power carried by surface waves; and for insulated antennas in a lossy medium, the present theory points out the limitations of the transmission line theory discussed in Chapter 5. A flat strip antenna in a dielectric slab and a cylindrical antenna in a dielectric cylinder are shown to exhibit quite similar impedance characteristics.

The first order effects of a nonuniform current are considered for insulated loops in a lossy dielectric or in dielectric layers in Chapter 10. Antenna parameters are discussed in situations where a loop with a thin insulation exhibits nearly the same impedance as the simpler computational model based on a bare loop with an assumed uniform current flow. A buried insulated horizontal loop exhibits nearly the same radiation efficiency as an insulated horizontal dipole of the same total wire length.

Chapter 11 treats antennas in a compressible isotropic plasma. The guided waves that depend on the resonances of the plasma waves in the transverse dimension of a plasma slab carry much less power than the waves guided along the boundaries of the slab. Only the presence of the latter waves is considered in the analysis of the finite antenna and even these are shown to have a small effect on the impedance of a linear antenna. However, they are significant when estimating the impedance of slots that are wide relative to the plasma wavelength.

Chapter 12 treats linear antennas in a cold plasma with a parallel or perpendicular static magnetic field. Simple expressions are shown to apply for the first order estimate of the wave numbers of the current distributions in these antennas, but such estimates do not apply to insulated antennas. Numerous comparisons are presented between the present calculations and earlier work on these problems.

Except for the first three chapters, this monograph is based largely on the author's work undertaken during the last several years. However, reference is made to many relevant results found by other investigators. It is hoped that the number of topics covered and the numerical examples given will make the monograph useful as a general reference. Also it attempts to illustrate the analytical procedures in sufficient detail to permit an interested reader to extend such techniques to related problem geometries.

1.2. Field Equations

The present considerations are restricted to harmonically time-varying fields with an implied $e^{-i\omega t}$ time variation of the field components. Using the rationalized MKS units, the electric and magnetic fields are related for source-free regions by

$$\nabla \times \mathbf{E} - i\omega\mu\, \mathbf{H} = 0, \tag{1.1}$$

$$\nabla \times \mathbf{H} + (i\omega\epsilon_r - \sigma)\, \mathbf{E} = 0, \tag{1.2}$$

where E, H, μ, ϵ_r, and σ are given in volts/meter, amperes/meter, Henries/meter, Farads/meter, and mho/meter, respectively. For free space, $\mu = \mu_0 = 4\pi \times 10^{-7}$ and $\epsilon_r = \epsilon_0 = (1/36\pi) \times 10^{-9}$. The real parameters ϵ_r and σ can be combined into a complex permittivity

$$\epsilon = \epsilon_r(1 + i \tan \delta), \tag{1.3}$$

where $\tan \delta = \sigma/(\omega\epsilon_r)$. The permittivity ϵ is a scalar quantity for isotropic media.

When currents are confined to the boundary between dielectric regions, the following boundary conditions apply to the interface between two media.

$$\mathbf{n} \times (\mathbf{E}_2 - \mathbf{E}_1) = 0, \tag{1.4}$$

$$\mathbf{n} \cdot (\epsilon_2\mathbf{E}_2 - \epsilon_1\mathbf{E}_1) = 0, \tag{1.5}$$

$$\mathbf{n} \cdot (\mu_2\mathbf{H}_2 - \mu_1\mathbf{H}_1) = 0, \tag{1.6}$$

$$\mathbf{n} \times (\mathbf{H}_2 - \mathbf{H}_1) = \mathbf{J}_{se}, \tag{1.7}$$

where the unit vector $\mathbf{n}$ is normal to the interface and to points from region 1 into region 2, and where $\mathbf{J}_{se}$ is the electric surface current in amperes/meter. For a perfectly conducting surface

$$\mathbf{n} \times \mathbf{E}_2 = 0, \tag{1.8}$$

$$\mathbf{n} \cdot \mu_2\mathbf{H}_2 = 0, \tag{1.9}$$

$$\mathbf{n} \times \mathbf{H}_2 = \mathbf{J}_{se}. \tag{1.10}$$

By analogy with (1.10) a magnetic surface current $\mathbf{J}_{sm}$ can be defined by

$$\mathbf{J}_{sm} = -\mathbf{n} \times \mathbf{E}_2 \tag{1.11}$$

which is assumed to flow on the surface of a perfect electric conductor. The latter relation is useful in the analysis of apertures in a conducting plane. The aperture can be replaced by an equivalent magnetic surface current that flows on the surface of the electrically conducting plane.

1.3. Hertz Potential

The fields of a homogeneous region can be conveniently derived from the Hertz potential (or vector) $\mathbf{\Pi}_j$, which satisfies the wave equation

$$\nabla^2\mathbf{\Pi}_j + k^2\mathbf{\Pi}_j = \nabla\nabla\cdot\mathbf{\Pi}_j - \nabla\times\nabla\times\mathbf{\Pi}_j + k^2\mathbf{\Pi}_j = 0, \tag{1.12}$$

where $k^2 = \omega^2\mu\epsilon$, and where $j = e$ or m designates the Hertz potential of the electric or magnetic type respectively. The electric and magnetic fields are related to $\mathbf{\Pi}_j$ by

$$\mathbf{E} = \nabla\nabla\cdot\mathbf{\Pi}_e + k^2\mathbf{\Pi}_e + i\omega\mu\,\nabla\times\mathbf{\Pi}_m, \tag{1.13}$$

$$\mathbf{H} = \nabla\nabla\cdot\mathbf{\Pi}_m + k^2\mathbf{\Pi}_m - i\omega\epsilon\,\nabla\times\mathbf{\Pi}_e. \tag{1.14}$$

The potentials $\mathbf{\Pi}_j$ are related to elementary electric or magnetic currents by

$$\mathbf{\Pi}_e = \frac{i\mathbf{I}\mathbf{dl}}{4\pi\omega\epsilon}\frac{e^{ikr}}{r}; \quad \mathbf{\Pi}_m = \frac{i\mathbf{K}\mathbf{dl}}{4\pi\omega\mu}\frac{e^{ikr}}{r}, \tag{1.15}$$

where the line element dl designates the length of the current, and r is the distance from the source to the observation point. The electric and magnetic currents I and K may be computed by integrating the corresponding surface current densities J_{sj} along the surface of the conductors in a direction transverse to the current flow as $\int J_{sj}ds$. The magnetic dipole moment **Kdl** of a volume distribution of electric current $\mathbf{i}_e$ is computed by

$$\mathbf{K}\mathbf{dl} = -\frac{i\omega\mu}{2}\iiint \mathbf{r}\times\mathbf{i}_e\,dv, \tag{1.16}$$

where $\mathbf{r}$ is the radius vector from the selected origin to the current element $\mathbf{i}_e$, and dv is a volume element. For a small loop of radius r_0 carrying a current I_0

$$\mathbf{K}\mathbf{dl} = -i\omega\mu_0\pi r_0^2 I_0\mathbf{i}_z = -i\omega\mu_0 I_0\,\mathbf{da}, \tag{1.17}$$

where $\mathbf{da} = \pi r_0^2\mathbf{i}_z$ is in the direction of the loop axis, and $\mathbf{i}_z$ is a unit vector.

Dipole fields of the homogeneous space are obtained by substituting (1.15) in (1.13) and (1.14), and by carrying out the indicated differentiations.

$\mathbf{\Pi}_e$ of (1.15) should be doubled when computing the field of vertical electric dipole situated on a conducting plane. Also $\mathbf{\Pi}_m$ of (1.15) should be doubled when computing the field of a loop of a horizontal axis or the field of a horizontal magnetic dipole above a conducting plane. For slots in a conducting plane,

$$\mathbf{K}d\mathbf{l} = \iint \mathbf{J}_{sm}\, da = \iint \mathbf{E}\times d\mathbf{a}, \tag{1.18}$$

where the tangential electric field of the aperture $\mathbf{E}\times\mathbf{n} = \mathbf{J}_{sm}$ is integrated over the area of the slot. $\mathbf{\Pi}_m$ of (1.15) should also be doubled in this case.

1.4. TE and TM Modes

In waveguide problems and for propagation in stratified media in unbounded space, it is convenient to express the fields as a superposition of transverse electric (TE) and transverse magnetic (TM) modes. TE modes are characterized by zero electrical fields in the selected direction of the propagation; TM modes have no magnetic field components in the same direction. Such modes can be derived from the electric and magnetic terms of Hertz potential (1.12) by assuming that $\mathbf{\Pi}_m$ and $\mathbf{\Pi}_e$ have components only in this direction of propagation. $\mathbf{\Pi}_j$ is selected to be of the form

$$\mathbf{\Pi}_m = \mathbf{i}_z \Psi(\xi, \eta, z), \tag{1.19}$$

$$\mathbf{\Pi}_e = \mathbf{i}_z \Phi(\xi, \eta, z), \tag{1.20}$$

where ξ and η are arbitrary orthogonal coordinates in the direction transverse to z, and where the z dependence is of the form $\exp(\pm\gamma z)$. A substitution of (1.19) and (1.20) into (1.13) and (1.14) shows that the fields derived from $\mathbf{\Pi}_m$ and $\mathbf{\Pi}_e$ exhibit $E_z = 0$ and $H_z = 0$ respectively, and can be therefore designated as TE and TM modes. The metric coefficients of the transverse coordinates are designated as h_ξ and h_η, and the field components of the electric and magnetic field are computed by substituting (1.19) and (1.20) into (1.13) and (1.14). This results in

$$E_\xi = \frac{i\omega\mu_0}{h_\eta}\frac{\partial}{\partial\eta}\Psi + \frac{1}{h_\xi}\frac{\partial^2}{\partial\xi\partial z}\Phi, \tag{1.21}$$

$$E_\eta = -\frac{i\omega\mu_0}{h_\xi}\frac{\partial}{\partial\xi}\Psi + \frac{1}{h_\eta}\frac{\partial^2}{\partial\eta\partial z}\Phi, \tag{1.22}$$

$$E_z = \left(\frac{\partial^2}{\partial z^2} + k^2\right)\Phi = -\nabla_t^2\Phi, \tag{1.23}$$

$$H_\xi = \frac{1}{h_\xi}\frac{\partial^2}{\partial\xi\partial z}\Psi - \frac{i\omega\epsilon}{h_\eta}\frac{\partial}{\partial\eta}\Phi, \tag{1.24}$$

$$H_\eta = \frac{1}{h_\eta}\frac{\partial^2}{\partial\eta\partial z}\Psi + \frac{i\omega\epsilon}{h_\xi}\frac{\partial}{\partial\xi}\Phi, \tag{1.25}$$

$$H_z = \left(\frac{\partial^2}{\partial z^2} + k^2\right)\Psi = -\nabla_t^2\Psi, \tag{1.26}$$

where ∇_t^2 designates the transverse part of the ∇^2 operator. The functions Ψ and Φ satisfy wave equations

$$\nabla_t^2\Psi + k_t^2\Psi = 0, \tag{1.27}$$

$$\nabla_t^2\Phi + k_t^2\Phi = 0, \tag{1.28}$$

where the transverse wave numbers $k_t^2 = k^2 + \gamma^2$, determined from boundary conditions, distinguish the different modes. These modes are orthogonal, which implies that the product of two axial components integrated over the cross-section of the guide is zero. Similarly the dot product between the TE field components of two different modes vanishes when integrated over the guide cross-section. This applies to the transverse field components of two different TE modes or two different TM modes, as well as to the transverse field components of one TE and one TM mode, as long as the guide boundary is nonabsorbing. A proof of this orthogonality is not given in the present text.

1.5. Definitions of Antenna Impedance or Admittance

The antenna impedance Z is defined as the ratio between the voltage V at the driving point to the antenna current I at this same point. Multiplying the numerator and the denominator of this expression by I, the impedance becomes equal to the complex power VI divided by the current at its driving point squared. For a lossless antenna, this power is equal to the complex power radiated by the antenna which is the sum of real power and the time rate of change of the electric and magnetic energy stored near the antenna. The Poynting's vector $\mathbf{S} = \mathbf{E} \times \mathbf{H}$ represents the intensity of energy flow at a point in the field, or the energy flow crossing a unit area whose normal is oriented in the direction of the vector $\mathbf{S}$. ($\mathbf{S}$ does not characterize the average intensity of the energy flow which is given as 0.5 Re($\mathbf{E} \times \mathbf{H}^*$) for fields of harmonic time variation.) The complex power radiated by the antenna can be computed by integrating the Poynting's vector $\mathbf{S}$ over a surface that encloses the antenna. The impedance is therefore given by

$$Z = \frac{V}{I} = \frac{VI}{I^2} = \frac{1}{I^2}\iint \mathbf{E} \times \mathbf{H} \cdot \mathbf{da}, \tag{1.29}$$

where the vector $\mathbf{da}$ is oriented in a direction perpendicular to the antenna surface. For a perfectly conducting antenna the tangential magnetic field is related to the surface current density by (1.10). After interchanging the dot and cross product signs in (1.29), a substitution of (1.10) shows that

$$Z = -\frac{1}{I^2}\iint \mathbf{E}\cdot\mathbf{J}_{se}\,da. \tag{1.30}$$

The impedance of a perfectly conducting antenna is therefore determined by integrating the tangential electric field and the surface current density over a surface surrounding the antenna.

The same expression for the antenna impedance can be derived without making explicit use of the Poynting's vector. The electric field near the antenna is related by (1.13) and (1.15) to the current distribution along the antenna, and the tangential electric field $\mathbf{E}$ on the antenna is formally related to its surface current density $\mathbf{J}_{se}$ by

$$\mathbf{E}\,(\xi,\eta) = -\iint \mathfrak{G}\,.\mathbf{J}'_{se}\,da', \tag{1.31}$$

where

$$\mathfrak{G} = \mathfrak{G}(\xi,\eta,\xi',\eta') = \mathfrak{G}(\xi',\eta',\xi,\eta),$$

$$\mathbf{J}'_{se} = \mathbf{J}_{se}(\xi',\eta'),$$

$$da' = d\xi'\,d\eta'.$$

For antennas in an unbounded homogeneous medium, the dyadic function $\mathfrak{G}$, which is also referred to as the tensor Green's function [Harrington, 1961; Levine and Schwinger, 1950], may be computed from (1.13) and (1.15) as

$$\mathfrak{G} = \frac{1}{4\pi i\omega\epsilon}(\nabla\nabla + k^2\mathfrak{I}),\,[\exp(ikr)/r], \tag{1.32}$$

where $\mathfrak{I}$ is a unit dyad, but in general $\mathfrak{G}$ is not restricted to this form. For a perfectly conducting antenna excited by voltage impressed across an infinitesimal gap at $\xi = 0$

$$\mathbf{E}(\xi,\eta) = -V(0)\,\delta(\xi)\,i_\xi, \tag{1.33}$$

where i_ξ denotes a unit vector. Substituting (1.33) in (1.31) and integrating along the antenna gives an integral equation for the antenna current density J'_{se}. The antenna impedance $Z = V(0)/I(0)$ can then be computed for various approximate solutions of J'_{se}. An example of such calculations is shown in Section 2.2.

Alternately, (1.31) and (1.33) can be multiplied by $J_{se} = J_{se}\ (\xi,\ \eta)$ and integrated over the antenna surface, giving

$$\begin{aligned}\iint \mathbf{E}\cdot\mathbf{J}_{se}\,da &= -\iiiint \mathbf{J}_{se}\cdot\mathfrak{G}\cdot\mathbf{J}'_{se}\,da'\,da \\ &= -V(0)\,I(0)\end{aligned} \tag{1.34}$$

where $\mathrm{d}a = \mathrm{d}\xi\,\mathrm{d}\eta$. Division of (1.34) by $[I(0)]^2$ gives the impedance Z as

$$Z = \frac{V(0)\,I(0)}{[I(0)]^2} = -\frac{\iint \mathbf{E}\cdot\mathbf{J}_{se}\,\mathrm{d}a}{[I(0)]^2}$$
$$= \frac{1}{[I(0)]^2}\iiiint \mathbf{J}_{se}\cdot\mathfrak{G}\cdot\mathbf{J}'_{se}\,\mathrm{d}a'\,\mathrm{d}a. \tag{1.35}$$

This impedance expression is stationary in the sense that a small change of $\delta\mathbf{J}_{se}$ about the correct surface current density $\mathbf{J}_{se}$ gives a zero first order change δZ in the impedance Z. It follows that

$$\delta Z = \frac{1}{[I(0)]^2}\left\{\iiiint (\delta\mathbf{J}_{se}\cdot\mathfrak{G}\cdot\mathbf{J}'_{se}+\mathbf{J}_{se}\cdot\mathfrak{G}\cdot\delta\mathbf{J}'_{se})\,\mathrm{d}a\,\mathrm{d}a' - 2Z\,\delta I(0)\,I(0)\right\}. \tag{1.36}$$

Multiplying (1.31) and (1.33) by $\delta\mathbf{J}_{se}$ and integrating over the antenna surface gives

$$V(0)\,\delta I(0) = \iiiint \delta\mathbf{J}_{se}\cdot\mathbf{J}'_{se}\,\mathrm{d}a\,\mathrm{d}a'$$
$$= Z\,\delta I(0)\,I(0). \tag{1.37}$$

Under conditions where $\mathfrak{G}$ is a symmetrical dyad or where the surface current $\mathbf{J}_{se}$ can be expressed as a single vector component $\mathbf{J}_{se} = i_\xi J_{se}$, it follows that

$$\iiiint \delta\mathbf{J}_{se}\cdot\mathfrak{G}\cdot\mathbf{J}'_{se}\,\mathrm{d}a\,\mathrm{d}a' = \iiiint \mathbf{J}_{se}\cdot\mathfrak{G}\cdot\delta\mathbf{J}'_{se}\,\mathrm{d}a\,\mathrm{d}a'. \tag{1.38}$$

The insertion of (1.37) and (1.38) in (1.36) shows that $\delta Z = 0$ and that Z of (1.35) is stationary. This stationary character of Z implies that the first order change in $I(\xi)$ can give only a second order change in Z, where $I(\xi) = \int J_{se}(\xi,\eta)\,\mathrm{d}\eta$. This property of the impedance will be used later for determining approximations to the antenna current $I(\xi)$. The impedance is stationary only with respect to small changes $\delta I(\xi)$ about the correct current $I_c(\xi)$; large changes of the impedance can be anticipated if $\delta I(\xi)$ differs significantly from $I_c(\xi)$.

The converse of this property is quite useful for estimating the computational accuracy. If antenna impedances that are computed using basically different trial functions for the antenna current remain nearly constant, it may be concluded that the assumed trial functions are sufficiently general. It can be also considered that the resulting current distributions do not depart significantly from the true current $I_c(\xi)$ for purposes of calculating the impedance, and that the resulting impedance approximates the correct antenna impedance. A strict proof of this converse characteristic can not be given, but its usefulness will be illustrated in several examples in Chapters 8 and 12.

The antenna impedance Z can be also defined using the complex conjugate of the current or current density as

$$Z = \frac{V(0)\, I(0)^*}{I(0)\, I(0)^*} = -\frac{\iint \mathbf{E} \cdot \mathbf{J}_{se}^* \, da}{I(0)\, I(0)^*}$$

$$= \frac{1}{I(0)\, I(0)^*} \iiiint \mathbf{J}_{se}^* \cdot \mathfrak{G} \cdot \mathbf{J}_{se}' \, da' \, da. \tag{1.39}$$

However, a development similar to the one shown in (1.36) to (1.38) indicates that $\delta Z = 0$ for real currents when $\mathbf{J}_{se}^* = \mathbf{J}_{se}$. The impedances Z of (1.35) and (1.39) are also equal for antenna current distributions of the form $\mathbf{J}_{se} = \sum_j A_j \mathbf{f}_j$, if the complex amplitudes of the individual real terms $\mathbf{f}_j$ are determined from the condition $\delta Z / \delta A_j = 0$, where Z is the impedance (1.35) [Galejs, 1967].

A similar development can be used for computing the admittance of a slot antenna: using (1.11) for the definition of a magnetic surface current $\mathbf{J}_{sm}$, it follows that

$$Y = \frac{I}{V} = \frac{IV}{V^2} = \frac{1}{V^2} \iint \mathbf{E} \times \mathbf{H} \cdot da$$

$$= -\frac{1}{V^2} \iint \mathbf{H} \cdot \mathbf{J}_{sm} \, da. \tag{1.40}$$

The admittance of a slot in a conducting plane is determined by integrating tangential magnetic fields and magnetic surface currents over the area of the aperture. The stationary character of the admittance in (1.40) can also be demonstrated similarly to the impedance (1.30) in the development of (1.34) to (1.38). For slot antennas excited by waveguides or by coaxial lines, the impedance is frequently referred to the principal mode of the exciting waveguide. In this case, I and V designate suitably defined current and voltage amplitudes of the principal waveguide mode.

It is also possible to formulate the antenna impedance in terms of the electric field (or magnetic current) on a surface which represents an extension of the antenna [Hurd, 1964] or the admittance of a slot antenna in terms of the current which flows on the ground plane surrounding the slot [Levine and Papas, 1951]. Such dual representations of the same physical quantity are particularly useful for determining lossless circuit parameters. The two formulations provide upper and lower bounds for the aperture susceptance in waveguides [Borgnis and Papas, Sections 12.10 to 13, 1955; Jones, Section 5.12, 1964], or the equivalent reactance in network representations of junction discontinuities [Primich, 1956]. In the two stationary formulations of a complex impedance the correct value is approached through different parts of the complex plane and it is possible to extrapolate the correct impedance by increasing the accuracy of the trial functions in the two formulations [fig. 8, Primich, 1956]. However, it is not possible to ascertain *a priori* if the resistance or the reactance of a given complex impedance formulation lies above or below its correct value. A similar conclusion is reached by Morgan [1963]; for integral equations of complex valued symmetrical kernels variational formulations may give approximate eigenvalues, but nothing can be said about the phase of the error. This conclusion is supported also by the subsequent numerical calculations which make use of a number of different one and two-term trial functions. The antenna resistance and reactance computed with less accurate trial functions can be either higher or lower than in computations based on more accurate representations (see Figs. 2.1, 2.2, and 2.4). In a given antenna problem it may be difficult to estimate the field or current with equal accuracy in the two different impedance formulations. For very long antennas it is simpler to estimate fields outside the antenna than the antenna current, and a better accuracy can be expected in the formulation based on the outside fields [Hurd, 1964]. However, for shorter antennas the formulations (1.35) or (1.40) are customary.

In the absence of experimental data or alternate impedance calculations one must rely on indirect accuracy estimates as was discussed following equation (1.38). Different trial functions may give nearly the same antenna impedance, but differences in antenna current or field distributions will be somewhat larger as is seen from the examples of Section 8.1.6 or 12.1.4.3.

Alternately the accuracy of an estimated current or field distribution can be checked by computing the electric field near the surface of the antenna. Based on (1.33) the tangential electric field should be equal to zero except near the center gap of the antenna. Such calculation of the antenna near fields has been carried out by King and Wu [1965, 1966] for antennas in free space, using approximations which force the tangential electric fields to vanish on the antenna surface for any non-sinusoidal current distribution.

Antenna surface fields calculated from (1.31) will not vanish for inaccurate current distributions, and minor changes of the current distribution will have significant effects on the surface fields. Current distributions varying linearly with the distance from the antenna ends violate the so-called edge conditions of diffraction theory [Bouwkamp, 1954] and excite strong electric fields tangential to the antenna surface. These field singularities can be avoided by making the antenna current to decay proportionately to the square root of the distance from the antenna ends. The more accurate current distributions produce the smaller surface fields and it is possible to distinguish the more accurate current distribution even in cases where the variationally determined antenna impedances differ insignificantly [Galejs, 1968].

1.6. References

BOOKER, H. G. (1946) Slot aerials and their relation to complementary wire aerials, *J. IEE (London)* **93**, Part IIIA, No. 4.

BORGNIS, F. E. and PAPAS, C. H. (1955) *Boundary Value Problems of Microwave Physics*, Springer-Verlag, Berlin.

BOUWKAMP, C. J., (1954) "Diffraction Theory," *Rep. Prog. Phys.*, **17**, 35–100.

CARTER, P. S. (1932) Circuit relations in radiating systems and applications to antenna problems, *Proc. IRE* **20** (6), 1004–41.

COLLIN, R. E. (1960) *Field Theory of Guided Waves*, McGraw-Hill, New York, NY.

EPSTEIN, P. S. (1956) Theory of wave propagation in a gyromagnetic medium, *Rev. Mod. Phys.* **28** (1), 3–17.

GALEJS, J., (1967) "Variationally computed antenna impedances and accuracy of resulting current distributions," *Electronic Letters*, **3** (8) 371.

GALEJS, J. (1968) "Surface fields on linear antennas", *Proc. IEE, 115*.

HALLEN, E. (1938) Theoretical investigations into transmitting and receiving qualities of antennae, *Nova Acta Regiae Soc. Sci. Upsaliensis* **4**, 1–44.

HARRINGTON, R. F. (1961) *Time-harmonic Electromagnetic Fields*, McGraw-Hill, New York NY.

HURD, R. A. (1964) Variational solutions for the admittance of the long cylindrical antenna, *Radio Science* **68D** (3), 311–16.

JONES, D. S. (1964) *The Theory of Electromagnetism*, Macmillan, New York, NY.

JORDAN, E. (1950) *Electromagnetic Waves and Radiating Systems*, Prentice-Hall, Englewood Cliffs, NJ.

KING, R. W. P. (1953) *Electromagnetic Eingineering*, McGraw-Hill, New York, NY.

KING, R. W. P. (1956) *Theory of Linear Antenna*, Harvard University Press, Cambridge, Mass.

KING, R. W. P. (1965) Currents, charges, and near fields of cylindrical antennas, *Radio Science* **69 D** (3), 429–46.

KING, R. W. P. and WU, T. T. (1966) The electric field very near a driven cylindrical antenna, *Radio Science* **1** (3), 353–9.

LEVINE, H. and PAPAS, C. H. (1951) Theory of circular diffraction antenna, *J. Appl. Phys.* **22**, 29–43.

LEVINE, H. and SCHWINGER, J. (1950) On the theory of electromagnetic wave diffraction by an aperture in an infinite plane conducting screen, *Comm. Pure and Appl. Math.* **3**, 355–91.

LEWIN, L. (1951) *Advanced Theory of Waveguides*, Iliffe & Sons, London.

MORGAN, S. P. (1963) On the integral equations of laser theory, *IEEE Trans. on Microwave Theory and Techniques* **MTT-4** (3), 156–66.

PRIMICH, R. I. (1956) A semi-infinite array of parallel metallic plates of finite thickness for microwave systems, *IRE Trans. on Microwave Theory and Techniques* **MTT-4** (3), 156–66.

SCHELKUNOFF, S. A. (1943) *Electromagnetic Waves*, D. Van Nostrand, Princeton, NJ.

SMYTHE, W. R. (1950) *Static and Dynamic Electricity*, 2nd ed., McGraw-Hill, New York, NY.

STORER, J. E. (1952) *Variational Solution to the Problem of the Symmetrical Cylindrical Antenna*, Cruft Lab. Rep. TR 101, Cambridge, Mass.

STORER, J. E. (1956) Impedance of thin-wire loop antennas, *Trans. Engrs.* **75**, 606; also Cruft Lab. Rep. No. 212 Harvard University, Cambridge, Mass.

STRATTON, J. A. (1941) *Electromagnetic Theory*, McGraw-Hill, New York, NY.

WAIT, J. R. (1959) *Electromagnetic Radiation from Cylindrical Structures*, Pergamon Press, New York.

CHAPTER 2

THIN WIRE ANTENNAS IN FREE SPACE

2.1. EMF Method

The impedance of a linear antenna can be determined by assuming a current distribution, then computing the electric field that it produces, and finally integrating this electric field and the current over the surface of the antenna as indicated in (1.30). Such calculations have been carried out by Carter [1932], and will be discussed here in the simple case of a half-wave antenna. The cylindrical antenna has a radius a and total length $2l$, with the origin of the coordinates in the antenna center. The surface current of (1.30) is assumed to exhibit no variations around the circumference of the antenna. With $l = \lambda/4$, the antenna current is of the form

$$I(z) = 2\pi a J_{se}(z) = I(0)\cos(kz). \tag{2.1}$$

$I(z)$ vanishes at the endpoints of the antenna $z = \pm l$. Substitution of (2.1) in (1.30) shows that

$$Z = -\frac{1}{I(0)}\int_{-l}^{l} E(z)\cos(kz)\,\mathrm{d}z. \tag{2.2}$$

$E(z)$ is related by (1.13) to the Hertz vector $\mathbf{\Pi}_{ze}$ of (1.15), which is computed for the assumed current distribution (2.1). It is shown in Section 2.5 that for calculating the fields outside the antenna, the Hertz vector $\mathbf{\Pi}_{ze}$ can be computed by assuming that the antenna current is concentrated at the antenna center ($\rho' = 0$). The distance r between the source at (ρ', ϕ', z') and the observation point at (a, ϕ, z) is therefore approximated by

$$r = [(z-z')^2 + a^2]^{1/2} \tag{2.3}$$

and the contributions of elementary currents in (1.15) can be integrated as

$$E_z(z) = \int_{-l}^{l} \mathrm{d}E_z(z) = \frac{iI(0)}{4\pi\omega\epsilon}\int_{-l}^{l}\cos(kz')\left(\frac{\partial^2}{\partial z^2} + k^2\right)\frac{\mathrm{e}^{ikr}}{r}\,\mathrm{d}z'. \tag{2.4}$$

It is noted that

$$\frac{\partial^2}{\partial z^2}\frac{\mathrm{e}^{ikr}}{r} = \frac{\partial^2}{\partial z'^2}\frac{\mathrm{e}^{ikr}}{r}, \tag{2.5}$$

and integrating (2.4) by parts twice, it follows that

$$E_z(z) = \frac{iI(0)k}{4\pi\omega\epsilon}\left[\sin(kz')\frac{e^{ikr}}{r}\right]_{z'=-l}^{l}. \tag{2.6}$$

Substituting (2.6) in (2.2), expressing cos (kz) as a sum of two exponentials, and by letting $a \to 0$ in r of (2.3),

$$Z = \frac{1}{4\pi}\left(\frac{\mu}{\epsilon}\right)^{1/2}\int_0^{2\pi}\frac{1-e^{iu}}{u}\,du = 30[Cin(2\pi) - iSi(2\pi)], \tag{2.7}$$

where $Si(z)$ is the sine integral, $Cin(z) = C + \log z - Ci(z)$, $C = 0.5772\ldots$ is the Euler's constant, and $Ci(z)$ is the cosine integral. Substituting the numerical values of the two functions in (2.7), $Z = 73.1 - i42.5$ ohms.

This impedance depends on the assumed form of the current distribution (2.1). Other work shows that this particular assumption is justified only for an extremely thin antenna having a length equal to one-half of the wavelength.

2.2. Hallen's Method

Hallen's method [1938] of antenna impedance calculations is based on an iterative solution of the integral equation for the antenna current. Using this method, King and Harrison [1943] have published extensive numerical results. The present development only illustrates the method and provides some illustrative examples.

The electric field outside the antenna can be computed by substituting (1.15) in (1.13) and by integrating the contributions of the individual current elements. It follows from Section 2.5 that the Hertz vector can be computed by assuming the current flow to be concentrated at the antenna center. Thus

$$E(z) = \left(\frac{\partial^2}{\partial z^2} + k^2\right)\int dz'\,\Pi_z'(z, z'), \tag{2.8}$$

where

$$dz'\,\Pi_z'(z, z') = \frac{iI(z')dz'}{4\pi\omega\epsilon}\frac{e^{ikr}}{r}, \tag{2.9}$$

$$r = [(z-z')^2 + a^2]^{1/2} \tag{2.10}$$

and a is the antenna radius.

The electric field along a perfectly conducting antenna will be zero, except at the driving point (antenna center), where the generator is assumed to maintain a voltage V across an infinitesimally thin gap. With the exception of the gap at $z = 0$, the left-hand side of (2.8) will be zero and (2.8) can be considered as a differential equation for the integral on its right-hand side. For a current that is symmetrical about $z = 0$, the integral of (2.8) should also be symmetrical about $z = 0$, and the formal solution of this differential equation can be written as

$$\int dz'\Pi_z(z, z') = C_1 \cos(kz) + C_2 \sin(k|z|). \tag{2.11}$$

Using $E(z) = -V(0)\delta(z)$, the integration of (2.8) yields

$$-V(0) = k^2 \int_{-l}^{l} dz \int_{-l}^{l} dz' \, \Pi_z'(z, z') + \frac{\partial}{\partial z} \int_{-l}^{l} dz' \, \Pi_z'(z, z')\Big|_{-l}^{l}. \tag{2.12}$$

Substituting (2.11) in (2.12) and evaluating the z-integrals, it follows that $V(0) = -2kC_2$. Using this value of the constant C_2, a substitution of (2.9) in (2.11) gives

$$\frac{i}{4\pi}\left[\frac{\mu_0}{\epsilon_0}\right]^{1/2} \int dz' \, I(z') \frac{e^{ikr}}{r} = C_1 k \cos(kz) - \frac{V}{2} \sin(k|z|) \tag{2.13}$$

as the integral equation for the unknown current distribution $I(z')$. $I(z)$ is added to and subtracted from $[I(z') \exp(ikr)]$ of the integral. One of the terms proportional to $I(z)$ is integrated and (2.13) is rearranged into

$$I(z) = \frac{1}{30i\Omega}\left[C_1 k \cos(kz) - \frac{V}{2}\sin(k|z|)\right] - \frac{1}{\Omega}\left\{I(z) \log\left[1 - \left(\frac{z}{l}\right)^2\right]\right.$$
$$\left. + I(z)\delta + \int_{-l}^{l} \frac{I(z') \, e^{ikr} - I(z)}{r} dz\right\}, \tag{2.14}$$

where

$$\Omega = 2 \log(2l/a), \tag{2.15}$$

and

$$\delta = \log\left\{\frac{1}{4}\left[\sqrt{1 + \left(\frac{a}{l-z}\right)^2} + 1\right]\left[\sqrt{1 + \left(\frac{a}{l+z}\right)^2} + 1\right]\right\}. \tag{2.16}$$

The parameter Ω will be of the order of 10 to 20 for thin antennas. The antenna current should be zero for $z = \pm l$. Substituting $z = l$ in (2.14) and subtracting $I(l)$ from $I(z)$ gives

$$I(z) = \frac{1}{30i\Omega}\left[C_1 k F_0(z) - \frac{V}{2} G_0(z)\right] - \frac{1}{\Omega}\left\{I(z) \log\left[1 - \left(\frac{z}{l}\right)^2\right] + I(z)\delta\right.$$
$$\left. + \int_{-l}^{l} \frac{I(z') \, e^{ikr} - I(z)}{r} dz' - \int_{-l}^{l} \frac{I(z')}{r_1} e^{ikr_1} dz'\right\}, \tag{2.17}$$

where

$$r_1 = [(l - z')^2 + a^2]^{1/2}, \tag{2.18}$$

$$F_0(z) = \cos kz - \cos kl, \tag{2.19}$$

$$G_0(z) = \sin k|z| - \sin kl. \tag{2.20}$$

For the initial estimate of the current $I(z)$, the terms involving $I(z)$ and $I(z')$ of the right-hand side of (2.17) are neglected because they are proportional to $1/\Omega$ and small relative to $I(z)$ or to the left-hand side of the same equation. Substituting

$$I_0(z) = \frac{1}{30i\Omega}\left[C_1kF_0(z) - \frac{V}{2}G_0(z)\right] \tag{2.21}$$

in (2.14) and setting $z = l$ results in

$$C_1k = \frac{V}{2}\frac{\sin kl + G_1(l)/\Omega}{\cos kl + F_1(l)/\Omega}, \tag{2.22}$$

where

$$F_1(l) = -\int dz' \frac{F_0(z')\, e^{ikr_1}}{r_1} \tag{2.23}$$

and $G_1(l)$ is obtained by substituting $G_0(z')$ in the right-hand side of (2.23). With C_1k related to V by (2.22), $I_0(z)$ of (2.21) is uniquely determined, and $I_0(0)$ can be used for computing the zero order antenna impedance as

$$Z_0 = \frac{V}{I_0(0)} = \frac{60i\Omega[\cos kl + (1/\Omega)F_1(l)]}{\sin kl + (1/\Omega)\,[1 - \cos kl)\, G_1(l) + \sin kl\, F_1(l)]}\cdot \tag{2.24}$$

The first order current $I_1(z)$ is obtained by substituting (2.21) and (2.22) in the right-hand side of (2.17), resulting in the first order impedance

$$Z_1 = \frac{V}{I_1(0)} = \frac{60i\Omega[\cos kl + (1/\Omega)\, F_1(l)]}{\sin kl + (1/\Omega)\,[G_1(l) - \cos kl\, G_1(0) + \sin kl\, F_1(0)]}, \tag{2.25}$$

where

$$F_1(0) = -\delta F_0(0) + \int dz' \frac{F_0(z')\, e^{ikr_1} - F_0(0)}{r}. \tag{2.26}$$

$G_1(0)$ is obtained by substituting $G_0(0)$ and $G_0(z')$ in the right-hand side of (2.26).

The functions $F_1(l)$, $G_1(l)$, $F_1(0)$, and $G_1(0)$ can be expressed in terms of sine and cosine integrals and are listed in a number of references [King and Harrison, 1943; Aharoni, 1946]. Table 2.1 lists values of Z_0 from (2.24), Z_1 from (2.25), and the more accurate second order impedance Z_2 of the theory by King and Middleton [King, 1956]. The examples in the table are restricted to half- and full-wave length antennas and Ω-values of 10, 20, and ∞.

TABLE 2.1. IMPEDANCE OF HALF-WAVE AND FULL-WAVE ANTENNAS

l	$\lambda/4$			$\lambda/2$		
Ω	10	20	∞	10	20	∞
Z_0	79.5–i40.8	76.1–i41.8	73.1–i42.5	1875 + i1405	8600 + i6030	∞
Z_1	64.8–i29.7	69.6–i35.7	73.1–i42.5	1260 + i645	5000 + 2900	∞
Z_2	86.5–i41.7	78.5–i43.6	73.1–i42.5	255 + i455	2700 + i2500	∞

It is seen that for a half-wave antenna ($l = \lambda/4$) the zero order impedance agrees more closely with the second order impedance Z_2 than with the first impedance Z_1. However, for a full-wave antenna ($l = \lambda/2$) the zero order calculations give too high impedance figures. Even the resistance of the first order calculations is in considerable disagreement with the resistance of Z_2. However, the reactance of Z_1 is the same order of magnitude as Z_2 of the second order theory by King and Middleton.

The zero and first order solutions of the iterative type are generally inaccurate unless the initial approximation of the antenna current is a good one. Higher order solutions are more accurate, but they are already very involved in the relatively simple problem of thin antennas in free space.

The previous development was carried out for a perfectly conducting antenna excited by a voltage source across an infinitesimal gap. The analysis can be extended to finitely conducting antennas and also to sources of a finite width [King, 1956]. These refinements are not considered in the present text.

2.3. Storer's Variational Solution

Schwinger [Saxon, 1945; Levine and Schwinger, 1950] applied variational methods to the solution of electromagnetic problems; several of the earlier applications have been discussed in the *Waveguide Handbook* [Marcuvitz, 1951]. Such a formulation has been applied by Storer [1952] for studying the impedance of a linear antenna. The essential steps of his development are summarized below. Further details of this can be seen in Section II. 39 of King [1956].

A perfectly conducting cylindrical antenna of length $2l$ and of radius a is excited by a voltage V, which is applied across an infinitesimal gap at the center of the antenna. The tangential electric field is zero along the antenna surface except for the gap at the antenna center, and by analogy with equations (2.8) to (2.10) can be written as

$$E_z(z) = -V(0)\,\delta(z) = -\int_{-l}^{l} G(z,z')\,I(z')\,dz', \tag{2.27}$$

where

$$G(z,z') = \frac{1}{4\pi i\omega\epsilon}\left(\frac{\partial^2}{\partial z^2} + k^2\right)\frac{e^{ikr}}{r}, \tag{2.28}$$

$$r = [(z-z')^2 + a^2]^{1/2}. \tag{2.29}$$

Both sides of (2.27) are multiplied with $I(z)$ and integrated with respect to z. After dividing both sides of the resulting expression by $[I(0)]^2$, the antenna impedance is computed as

$$Z = \frac{V(0)\, I(0)}{[I(0)]^2} = \frac{1}{[I(0)]^2} \int_{-l}^{l} \int_{-l}^{l} G(z, z')\, I(z)\, I(z')\, dz\, dz'. \tag{2.30}$$

The same expression can be also obtained by substituting (1.13) and (1.15) in the general impedance expression (1.30). Following (1.35) to (1.38) this impedance expression is stationary in the sense that small change $\delta I(z)$ about the correct antenna current $I(z)$ gives a zero first order change δZ in the impedance Z. The antenna current $I(z)$ is assumed to be of the form

$$I(z) = Af_A(z) + Bf_B(z), \tag{2.31}$$

where

$$f_A(z) = \sin [k(l - |z|)] \tag{2.32}$$

$$f_B(z) = 1 - \cos [k(l - |z|)] \tag{2.33}$$

and where A and B are the amplitudes of the as yet undetermined trial functions of the antenna current. Substituting (2.31) in the impedance expression (2.30) it follows that

$$Z = \frac{A^2\gamma_{AA} + 2AB\gamma_{AB} + B^2\gamma_{BB}}{(AF_A + BF_B)^2}, \tag{2.34}$$

where

$$F_N = f_N(0), \tag{2.35}$$

$$\gamma_{NM} = \iint dz\, dz'\, G(z, z')\, f_N(z)\, f_M(z'), \tag{2.36}$$

and where N, $M = A$ and/or B. After dividing the numerator and the denominator of (2.34) by B, the resulting impedance expression is a function of the ratio of trial function amplitudes (A/B). Because of the stationary character of Z, the optimum ratio (A/B) may be determined from the requirement that

$$\frac{\partial Z}{\partial (A/B)} = 0. \tag{2.37}$$

After carrying out the differentiation, it follows that

$$\frac{A}{B} = \frac{\gamma_{BB}F_A - \gamma_{AB}F_B}{\gamma_{AA}F_B - \gamma_{AB}F_A}. \tag{2.38}$$

Substituting (2.38) in (2.34), the impedance expression Z becomes

$$Z = \frac{\gamma_{AA}\gamma_{BB} - \gamma_{AB}^{2}}{\Delta} \tag{2.39}$$

with

$$\Delta = F_B^2\gamma_{AA} - 2F_AF_B\gamma_{AB} + F_A^2\gamma_{BB}. \tag{2.40}$$

The impedance can be computed for the sinusoidal field distribution term alone by setting $B = 0$ in (2.31). This gives

$$Z_s = \frac{\gamma_{AA}}{F_A^2}. \tag{2.41}$$

For a half-wave antenna ($l = \lambda/4$) the impedance in (2.41) can be seen to be the same as the impedance computed from (2.2) and (2.4). This indicates that for an assumed current distribution of the antenna the EMF method discussed in Section 2.1 gives the same impedance as the variational computations. However, in the variational method it is possible to assume a more general form of the antenna current distribution, and subsequently to optimize the amplitudes of the individual terms of trial functions for the antenna current. Of course, it is possible to assume an antenna current consisting of more than a two-term trial function. The rélative amplitudes of the subsequent terms of the antenna current can be again considered as variables in an impedance expression analogous to (2.34). Since the impedance is stationary with respect to small changes in the amplitudes of the currents, equations similar to (2.37) can be used for optimizing amplitudes of the various trial functions. The resulting impedance expression will be algebraically more complicated than Z of (2.39).

For the particular choice of the antenna-current trial functions shown in (2.32) and (2.33), the integration indicated in (2.36) has been carried out by Storer [1952]. Values of γ_{NM} are given by the following expressions:

$$\begin{aligned}\gamma_{AA} = K_\gamma[&4\cos^2 kl\,S(kl) - \cos 2kl\,S(2kl)\\ &- \sin 2kl[2C(kl) - C(2kl)]],\end{aligned} \tag{2.42}$$

$$\begin{aligned}\gamma_{AB} = K_\gamma[&-2\cos kl\,E(kl) + E(2kl)\\ &+ \sin 2kl[2S(kl) - S(2kl)]\\ &+ \cos 2kl[2C(kl) - C(2kl)]],\end{aligned} \tag{2.43}$$

$$\begin{aligned}\gamma_{BB} = K_\gamma[&-4\sin kl\,E(kl) + 2kl\,E(2kl)\\ &+ 4\sin^2 kl\,S(kl) + \cos 2kl\,S(2kl)\\ &+ \sin 2kl[2C(kl) - C(2kl)]\\ &+ i(e^{i2kl} - 1 - ka)],\end{aligned} \tag{2.44}$$

where $$K_\gamma = (\mu/\epsilon)^{1/2}/(2\pi i)$$

$$E(kx) = \log\frac{2x}{a} - Cin(kx) + iSi(kx), \tag{2.45}$$

$$S(kx) = \frac{1}{2}Si(2kx) + \frac{i}{2}Cin(2kx) - ka, \tag{2.45}$$

$$C(kx) = \log\frac{2x}{a} - \frac{1}{2}Cin(2kx) + \frac{i}{2}Si(2kx), \tag{2.47}$$

and where $Si(z)$ is the sine integral, $Cin(z) = C + \log z - Ci(z)$, $C = 0.5772\ldots$ is the Euler's constant, and $Ci(z)$ is the cosine integral.

For $kl \ll 1$, γ_{AA} of (2.42) can be expanded in a power series, and Z_s of (2.41) is approximated by the leading terms of this expansion as

$$Z_s = \frac{1}{\pi}\left(\frac{\mu}{\epsilon}\right)^{1/2}\left[\frac{1}{6}(kl)^2 + \frac{i}{kl}\left(\log\frac{l}{a} - 1\right)\right]. \tag{2.48}$$

The leading term of the function (2.32) is linear. Therefore Z_s of (2.48) is based on an assumed triangular current distribution along a short, linear antenna.

The above equations can be modified for a flat strip antenna of width $w = 2\epsilon$ (where the half-width ϵ should not be confused with the permittivity designated by the same symbol). A development that is summarized in Section 7.1 shows that (2.42) to (2.44) apply to flat strip antennas if $\log 2x/a$ is replaced by $\log 4x/\epsilon$ and ka by $8k\epsilon/\pi^2$ in (2.45) to (2.47). The impedance of a flat strip antenna of width $w = 2\epsilon$ will differ negligibly from the impedance of the cylindrical antenna of radius a if $\epsilon = 2a$ and $ka \ll 1$. For strip antennas, the parameter Ω is therefore defined as $\Omega = 2\log(4l/\epsilon)$.

A half-wave antenna with an assumed sinusoidal current distribution may be considered as the simplest example. Letting $l = \lambda/4$, (2.42) gives $\gamma_{AA} = K_\gamma S(\pi)$ and, in the limit of a approaching zero, Z_s of (2.41) is the same as Z of (2.7). For antennas of a finite radius a it is necessary to consider the complete impedance expression (2.39). The results of several such calculations are indicated in Fig. 2.1.

The calculations for the strip antenna are made by numerically evaluating integrals similar to (2.36), which will be discussed later in Section 9.1. The admittance Y of a slot antenna of comparable dimensions is computed using equations equivalent to (2.42) to (2.44). This admittance Y is related to the impedance of a complementary dipole antenna Z using the relation (3.6). The impedance Z computed from the admittance Y of the complementary slot antenna is also indicated in Fig. 2.1. The second and third order impedance data of King and Middleton (tables 11.30.1 and 15 by King [1956]),

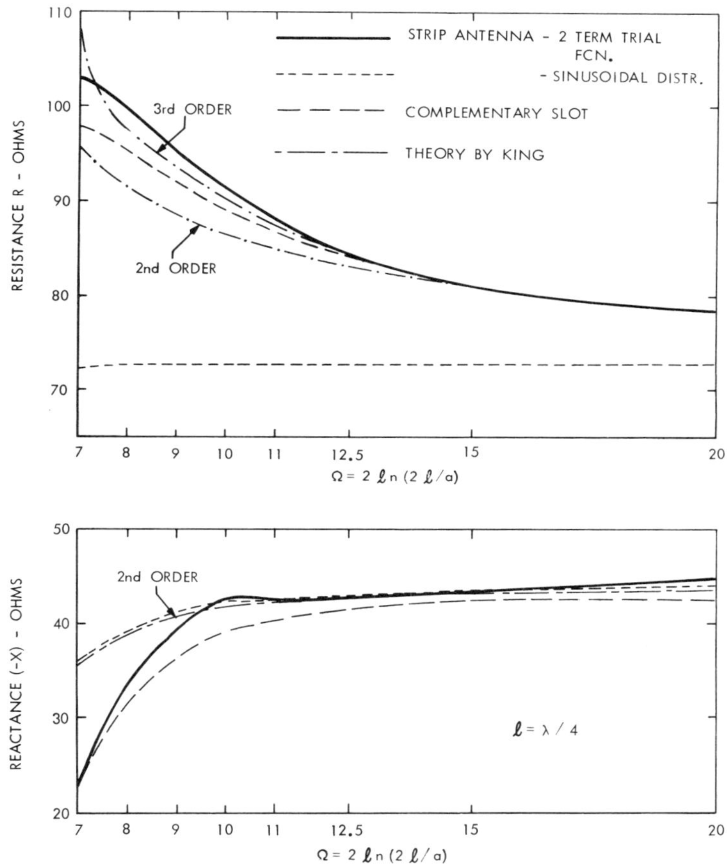

FIG. 2.1. Free space antenna impedance $Z = R + iX$ of a half-wave antenna.

are shown in the same figure. The assumption of the sinusoidal current distribution is seen to give an incorrect resistance value for antennas of finite transverse dimensions. The computations for strip antennas using the two-term trial function of (2.31) through (2.33) give results comparable to the third order theory by King and Middleton. The impedance of the complementary slot is nearly the same. For $\Omega < 10$, the reactance computed by King [1956] is nearly the same as with an assumed sinusoidal current distribution; it differs from the other computations. There is good agreement between the various computations for antennas of $\Omega > 11$.

Further calculations are shown in Fig. 2.2 for antennas of $\Omega = 8$ and 12.5, and varying antenna lengths l. In the resistance plot the left-hand scale applies to $l/\lambda \leqslant 0.2$ and the right-hand scale to $l/\lambda \geqslant 0.2$. For $\Omega = 8$ the

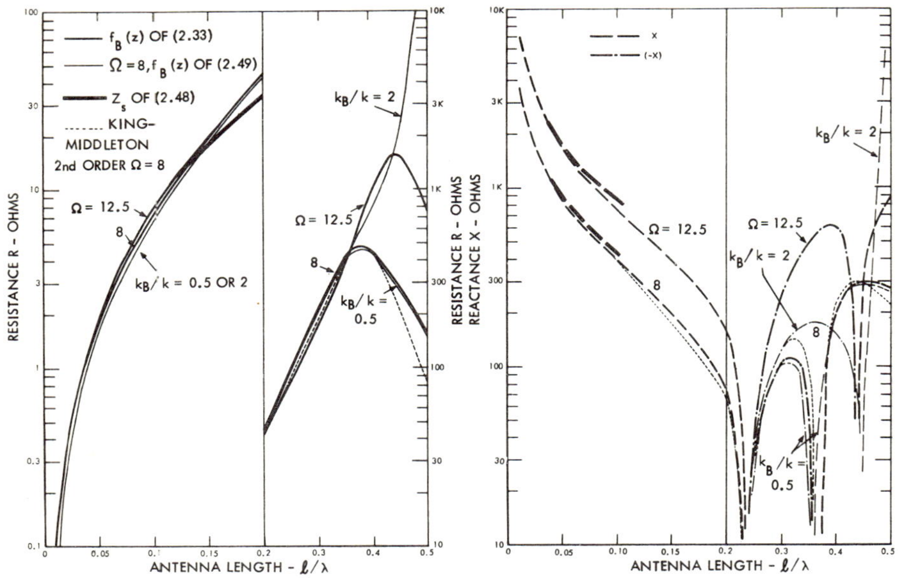

FIG. 2.2. Free space impedance $Z = R + iX$ linear antenna.

solutions based on the trial functions (2.32) and (2.33) are also compared with solutions using

$$f_B(z) = \sin k_B(l - |z|) \tag{2.49}$$

instead of $f_B(z)$ of (2.33). The wave number k_B is selected as $k/2$ and $2k$ in the present computations. These curves are indicated by thin lines in Fig. 2.2. Also shown by dotted curves are the second order King and Middleton impedance data. The approximate impedance Z_s of short antennas from (2.48) is indicated by heavy lines only for those antenna lengths where Z_s starts deviating from the more accurate calculations.

All of the calculations give nearly the same resistance figures for small values of l except that calculations using $f_B(z)$ of (2.49) give a slightly lower resistance. For larger values of l the resonance peak of the resistance is increased in magnitude for increasing values of Ω. The computations that make use of $k_B/k = 0.5$ give nearly the same resistance values as computations using $f_B(z)$ of (2.33). However, the calculations based on $k_B/k = 2$ give an exceedingly high resistance as l approaches $\lambda/2$. The reactance data of the various computations differ only to a small degree for $l/\lambda < 0.3$. The reactance based on $k_B/k = 2$ becomes inaccurate for larger values of l, and approaches infinity as l approaches $\lambda/2$. Examination of the impedance expression (2.34) or (2.39) shows that the denominator approaches zero for $k_B/k = 2$ and l approaching $\lambda/2$, which makes the impedance approach infinity. The variational impedance formulation cannot be expected to be accurate if F_A and F_B are both equal to zero simultaneously. The calculations based on the trial functions (2.32) and (2.33) will fail for l approaching λ. The calculations made using $f_B(z)$ of (2.49) and $k_B = 0.5\,k$ will also fail for l approaching λ. For $\Omega = 12.5$ the differences between the various sets of computations are much smaller, and the variational calculations are relatively insensitive to the choice of the various trial functions as long as F_A and F_B do not approach zero simultaneously. There is a relatively wide choice of the antenna trial function for shorter antennas.

C. T. Tai [King, 1956] has introduced an alternate form $f_B(z) = (l - |z|) \cos k(l - |z|)$ which gives non-vanishing values of F_B for $l \neq 0$ and accurate impedance data also for $l \approx \lambda$. Several comparisons between impedance computation based on variational and iterative procedures have been presented by King [pages 251 and 843, 1956]. The antenna impedances based on the King–Middleton second order theory compare closely with calculations by Tai and with a number of measurements. Calculations based on (2.32) and (2.33) are not accurate when l is near an integer multiple of λ for reasons indicated earlier.

The above comparisons between variational and iterative solutions have been limited to second and third order iterative solutions. More recently the admittance of a cylindrical antenna has been examined for one to 30 iterations [King *et al.*, 1966a]. The conductance is shown to converge rapidly but the susceptance increases with added iterations and it tends to approach the infinite value of the knife-edge susceptance of the delta function generator. This knife-edge susceptance should be subtracted to obtain an approximate theoretical admittance value. The knife-edge susceptance does not appear in the variational solutions discussed earlier, and there is no need to correct the solutions for it.

For long antennas it may be more appropriate to apply the recently developed theory of Wu [1961], Wu and Chen [1967] or the variational formulation of Hurd [1964]. Hurd [1964] does not specify the antenna current, but assumes that the electric field along the antenna axis is approximated by

$$E_z(z) = |z|^{-1}(z^2 - l^2)^{-1/2} \exp(ik|z|)$$

for $|z| > l$. His numerical results compare closely with the computations by Wu [1961].

The electric fields very near a circular cylindrical antenna have been recently investigated by King and Wu [1965, 1966a]. This work has been extended to imperfectly conducting cylindrical transmitting antennas [King and Wu, 1966b], and extensive numerical results are also presented [King *et al.*, 1966b]. The surface fields on a flat strip antenna have been studied by Galejs [1968].

2.4. Thin Wire Loop Antenna

Loop antennas are simplest to analyze for an assumed uniform current flow. In the limit of a small loop the fields will approach those of elementary magnetic dipoles, which can be computed from (1.13), (1.14), (1.15), and (1.17). For loops of a circumference that is of the order of $\lambda/5$ assumption of a uniform current flow is no longer justified. Such a loop exhibits also an electric dipole moment, and it is analyzed by considering the electric current to flow along the curved antenna surface. Similarly as for the linear antenna, the antenna impedance can be determined only after establishing the current distribution along the antenna, which requires the solution of an integral equation or the use of variational approximations. It will be shown later that both yield identical results for this problem. In the integral equation approach, Storer [1956] expresses the antenna current as a Fourier series; he has also presented extensive numerical results.

The fields of the circular loop antennas can be derived from the Hertz vector of elementary electric currents (1.15). For a source located at (ρ', ϕ'), the Hertz potential is in the direction of the current flow, which is in the ϕ' direction. In terms of the coordinates of the observation points at (ρ, ϕ), the Hertz potential Π_e will exhibit components both parallel and transverse to the current flow. For the geometry indicated in Fig. 2.3, Π_e is computed as

$$\mathbf{\Pi}_e = \mathbf{i}_{\phi'} \Pi_e = \Pi_e [\mathbf{i}_\rho \sin(\phi-\phi') + \mathbf{i}_\phi \cos(\phi-\phi')], \tag{2.50}$$

where

$$\Pi_e = \frac{iI(\phi')\,\rho'\,\mathrm{d}\phi'}{4\pi\omega\epsilon} \cdot \frac{\mathrm{e}^{ikr}}{r}, \tag{2.51}$$

$$r = [\rho^2 + \rho'^2 - 2\rho\rho' \cos(\phi-\phi') + (z-z')^2]^{1/2}, \tag{2.52}$$

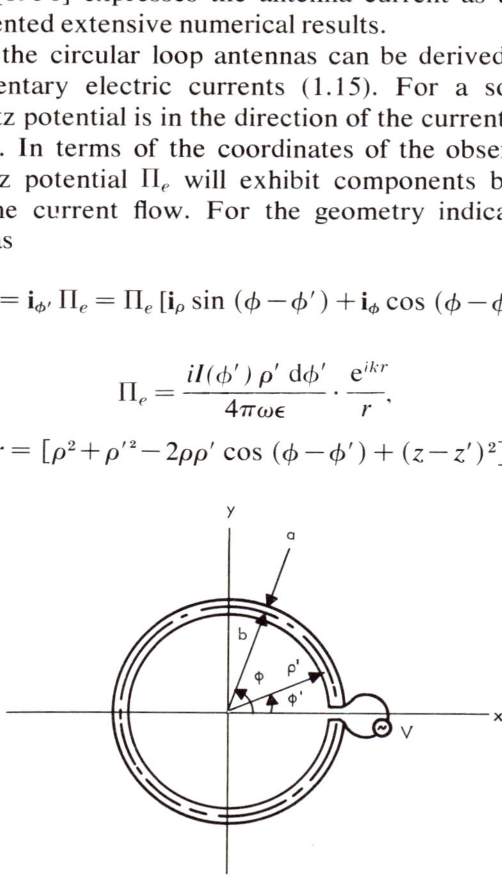

FIG. 2.3. Geometry of the loop antenna.

and where $\mathbf{i}_{\phi'}$, $\mathbf{i}_\rho$, and $\mathbf{i}_\phi$ represent unit vectors. Substituting (2.50) in (1.15) and carrying out the indicated operations in circular cylindrical coordinates it follows that

$$dE_\phi(\rho,\phi) = \frac{1}{\rho^2}\left\{\cos(\phi-\phi')\left[\rho\frac{\partial}{\partial\rho}\Pi_e + \frac{\partial}{\partial\phi^2}\Pi_e\right] + \sin(\phi-\phi')\right.$$

$$\left.\left[-\frac{\partial}{\partial\phi}\Pi_e + \rho\frac{\partial^2}{\partial\phi\partial\rho}\Pi_e\right]\right\} + k^2\cos(\phi-\phi')\Pi_e. \tag{2.53}$$

This represents the contribution to $E_\phi(\rho,\phi)$ by the current flowing at (ρ',ϕ'). The ρ and ϕ coordinates occur only in the second factor of (2.51). The derivatives indicated in (2.53) can be expressed in terms of the r derivatives of this factor and in terms of derivatives of r with respect to ρ and ϕ. Carrying out these differentiations, and noting that $|\rho-\rho'| \ll \rho$ or ρ', results in

$$dE_\phi(\rho,\phi) = \left\{\frac{\sin^2(\phi-\phi')\rho'^2}{r^2}\frac{\partial^2}{\partial r^2}\Pi_e + \left[\frac{\rho'\cos(\phi-\phi')}{r\rho} - \frac{\rho'^2\sin^2(\phi-\phi')}{r}\right]\right.$$

$$\left.\cdot\frac{\partial}{\partial r}\Pi_e\right\} + k^2\cos(\phi-\phi')\Pi e, \tag{2.54}$$

which can be expressed more compactly as

$$dE_\phi(\rho,\phi) = \frac{1}{\rho^2}\frac{\partial^2}{\partial\phi^2}\Pi_e + k^2\cos(\phi-\phi')\Pi_e. \tag{2.55}$$

By analogy with the linear antenna (see Section 2.5), the antenna fields are computed by assuming the current $I(\phi')$ is concentrated in the antenna center ($\rho' = b$). The radius ρ designates the distance from the loop center to the wire surface and is replaced by the average value $(a^2+b^2)^{1/2}$. For $\phi = \phi'$, the correct distance $r = a$ is given, but for ϕ different from ϕ', only negligible errors are introduced for thin antennas such that $a \ll b$. The distance r of (2.52) is therefore replaced by

$$r = [a^2 + 2b^2 - 2b^2\cos(\phi-\phi')]^{1/2} \tag{2.56}$$

Substituting Π_e of (2.51) in (2.55) and integrating along the circumference of the loop gives

$$E_\phi(\phi) = \frac{i}{4\pi\omega\epsilon}\int_{-\pi}^{\pi} d\phi'\, I(\phi')$$

$$\left[\frac{1}{b}\frac{\partial^2}{\partial\phi^2} + k^2 b\cos(\phi-\phi')\right]\cdot\frac{e^{ikr}}{r}. \tag{2.57}$$

The perfectly conducting loop is excited by a voltage V impressed across an infinitesimally narrow gap at $\phi = 0$. The resulting tangential electric field $E_\phi(\phi)$ can be expressed as

$$E_\phi(\phi) = -V\delta(\phi)/b, \tag{2.58}$$

where $\delta(\phi)$ represents a delta function. A substitution of (2.58) in (2.57) gives an integral equation for determining $I(\phi')$. Equations (2.57) and (2.58) are rewritten as

$$V\delta(\phi) = \int_{-\pi}^{\pi} d\phi' \, I(\phi') \, G(\phi - \phi'), \tag{2.59}$$

where

$$G(\phi - \phi') = \frac{1}{4\pi i\omega\epsilon}\left[\frac{\partial^2}{\partial\phi^2} + (kb)^2 \cos(\phi - \phi')\right]\frac{e^{ikr}}{r}. \tag{2.60}$$

$G(\phi - \phi')$ is a bounded, periodic, and even function of $(\phi - \phi')$, and is expanded into a Fourier series

$$G(\phi - \phi') = \sum_{-\infty}^{\infty} \alpha_n e^{-in(\phi - \phi')}, \tag{2.61}$$

$$\alpha_n = \alpha_{-n} = \frac{1}{2\pi}\int_{-\pi}^{\pi} d\psi \, G(\psi) \, e^{in\psi}. \tag{2.62}$$

Inserting (2.61) in (2.59) yields

$$V\delta(\phi) = \sum_{-\infty}^{\infty} \alpha_n \, e^{in\phi} \int_{-\pi}^{\pi} d\phi' \, I(\phi') \, e^{in\phi'}. \tag{2.63}$$

The integral is equal to 2π times the coefficient I_n in the Fourier series expansion of the current $I(\phi')$, i.e.

$$I_n = \frac{1}{2\pi}\int_{-\pi}^{\pi} d\phi \, I(\phi) \, e^{in\phi}, \tag{2.64}$$

$$I(\phi) = \sum_{-\infty}^{\infty} I_n \, e^{-in\phi}. \tag{2.65}$$

Substituting (2.64) in (2.63)

$$V\delta(\phi) = 2\pi \sum_{-\infty}^{\infty} \alpha_n I_n \, e^{-in\phi}. \tag{2.66}$$

After multiplying both sides of (2.66) with $e^{in\phi}$ and integrating between the limits of $(-\pi)$ and π, it can be seen that

$$I_n = \frac{1}{(2\pi)^2\alpha_n}\int_{-\pi}^{\pi} d\phi \, V\delta(\phi) \, e^{in\phi} = \frac{V}{(2\pi)^2\alpha_n}. \tag{2.67}$$

Substituting (2.67) in (2.65)

$$I(\phi) = \frac{V}{(2\pi)^2} \sum_{-\infty}^{\infty} \frac{e^{-in\phi}}{\alpha_n} = \frac{V}{(2\pi)^2} \cdot \left[\frac{1}{\alpha_0} + 2 \sum_{1}^{\infty} \frac{\cos n\phi}{\alpha_n}\right], \tag{2.68}$$

and the impedance of the loop is found to be

$$Z = \frac{V}{I(0)} = (2\pi)^2 \left[\frac{1}{\alpha_0} + 2 \sum_{1}^{\infty} \frac{1}{\alpha_n}\right]^{-1}. \tag{2.69}$$

The impedance expression (2.69) was obtained from a Fourier solution of the integral equation (2.59). A similar impedance expression can be also obtained from a variational formulation. After multiplying both sides of (2.59) with $I(\phi)$ and integrating with respect to ϕ, the antenna impedance is computed as

$$Z = \frac{VI(0)}{[I(0)]^2} = \frac{1}{[I(0)]^2} \iint d\phi \, d\phi' \, I(\phi) \, I(\phi') \, G(\phi - \phi'), \tag{2.70}$$

where $G(\phi - \phi')$ is defined by (2.60). The same impedance expression is also obtained by substituting (2.57) in (1.30) and by replacing the indicated surface current density by the antenna current. The antenna current $I(\phi)$ is expanded in the Fourier series (2.64). $G(\phi - \phi')$ is represented by a Fourier series in (2.61), and it can be readily shown that

$$\int_{-\pi}^{\pi} \int_{-\pi}^{\pi} d\phi \, d\phi' \cos n\phi \cos m\phi' \, G(\phi - \phi') = 0 \quad \text{if} \quad n \neq m. \tag{2.71}$$

When (2.65) is substituted in (2.70) and (2.71) is applied, (2.70) is reduced to

$$Z = \frac{1}{[I(0)]^2} \iint d\phi \, d\phi' \cdot \left[I_0^2 + 4 \sum_{n=1}^{\infty} I_n^2 \cos n\phi \cos n\phi'\right] G(\phi - \phi'). \tag{2.72}$$

Inserting the Fourier series representation (2.61) of $G(\phi - \phi')$ and evaluating the integrals,

$$\frac{1}{Z} = \frac{1}{(2\pi)^2} \frac{\left[I_0 + 2 \sum_{n=1}^{\infty} I_n\right]^2}{\left[\alpha_0 I_0^2 + 2 \sum_{n=1}^{\infty} \alpha_n I_n^2\right]}. \tag{2.73}$$

This impedance expression is stationary, which can be demonstrated in the same way as for equation (1.35) in Section 1.5. The optimum values of the relative current amplitudes $i_n = I_n/I_0$ are determined from the requirement that $\partial Z/\partial i_n = 0$. This gives

$$\alpha_n i_n = \frac{\alpha_0 + 2 \sum_{m=1}^{\infty} \alpha_m i_m^2}{1 + 2 \sum_{m=1}^{\infty} i_m}, \tag{2.74}$$

which is valid for any value of n. The right-hand side of (2.74) does not depend on n, and $a_n i_n$ is therefore constant and independent of n. After dividing the right hand side of equation (2.74) by $a_n i_n$, it is seen that

$$\alpha_n i_n = \alpha_0. \tag{2.75}$$

The impedance expression (2.73) is rearranged into

$$\frac{1}{Z} = \frac{1}{(2\pi)^2} \frac{1 + 2\sum_{n=1}^{\infty} i_n}{\alpha_n i_n} = \frac{1}{(2\pi)^2}\left[\frac{1}{\alpha_0} + 2\sum_{n=1}^{\infty} \frac{1}{\alpha_n}\right], \tag{2.76}$$

which is in agreement with (2.69).

The coefficients a_n of (2.68) and (2.69) or (2.76) have been evaluated by Storer [1956] as

$$\alpha_n = \alpha_{-n} = \frac{1}{4\pi i}\left(\frac{\mu_0}{\epsilon_0}\right)^{1/2}\left[\frac{kb}{2}(K_{n+1} + K_{n-1}) - \frac{n^2}{kb} K_n\right]. \tag{2.77}$$

The coefficients $K_n = K_{-n}$ are computed from

$$K_0 = \frac{1}{\pi}\log\frac{8b}{a} - \frac{1}{2}\int_0^{2kb} \Omega_0(x)\, \mathrm{d}x + \frac{i}{2}\int_0^{2kb} J_0(x)\, \mathrm{d}x, \tag{2.78}$$

$$K_{n+1} - K_n = \Omega_{2n+1}(2kb) - iJ_{2n+1}(2kb), \tag{2.79}$$

where $J_n(x)$ is the Bessel function of the first kind and $\Omega_n(x)$ is the Lommel–Weber function

$$\Omega_n(x) = \frac{1}{\pi}\int_0^{\pi} \cos(x \sin\theta - n\theta)\, \mathrm{d}\theta. \tag{2.80}$$

It has also been shown that α_n becomes extremely small for values of n that are of the order of $2b/a$, but α_n is never equal to zero. The summations of the equations (2.68) or (2.69) converge very slowly for large values of n, and have been approximated by integrals.

Power series expansions of K_0, K_1 and K_2 have been listed by Chen and King [1963]. For loops that are small relative to the wavelength ($kb \ll 1$), the current is nearly uniform and independent of ϕ. The impedance Z of (2.69) or (2.76) is approximated by its leading term

$$Z \approx (2\pi)^2\alpha_0 = \left(\frac{\mu_0}{\epsilon_0}\right)^{1/2}\left\{\frac{\pi}{6}(kb)^4\left[1 - \frac{(kb)^2}{5} + \ldots\right]\right.$$

$$\left. - ikb\left[\log\left(\frac{8b}{a}\right) - 2 + \frac{2}{3}(kb)^2 + .\ \right]\right\}; \tag{2.81}$$

Loop antennas can be also analyzed without invoking the thin wire assumption. Such a development is outlined in Chapter 10 for a flat loop in a stratified dielectric. For a uniform current around the loop only I_0 is different from zero and $\alpha_n = \alpha_{-n} = \infty$ for $n \neq 0$ according to (2.67). The first order effects of a nonuniform current are considered by assuming I_0 and I_1 different from zero or $\alpha_n = \alpha_{-n} = \infty$ for $n = \geqslant 2$. These calculations, which consider either a uniform current or a first order effect of a nonuniform current, are compared for free space around the antenna with Storer's calculations, which use more accurate current distributions. A flat loop of an average radius ρ_0 and width $w = \rho_2 - \rho_1$ is compared with a thin wire loop of radius $b = \rho_0$ and wire radius $a = w/4$. The current distributions based on a two-term approximation (α_0 and α_1 finite) are compared with calculations of Storer [1956] in Fig. 2.4.

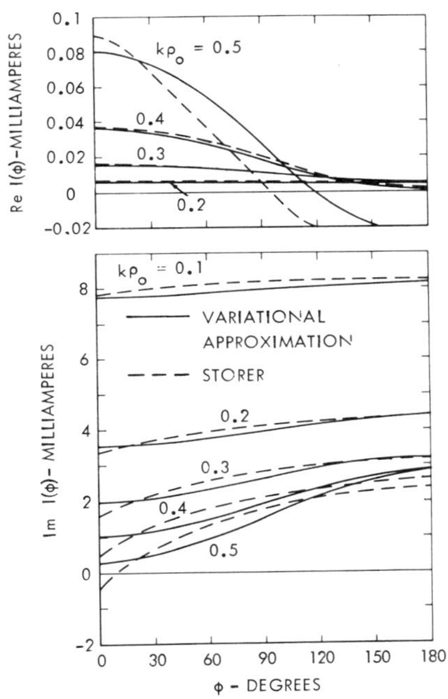

FIG. 2.4. Current distributions for loops in free space. $\Omega = 2 \log |8\pi\rho_0/(\rho_2 - \rho_1)| = 10$, $a = 2.27\ \lambda$.

There is a reasonable agreement for $k\rho_0 < 0.3$. The differences in the imaginary part of the current distribution become more pronounced for small values of ϕ with $k\rho_0 > 0.4$, and the two-term approximation of the current distribution is inadequate for loops of larger diameters. In Fig. 2.5, Y_0 represents the loop admittance computed for an assumed uniform current distribution (only α_0 if finite), and Y_1 considers the nonuniform current

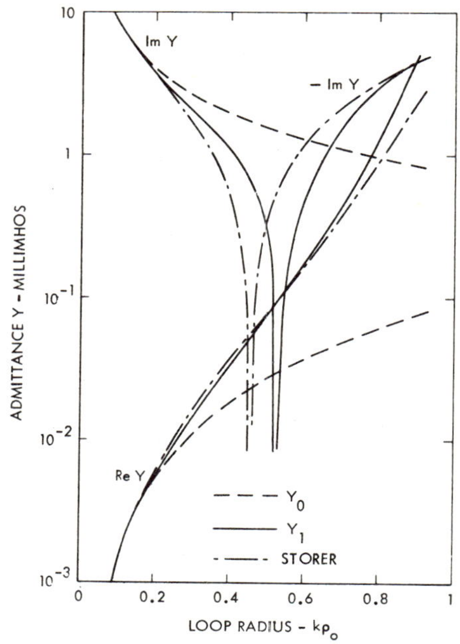

FIG. 2.5. Free space admittance of the loop.
$\Omega = 2 \log |8\pi\rho_0/(\rho_2 - \rho_1)| = 10$, $a = 2.27\ \lambda$.

(α_0 and α_1 finite). The conductance $\text{Re}Y_1$ is reasonably accurate for loops of $k\rho_0 < 0.7$, but the susceptance $\text{Im}Y_1$ is incorrect in the vicinity of the loop resonance that occurs for $k\rho_0 \approx 0.5$. The assumption of a uniform loop current is justified in the impedance computations only for loops of radii satisfying $k\rho_0 < 0.2$. The use of the two-term trial function is therefore limited to relatively small loops, and the numerical results of Storer [1956] should be used for loops of other parameters. The more recent theory of Wu [1962] applies in particular to larger loop antennas.

2.5. Appendix: Hertz Vector of a Cylindrical Wire

It will be shown that for thin cylindrical wires the fields outside the wire can be computed by assuming that the total current is concentrated at the center of the wire. The Hertz potential (or vector) $\Pi_z(a, \phi, z)$ due to a distribution of current elements $J_z(\rho', \phi', z')$ is computed by integrating (1.15), so that

$$\Pi_z = \frac{i}{4\pi\omega\epsilon} \int d\rho' \int d\phi' \int dz'\, J_z(\rho', \phi', z') \frac{e^{ikr}}{r}, \tag{2.82}$$

where

$$r = [(z - z')^2 + a^2 + \rho'^2 - 2a\rho' \cos(\phi - \phi')]^{1/2}, \tag{2.83}$$

and $\rho' \leqslant a$. Π_z can be computed by assuming the current to flow along the antenna axis if it can be shown that

$$\Delta\Pi_z = \Pi_z - \Pi_{z1} \tag{2.84}$$

is small. Π_{z1} is obtained from (2.82) by letting $\rho' \to 0$ in (2.83). $J_z(\rho', \phi', z')$ which is independent of ϕ', is assumed to vary relatively slowly with z', and is removed from the ϕ' and z' integrals. The principal contribution to Π_z and Π_{z1} occurs for small values of $|z-z'|$, and the integrals will be evaluated only in the vicinity of $z = z'$, where the exponential of (2.82) is near unity. The exponential is approximated by $\exp(ikr) = 1 + ikr$, with the terms proportional to ikr cancelling in the integrals representing $\Delta\Pi_z$. The z' integration is carried out between the limits $(z-b) < z' < (z+b)$, where b is large relative to a. This gives

$$\Delta\Pi_z \approx \frac{i}{2\pi\omega\epsilon} \int_0^{\rho 1} d\rho' \, J_z(\rho', \phi', z) \int_0^{2\pi} d\phi' \cdot \{ \log [y + (y^2 + a^2 + \rho'^2 - 2a\rho' \cos(\phi - \phi'))^{1/2}] - \log [y + (y^2 + a^2)^{1/2}] \}_{y=0}^{b}. \tag{2.85}$$

For $y = 0$, the ϕ' integrals of the two logarithmic terms cancel, which can be shown using the integral

$$\int_0^{2\pi} \log(\alpha - \beta \cos\phi) \, d\phi = 2\pi \log \frac{\alpha + (\alpha^2 - \beta^2)^{1/2}}{2}. \tag{2.86}$$

For $y = b$, the difference of the two logarithmic terms is expanded in a series, and the ρ integral is evaluated assuming an ρ-independent current density. This results in

$$\frac{\Delta\Pi_z}{\Pi_{z1}} = \frac{1}{12} \frac{(\rho_1/b)^2}{\log(2b/a)}, \tag{2.87}$$

which is 1.45×10^{-3} for $\rho_1 = a = b/5$, and becomes negligible for larger values of b/a. It is therefore legitimate to compute the Hertz vector by assuming that the current flow is confined to the center of the wire, if the computations consider the integrated effect of the current over a length exceeding several wire diameters. This result has been also derived by King and Harrison [1943].

2.6. Exercises

2.1. Derive the general impedance expression (2.39) starting out with the formulation (2.30).

2.2. Derive the coefficients γ_{NM} (2.42) to (2.44) of the cylindrical antenna.

2.3. Derive the integral equation of the loop current (2.57) starting out with the Hertz potential (2.50).

2.4. Derive the coefficients a_n of (2.77) of the loop antenna and expand the coefficients K_0 and K_1 in powers of kb.

2.7. References

AHARONI, J. (1946) *Antennae*, Clarendon Press, Oxford.

CARTER, P. S. (1932) Circuit relations in radioating systems and applications to antenna problems, *Proc. IRE* **20** (6), 1004–41.

CHEN, C. L. and KING, R. W. P. (1963) The small bare loop antenna immersed in a dissipative medium, *IEEE Trans. on Ant. and Prop.* **AP-11** (3), 266–9.

GALEJS, J., (1968), "Surface fields on linear antennas", *Proc. IEE, 115*.

HALLEN, E. (1938) Theoretical investigations into transmitting and receiving qualities of antennae, *Nova Acta Regiae Soc. Sci. Upsaliensis,* **4**, 1–44.

HURD, R. A. (1964) Variational solutions for the admittance of the long cylindrical antenna, *Radio Science* **68D** (3), 311–16.

KING, R. W. P. (1956) *Theory of Linear Antennas*, Harvard University Press, Cambridge, Mass.

KING, R. W. P., ARONSON, E. A. and HARRISON, C. W. (1966a) Determination of the admittance and effective length of cylindrical antennas, *Radio Science* **1** (7), 835–50.

KING, R. W. P., HARRISON, C. W. and ARONSON, E. A. (1966b) The imperfectly conducting cylindrical transmitting antenna: numerical results, *IEEE Trans. on Ant. and Prop.* **AP–14** (5), 535–42.

KING, R. W. P. and HARRISON Jr., C. W. (1943) The distribution of current along a symmetrical center-driven antenna, *Proc. IRE* **31**, 548–67.

KING, R. W. P. and WU, T. T. (1965) Currents, charges, and near fields of cylindrical antenna, *Radio Science* **69D** (3), 429–46.

KING, R. W. P. and WU, T. T. (1966a) The electric field very near a driven cylindrical antenna, *Radio Science* **1** (3), 353–9.

KING, R. W. P. and WU, T. T. (1966b) The imperfectly conducting cylindrical transmitting antenna, *IEEE Trans. on Ant. and Prop.* **AP–14** (5), 524–34.

KRAUS, J. D. (1950) *Antennas*, McGraw-Hill, New York, NY.

LEVINE, H. and SCHWINGER, J. (1950) On the theory of electromagnetic wave diffraction by an aperture in an infinite plane conducting screen, *Comm. Pure Appl. Math.* **3**, 355–91.

MARCUVITZ, N. (1951) *Waveguide Handbook*, McGraw-Hill, New York, NY, sections 2.3 and 2.4.

SAXON, D. S. (1945) *Notes on Lectures by J. Schwinger*, MIT.

STORER, J. E. (1952) *Variational Solution to the Problem of the Symmetrical Cylindrical Antenna*, Cruft Lab. Rep. TR 101, Cambridge, Mass.

STORER, J. E. (1956) Impedance of thin-wire loop antennas, *Trans. Am. Inst. Engrs.* 75, 606; also Cruft Lab. Rep. No. 212, Harvard University, Cambridge, Mass.

WU, T. T. (1961) Theory of the dipole antenna and the two-wire transmission Line, *J. Math. Physics* **2** (4), 550–74.

WU, T. T. (1962) Theory of thin circular loop antenna, *J. Math. Physics* **3**, 1301–4.

WU, T. T. and CHEN, C. L. (1967) Theory of the long dipole antenna, *Antenna Theory*, ed. by R. E. Collin and F. A. Zucker, McGraw-Hill, New York, NY.

CHAPTER 3

SLOT ANTENNAS IN FREE SPACE

3.1. Concept of Complementary Antennas

A set of complementary antennas can be defined in the following manner. A slot of width w and length $2l$ is cut in a conducting ground plane, is excited by a voltage V impressed across its center portion, and radiates into the upper half space. An ideal open circuit (or zero admittance) is reflected into the slot plane from the lower half space. A complementary dipole antenna is made of a thin conducting metal strip of identical size to the slot, and is excited by a current I at its center. A simple relation will be derived between the admittance $Y = I/V$ of the slot antenna and the impedance $Z = V/I$ of the complementary dipole. This relation was obtained first by Booker [1946].

The impedance of the electric dipole antenna can be written following equations (1.32) and (1.35) as

$$Z = \frac{1}{[I(0)]^2} \iint \mathbf{J}_{se} \cdot \mathfrak{G}_e \cdot \mathbf{J}'_{se} \, da' \, da, \tag{3.1}$$

where

$$\mathfrak{G}_e = \frac{1}{4\pi i \omega \epsilon} (\nabla\nabla + k^2 \mathfrak{J}), \tag{3.2}$$

$\mathfrak{J}$ is a unit dyad, and the notations $\mathbf{J}_{se}$, $\mathbf{J}'_{se}$, da', and da are defined in Section 1.4. Equations (3.1) and (3.2) can be derived also from (1.13), (1.15), and (1.30). For a slot antenna, (1.14), (1.15), and (1.40) are similarly manipulated to establish

$$Y = \frac{1}{[V(0)]^2} \iint \mathbf{J}_{sm} \cdot \mathfrak{G}_m \cdot \mathbf{J}'_{sm} \, da' \, da \tag{3.3}$$

with

$$\mathfrak{G}_m = \frac{2}{4\pi i \omega \mu} (\nabla\nabla + k^2 \mathfrak{J}). \tag{3.4}$$

The factor 2 of (3.4) accounts for the presence of the ground plane, which was pointed out in Section 1.3. A comparison of (3.2) and (3.4) shows that

$$\mathfrak{G}_m = 2\frac{\epsilon}{\mu} \mathfrak{G}_e. \tag{3.5}$$

In a complementary antenna geometry the distribution of the magnetic current density $J_{sm}(\xi, \eta)/V(0)$ will be the same as the electric current distribution $J_{se}(\xi, \eta)/I(0)$. Substituting this and the relation (3.5) into (3.3), Y of (3.3) is related to Z of (3.1) by

$$Y = 2\frac{\epsilon}{\mu}Z. \tag{3.6}$$

For free space, $(\mu_0/\epsilon_0)^{1/2} = 120\pi$ and $(\mu_0/2\epsilon_0) = 7200\pi^2$. The validity of the relation (3.6) is not restricted to the rectangular geometry shown in Fig. 3.1. This equation can be also used to relate the impedance of a circular loop to the admittance of an annular slot as shown in Section 3.3.

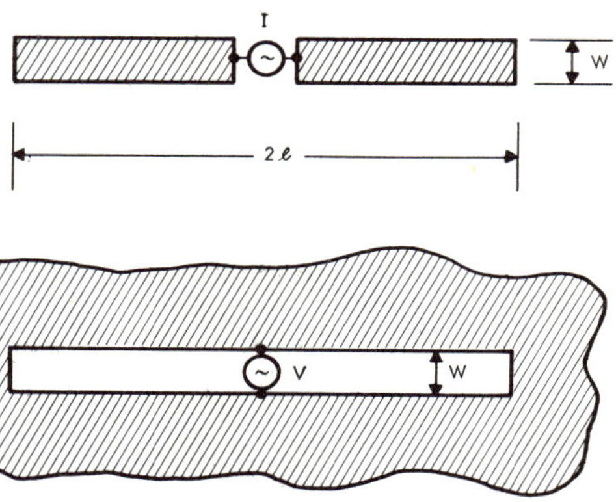

FIG. 3.1. A flat dipole antenna and a complementary slot.

A slot in a thin screen radiates the same amount of power into either of the two half spaces. Its admittance is two times larger than indicated in (3.3). $\mathfrak{G}_m$ of (3.5) and Y of (3.6) should be doubled for slots which radiate in both directions.

An experimental study of slot antennas and approximately complementary cylindrical antennas has been reported by King and Owyang [1960]. The exact quantitative equivalence between complementary quantities of infinitely thin strip antennas and slots in ground planes of zero thickness and infinite extent is changed into a qualitative agreement between the complementary quantities when cylindrical dipoles are compared with slot antennas in ground screens of a finite thickness. Because of the additional surfaces that may be charged or may carry currents, the equivalent length of cylindrical dipoles is greater than that of an infinitely thin strip of the same length. The equivalent length of a slot in a ground screen of finite thickness is shorter than that of the same slot cut in an ideal ground plane of zero thickness.

3.2. Rectangular Waveguide with an Infinite Flange

The impedance of the rectangular waveguide shown in Fig. 3.2 will be formulated after expressing the field in the waveguide and in the semi-infinite free space region in terms of the aperture fields at $z = 0$. For purposes of impedance calculations it is assumed that the principal waveguide mode is adequate for representing the aperture fields. This assumption is justified by

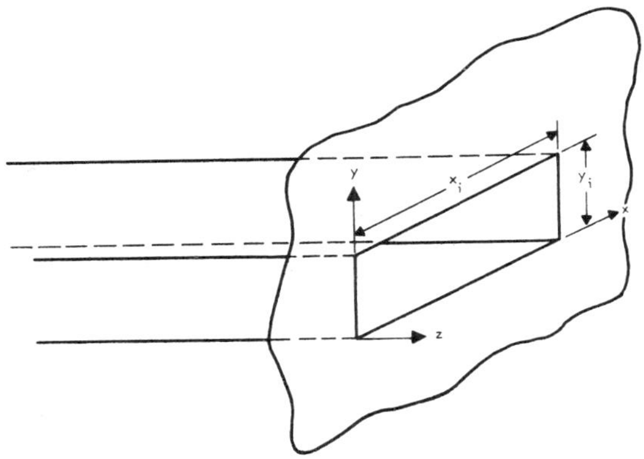

FIG. 3.2. Rectangular waveguide with an infinite flange.

more exact calculations reported in Chapter 8. The principal waveguide mode exhibits only an E_y electric field component, and the substitution of (1.18) into (1.14) shows that the magnetic field component H_x is given in the outside space ($z \geqslant 0$) by

$$H_x(x, y) = \frac{2i}{4\pi\omega\mu_0} \int_0^a \int_0^b \mathrm{d}x' \, \mathrm{d}y' \, E_y(x', y') \left(k^2 + \frac{\partial^2}{\partial x^2}\right) \frac{\mathrm{e}^{ikr}}{r} \tag{3.7}$$

with

$$r = [(x - x')^2 + (y - y')^2 + z^2]^{1/2}.$$

The waveguide ($z \leqslant 0$) is excited in its principal TE_{10} mode by waves that propagate in the positive z direction. There are also reflected waves in this guide. The fields in the guide consist of a superposition of TE and TM modes that can be derived from the scalar functions Ψ_i and Φ_i, respectively, as shown in Section 1.4. For an assumed exp $(-i\omega t)$, time dependences of the fields Ψ_i and Φ_i are expressed following Collin [1960] as

$$\Psi_i = a_{10} \cos \beta_{xil} x (\mathrm{e}^{\gamma_{il} z} + R \mathrm{e}^{-\gamma_{il} z}) + \sum_n \sum_m{}' a_{nm} \cos \beta_{xi} x \cos \beta_{yi} y \, \mathrm{e}^{-\gamma_i z}, \tag{3.8}$$

$$\Phi_i = \sum_n \sum_m b_{nm} \sin \beta_{xi} x \sin \beta_{yi} y \, \mathrm{e}^{-\gamma_i z}, \tag{3.9}$$

where

$$\beta_{xi}=\frac{n\pi}{x_i},\quad \beta_{yi}=\frac{m\pi}{y_i},\quad \beta_{xil}=\frac{\pi}{x_i},$$

$$\gamma_i=i[k_i^2-\beta_{xi}^2-\beta_{yi}^2]^{1/2},\quad \gamma_{il}=i[k_i^2-\beta_{xil}^2]^{1/2},$$

$$k_i=[\omega^2\mu_0\epsilon_i+i\omega\mu_0\sigma_i]^{1/2},$$

and R is a reflection coefficient. The prime on the double summation (3.8) designates the omission of its $n=1$, $m=0$ term. The field components are related to Ψ_i and Φ_i by (1.21) to 1.26). For rectangular coordinates, this results in

$$E_{xi}=i\omega\mu_0\frac{\partial}{\partial y}\Psi_i+\frac{\partial^2}{\partial z\partial x}\Phi_i, \tag{3.10}$$

$$E_{yi}=-i\omega\mu_0\frac{\partial}{\partial x}\Psi_i+\frac{\partial^2}{\partial z\partial y}\Phi_i, \tag{3.11}$$

$$E_{zi}=-\left(\frac{\partial^2}{\partial x^2}+\frac{\partial^2}{\partial y^2}\right)\Phi_i, \tag{3.12}$$

$$H_{xi}=\frac{\partial^2}{\partial z\partial x}\Psi_i+\frac{1}{i\omega\mu_0}k_i^2\frac{\partial}{\partial y}\Phi_i, \tag{3.13}$$

$$H_{yi}=\frac{\partial^2}{\partial z\partial y}\Psi_i-\frac{1}{i\omega\mu_0}k_i^2\frac{\partial}{\partial x}\Phi_i, \tag{3.14}$$

$$H_{zi}=-\left(\frac{\partial^2}{\partial x^2}+\frac{\partial^2}{\partial y^2}\right)\Psi_i. \tag{3.15}$$

After multiplying (3.11) with $\sin\beta_{xil}x'$ and integrating over the aperture at $z=0$, it follows that

$$a_{10}=\frac{2/(i\omega\mu_0)}{\beta_{xil}x_1y_1(1+R)}\iint E_y\sin\beta_{xil}x'\,\mathrm{d}x'\,\mathrm{d}y' \tag{3.16}$$

where E_y refers to the aperture fields $z=0$.

Equation (3.10) is multiplied by $\cos\beta_{xj}x\sin\beta_{yj}y$ and (3.11) by $\sin\beta_{xj}x\cos\beta_{yj}y$. Integrating over the aperture gives two equations with unknowns a_{nm} and b_{nm}. E_x was assumed to be zero in the aperture, and the coefficients a_{nm} and b_{nm} may be computed as

$$a_{nm}=\frac{2\epsilon_m\beta_{xi}/(i\omega\mu_0)}{x_iy_i(\beta_{xi}^2+\beta_{yi}^2)}\iint E_y\sin\beta_{xil}x'\,\mathrm{d}x'\,\mathrm{d}y' \tag{3.17}$$

$$b_{nm}=\frac{2\epsilon_m\beta_{yi}}{x_iy_i(\beta_{xi}^2+\beta_{yi}^2)\,\gamma_i}\iint E_y\sin\beta_{xi}x'\cos\beta_{yi}y'\,\mathrm{d}x'\,\mathrm{d}y' \tag{3.18}$$

where

$$\epsilon_m=1\quad\text{for}\quad m=0$$

and

$$\epsilon_m = 2 \quad \text{for} \quad m \neq 0.$$

After substituting a_{nm} and b_{nm} in (3.13), the magnetic field components H_x of (3.7) and (3.13) are both related to the integrals of the electric aperture field E_y. After equating the tangential magnetic field components in the aperture and letting

$$D = \iint E_y \sin \beta_{xil} x' \, dx' \, dy', \tag{3.19}$$

it follows that

$$\frac{2\gamma_{il}}{x_i y_i} \frac{1-R}{1+R} D \sin \beta_{xil} x = \iint E_y(x', y') G_1 \, dx' \, dy', \tag{3.20}$$

where

$$G_1 = \sum_n \sum_m{}' \frac{2\epsilon_m}{x_i y_i \gamma_i} (\beta_{xi}^2 - k_i^2) \sin \beta_{xi} x' \cos \beta_{yi} y' \cdot \sin \beta_{xi} x \cos \beta_{yi} y + \frac{1}{2\pi}\left(k^2 + \frac{\partial^2}{\partial x^2}\right)\frac{e^{ikr}}{r}. \tag{3.21}$$

The normalized slot admittance of the waveguide is

$$y_s = \frac{Y_s}{Y_0} = \frac{1-R}{1+R}, \tag{3.22}$$

where the characteristic admittance $Y_0 = \gamma_{il}/(i\omega\mu_0)$.

A variational principle for computing y_s may be constructed by multiplying (3.20) by $E_y(x,y)$ and by integrating over the slot. It follows that

$$y_s = \frac{\iiiint E_y(x, y) E_y(x', y') G_1 \, dx \, dy \, dx' \, dy'}{2\dfrac{\gamma_{il}}{x_i y_i} D^2}. \tag{3.23}$$

The stationarity of (3.23) can be demonstrated using the method shown in Section 1.5. The principal waveguide mode exhibits the cross-sectional variation

$$E_y(x, y) = E_0 \sin \beta_{xil} x. \tag{3.24}$$

Substituting this for the aperture fields in (3.23), it is seen that all terms of the double summation of (3.21) vanish. The integral of (3.19) is readily evaluated and (3.23) simplifies to

$$y_s = \frac{1}{\pi x_i y_i \gamma_{il}} \iiiint \sin(\pi x/x_i) \sin(\pi x'/x_i) \left(k^2 + \frac{\partial^2}{\partial x^2}\right) \cdot \frac{e^{ikr}}{r} \, dx \, dy \, dx' \, dy'. \tag{3.25}$$

The term $\partial^2/\partial x^2$ is replaced by $-\partial^2/\partial x \partial x'$ and can be integrated by parts. After defining new variables of integration by $x-x' = \xi$, $y-y' = \eta$, $x+x' = x_i+\nu$, and $y+y' = y_i+\mu$, the μ and ν integrals can be readily evaluated, and (3.25) reduced to a double integral. One further integration of this equation can be made in terms of elementary functions, and y_s can be computed by a single numerical integration. Such calculations have been carried out by Lewin [1951] and representative admittance data are shown in Fig. 3.3. The normalized admittance has a negative imaginary part, which corresponds to capacitive susceptance.

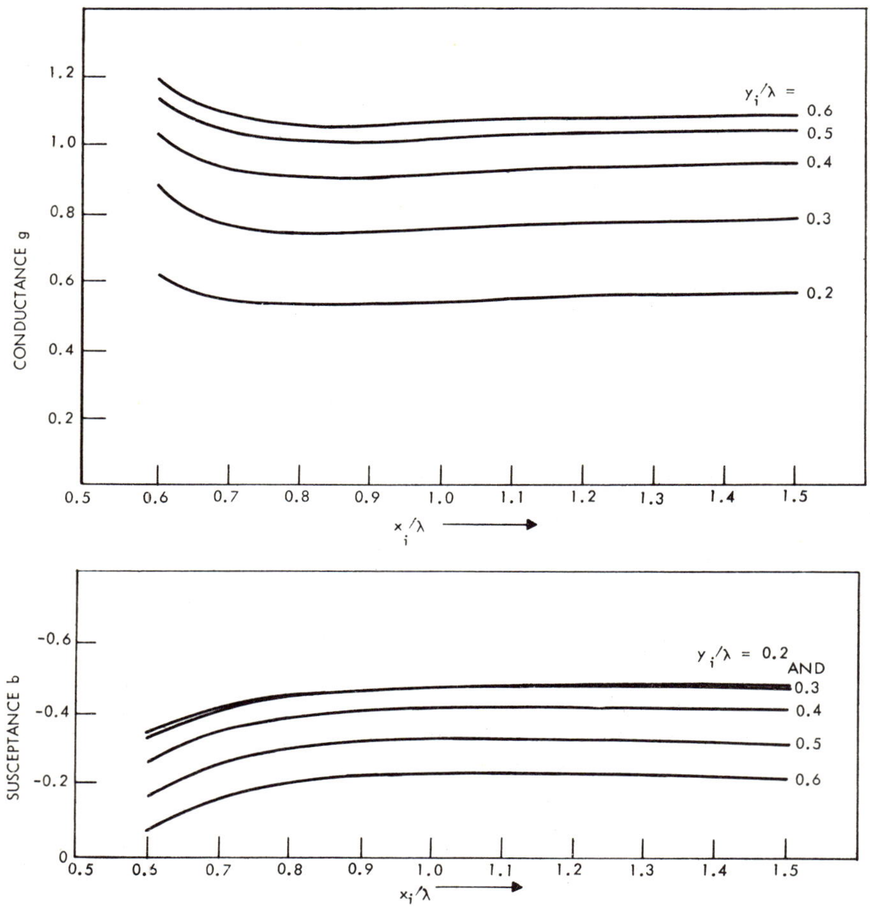

FIG. 3.3. The normalized admittance $y_s = g + ib$ of a rectangular waveguide with an infinite flange.

3.3. Coaxial Line with an Infinite Flange

A coaxial line of radius b has a center conductor of radius a and is terminated in an infinite flange as shown in Fig. 3.4. The coaxial line is excited in its dominant TEM mode, which exhibits no angular variations and has only the H_ϕ and E_ρ field components. The discontinuity at $z = 0$ also excites higher order TM modes, which still exhibit no angular variations. Also the fields in the outside space ($z \geqslant 0$) exhibit no variations in the ϕ direction.

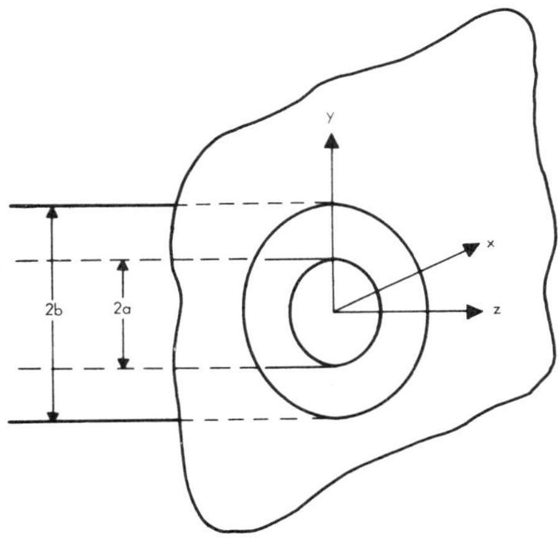

FIG. 3.4. Coaxial line with an infinite flange.

The fields inside the coaxial line and in the outside space can be related to the tangential electric field of the aperture at $z = 0$. The continuity of the tangential magnetic field in the aperture plane gives an integral equation for the electric fields in the aperture. This integral equation can be used for computing approximations to the aperture fields and to the antenna impedance. Such a formulation has been developed by Levine and Papas [1951] but their numerical calculations have been carried out by assuming that the principal mode of the coaxial line is the aperture field distribution. This principal mode represents the leading term of a more general aperture distribution, and the principal mode of the coaxial line is assumed to represent the aperture fields also in the present calculations.

The magnetic field in the outside space is related to the tangential electric field $E_\rho(\rho', \phi')$ of the aperture by (1.14), (1.15), and (1.18) as

$$d\mathbf{H} = \frac{2i}{4\pi\omega\mu} E_\rho(\rho', \phi')(\nabla\nabla\cdot + k^2)\frac{e^{ikr}}{r}(\mathbf{i}_{\rho'} \times \mathbf{i}_z)\,\rho'\,d\rho'\,d\phi', \tag{3.26}$$

where

$$r = [\rho^2 + \rho'^2 - 2\rho\rho' \cos(\phi - \phi') + z^2]^{1/2}. \tag{3.27}$$

The cross product of the two unit vectors is equal to $(-i_{\phi'})$, which has components in ρ and ϕ directions in the unprimed coordinates, as was shown by (2.50). $E_\rho(\rho', \phi') = E_\rho(\rho')$ is taken to be the principal mode of the coaxial line, and exhibits no variations in the ϕ' direction. The operator $(\nabla\nabla\cdot)$ involves differentiations with respect to unprimed coordinates, and (3.26) can be integrated with respect to ϕ' prior to the evaluation of $(\nabla\nabla\cdot)$. The resulting expression has only a ϕ component, and is independent of ϕ. The operator $\nabla = \mathbf{i}_\phi\ \partial/\partial\phi$ is zero, and the term proportional to $(\nabla\nabla\cdot)$ is consequently zero. The magnetic field of the free space region is therefore related to $E_\rho(\rho')$ as

$$H_\phi(\rho, z) = \frac{k^2}{2\pi i\omega\mu_0}\int_a^b E_\rho(\rho')\rho'\,\mathrm{d}\rho'\int_0^{2\pi}\frac{\mathrm{e}^{ikr}}{r}\cos\psi\,\mathrm{d}\psi, \tag{3.28}$$

where $\psi = \phi - \phi'$. The last integral can be expressed in terms of Bessel functions [Sommerfeld, 1949, equations (31.14) and 21.3a)]. Evaluating the ϕ integral gives

$$H_\phi(\rho, z) = -i\omega\epsilon_0\int_a^b E_\rho(\rho')\rho'\,\mathrm{d}\rho'$$
$$\cdot\int_0^\infty \frac{\exp[-(\zeta^2-k^2)^{1/2}|z|]J_1(\zeta\rho)J_1(\zeta\rho')\zeta\,\mathrm{d}\zeta}{(\zeta^2-k^2)^{1/2}}, \tag{3.29}$$

where $\arg(\zeta^2-k^2)^{1/2} = 0$ and $-\pi/2$ for $\zeta > k$ and $\zeta < k$, respectively.

The ϕ symmetrical magnetic fields of the coaxial line are expressed as [Marcuvitz, 1951]

$$H_\phi(\rho, z) = \frac{A_0}{\rho}\left(\mathrm{e}^{ik_iz} - R_0\mathrm{e}^{-ik_iz}\right) + \sum_{n=1}^{\infty} A_nR_n(\rho)\exp(\gamma_nz) \tag{3.30}$$

Where $k_i = \omega(\mu_0\epsilon_i)^{1/2}$, $\gamma_n = i(k_i^2-\lambda_n^2)^{1/2}$, $\operatorname{Re}\gamma_n \leqslant 0$, and R is the reflection coefficient of the principal (TEM) mode. The radial function $R_n(\rho)$ is defined by

$$R_n(\rho) = N_n[J_1(\lambda_n\rho)Y_0(\lambda_na) - J_0(\lambda_na)Y_1(\lambda_n\rho)], \tag{3.31}$$

where $J_m(x)$ is the Bessel function of the first kind and $Y_m(x)$ is the Bessel function of second kind of order m. The eigenvalues λ_n are computed from

$$Y_0(\lambda_na)J_0(\lambda_nb) = J_0(\lambda_na)Y_0(\lambda_nb), \tag{3.32}$$

and the normalizing factor N_n is defined by

$$N_n^{-1} = \frac{(2)^{1/2}}{\pi\lambda_n}\left[\frac{J_0^2(\lambda_na)}{J_0^2(\lambda_nb)} - 1\right]^{1/2}. \tag{3.33}$$

The radial electric field component E_ρ is computed from (3.30) as

$$i\omega\epsilon E_\rho(\rho, z) = \frac{\partial}{\partial z} H_\phi(\rho, z). \tag{3.34}$$

The amplitudes A_n are related to the aperture field $E_\rho(\rho, 0) = E_\rho(\rho)$. Multiplying $E_\rho(\rho, 0)$, as computed from (3.34), with $R_n(\rho)\rho\, d\rho$ and integrating from $\rho = a$ to b gives

$$A_n = \frac{i\omega\epsilon_i \int_a^b E_\rho(\rho) R_n(\rho)\rho\, d\rho}{\gamma_n}. \tag{3.35}$$

Multiplying $E_\rho(\rho, 0)$ with $d\rho$ and integrating from $\rho = a$ to b gives

$$A_0 = \frac{\omega\epsilon_i}{k_i} \frac{\int_a^b E_\rho(\rho)\, d\rho}{[1+R] \log(b/a)}. \tag{3.36}$$

The tangential magnetic fields are continuous in the aperture at $z = 0$, and $H_\phi(\rho, 0)$ of (3.29) is equal (3.30) in the aperture $a \leqslant \rho \leqslant b$. Inserting (3.35) and (3.36) in (3.30) gives

$$\frac{\omega\epsilon_i}{k_i} \frac{1-R}{1+R} \frac{\int_a^b E_\rho(\rho')\, d\rho'}{\rho \log(b/a)} + \sum_{n=1}^{\infty} \frac{i\omega\epsilon_i R_n(\rho) \int_a^b E_\rho(\rho') R_n(\rho')\, d\rho'}{\gamma_n}$$

$$= -i\omega\epsilon_0 \int_a^b E_\rho(\rho')\rho'\, d\rho' \int_0^\infty \frac{J_1(\zeta\rho) J_1(\zeta\rho')\zeta\, d\zeta}{[\zeta^2 - k^2]^{1/2}}. \tag{3.37}$$

This expression is brought into stationary form by multiplying both sides of (3.37) by $\rho E_\rho(\rho)$, and then integrating the resulting equation from $\rho = a$ to $\rho = b$. $E_\rho(\rho)$ is assumed of the form

$$E_\rho(\rho) = E_0/\rho, \tag{3.38}$$

which results in

$$\frac{\omega\epsilon_i}{k_i \log(b/a)} \frac{1-R}{1+R} \left[\int_a^b \frac{d\rho}{\rho}\right]^2 + \sum_{n=1}^{\infty} \frac{i\omega\epsilon_i}{\gamma_n} \left[\int_a^b R_n(\rho)\, d\rho\right]^2$$

$$= -i\omega\epsilon_0 \int_0^\infty \frac{\zeta\, d\zeta}{(\zeta^2 - k^2)^{1/2}} \left[\int_a^b J_1(\zeta\rho)\, d\rho\right]^2. \tag{3.39}$$

The ρ integrals of the n summation are equal to zero, and the other ρ integrals are readily evaluated. The normalized admittance y_s is computed by analogy with (3.22) as

$$y_s = \frac{1-R}{1+R} = -\frac{ik}{(\epsilon_i/\epsilon_0)^{1/2}\log(b/a)} \int_0^\infty \frac{d\zeta}{\zeta(\zeta^2-k^2)^{1/2}} \cdot [J_0(\zeta a) - J_0(\zeta b)]^2. \qquad (3.40)$$

The integral is considered separately over the ranges $0 < \zeta < k$ and $k < \zeta < \infty$, the integrand changes from imaginary to real at $\zeta = k$, and the integration over these two ranges gives the slot conductance and susceptance respectively. For $0 < \zeta < k$, the variable of integration is changed to $\zeta = k \sin\theta$; for $k < \zeta < \infty$, the present infinite integral can be expressed as a finite integral of integral sines. It follows that

$$y_s = \frac{1}{(\epsilon_i/\epsilon_0)^{1/2}\log(b/a)} \left\{ \int_0^{\pi/2} \frac{d\theta}{\sin\theta}[J_0(ka\sin\theta) - J_0(kb\sin\theta)]^2 \right.$$
$$- \frac{i}{\pi}\int_0^\pi [2Si(ka\sqrt{1+(b/a)^2-2(b/a)\cos\psi}) - Si(2ka\sin\psi/2)$$
$$\left. - Si(2kb\sin\psi/2)]\,d\psi \right\}. \qquad (3.41)$$

These integrals have been evaluated numerically. For small apertures of $kb \ll 1$, the functions can be replaced by series expansions and the individual terms readily integrated. The slot admittance Y_s is related to the normalized admittance y_s by the characteristic admittance of the coaxial line Y_0 as

$$\frac{Y_s}{y_s} = Y_0 = 2\pi[(\mu/\epsilon_i)^{1/2}\log(b/a)]^{-1}, \qquad (3.42)$$

and Y_s is approximated by

$$Y_s = \frac{(k\rho)^4}{360}\left[1 - \frac{(k\rho)^2}{5} + \frac{(k\rho)^4}{56} + \ldots\right]$$
$$-i\frac{k\rho}{60\pi}\left[\log\frac{8\rho}{b-a} - \frac{1}{2} + \frac{2}{3}(k\rho)^2 + \ldots\right], \qquad (3.43)$$

where $\rho = (a+b)/2$.

An example of admittance calculations is shown in Fig. 3.5. The closed form approximation (3.43) (indicated by thin lines) gives nearly the same admittance $Y_s = G + iB$ as the numerical evaluation of the integrals (3.41) (indicated by heavy lines) for $k\rho < 1$. However, there are significant

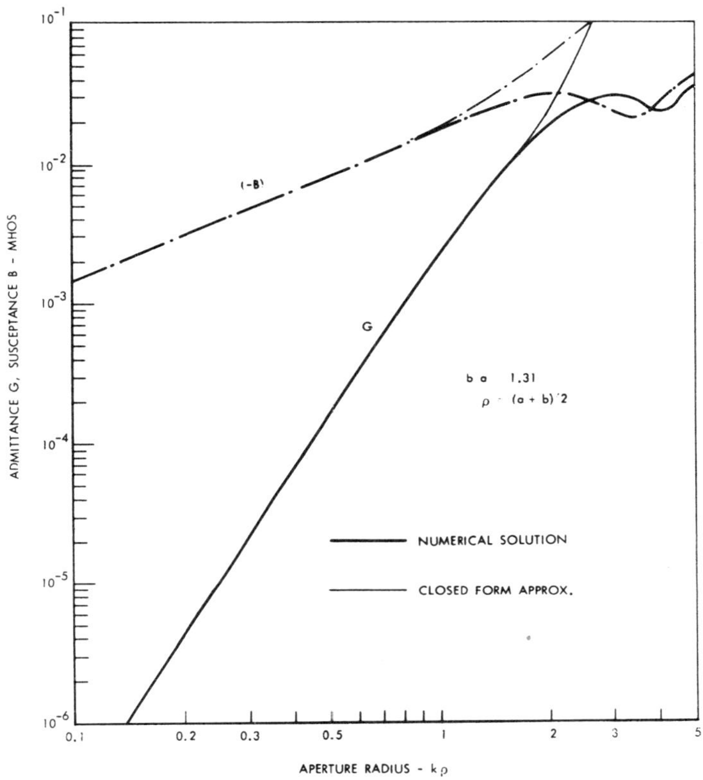

FIG. 3.5. Admittance of a coaxial line.

differences for larger values of $k\rho$. The slot susceptance is capacitive ($B < 0$).

The slot susceptance (3.43) can be related by (3.6) to the impedance of a complementary flat loop antenna. Substitution of the leading terms of (3.43) in (3.6) gives the loop impedance as

$$Z_{\text{loop}} = 7200\pi^2 Y_s = 120\pi\left[\frac{\pi}{6}(k\rho)^4 - ik\rho\left(\log\frac{8\rho}{b-a} - \frac{1}{2}\right)\right]. \qquad (3.44)$$

A flat loop of width $w = (b-a)$ corresponds to a loop made of a cylindrical wire if its radius is $w/4$. Under these conditions, (3.44) is nearly the same as the impedance Z of (2.81) except that a term 2 of the reactance (2.81) corresponds to a term $2 \log 2 + \frac{1}{2} \simeq 1.8863$ in (3.44). However, the complementary relationship (3.6) does not apply to flat loops and apertures of a coaxial line for larger radii ρ. The fields of the coaxial line do not vary with ϕ in the TEM mode, but the current distribution along the loop will exhibit ϕ-variations, and the basis for deriving (3.6) will be violated. Equation (3.6) applies only to antennas with complementary field or current distributions (or complementary sources).

3.4 Exercises

3.1. Reduce the fourfold integral of the admittance formulation (3.25) to a single integral.

3.2. Show that

$$\int_a^b R_n(\rho) R_m(\rho) \rho \, d\rho = \begin{matrix} 1 & \text{if} & n = m \\ 0 & \text{if} & n \neq m \end{matrix}$$

where $R_n(\rho)$ is defined by (3.31).

3.3. Transform the integral of the admittance expression (3.40) into the integral (3.41) and expand it in powers of $\rho = (a+b)/2$ by assuming that $kb \ll 1$.

3.5. References

BOOKER, H. G. (1946) Slot aerials and their relation to complementary wire aerials, *J. IEE (London)* **93**, Part IIIA, No. 4.

COLLIN, R. E. (1960) *Field Theory of Guided Waves*, McGraw-Hill, New York, NY.

KING, R. W. P. and OWYANG, G. H. (1960) An experimental study of the slot aerial and the three elements colinear array of slot aerials, *Proc. IEE* **107C**, 216–27.

KRAUS, J. D. (1950) *Antennas*, McGraw-Hill, New York, NY.

LEVINE, H. and PAPAS, C. H. (1951) Theory of circular diffraction antenna, *J. Appl. Phys.* **22**, 29–43.

LEWIN, L. (1951) *Advanced Theory of Waveguides*, Iliffe & Sons, London.

MARCUVITZ, N. (1951) *Waveguide Handbook*, McGraw-Hill, New York, NY.

SOMMERFIELD, A. (1949) *Partial Differential Equations*, Academic Press Inc., New York, NY.

CHAPTER 4

DIELECTRIC LOADING OF SMALL ANTENNAS

IN THIS chapter the characteristics of small electric dipole antennas and loops will be determined in the presence of spherically shaped dielectric regions. An elementary electric dipole in the center of a dielectric sphere will be treated first in Section 4.1, and loops wound on a spherical core will be compared with capacitor type antennas in Section 4.2. The effects of spherical shells on radiation characteristics of elementary dipoles and loops are discussed in Section 4.3. The expressions of the Hertz potential of elementary dipoles or loops of Section 1.3 can be used to deduce the characteristics of such antennas in the presence of a concentric dielectric region, but for finite antennas, it is more convenient to express the field in terms of spherical wave functions. For small antennas in free space, it is permissible to compute the antenna characteristics by assuming that the current distributions of linear antennas is triangular (Section 2.3), and is uniform along a circular loop antenna (Section 2.4). The impedance characteristics of dielectrically loaded electrically small antennas will also be calculated by making use of similar assumptions about antenna current distributions.

Concentric dielectric regions can also be treated with finite biconically shaped antennas. The problem of a small-angle biconical antenna embedded in a dielectric sphere has been solved by Tai [1948]. Polk [1959] has provided numerical results for antennas with dimensions comparable to a wave length, and Galejs [1962] has examined electrically small antennas. For wide angle cones of small size, the variational formulation of Tai [1949] has been adapted to dielectric antenna loading by Galejs [1962]. Galejs [1964] has also considered top-loaded and half-buried biconical structures of spherical cores of lossy dielectric. Antennas of conical shape have been recently discussed by Wait [1967], who also lists numerous references in this problem area. Biconically shaped antennas are not considered in the present text.

4.1. Dielectric Loading of Electric Dipole Antennas

An electric dipole is located in the center of a dielectric sphere having a radius a, a complex dielectric constant ϵ_1, and a magnetic permeability

μ_0. The outer medium is characterized by ϵ_2 and μ_0. For an unbounded space, the fields excited by an elementary dipole are computed directly from equations (1.13) to (1.15). The fields of the dipole located at the center of the coordinates in the z-direction can be expressed in the spherical (r, θ, ϕ) coordinates. The unit vector, $\mathbf{i}_z$, in the z direction is related to the unit vectors, $\mathbf{i}_r$ and $\mathbf{i}_\theta$, in the r and θ directions by

$$\mathbf{i}_z = \mathbf{i}_r \cos\theta - \mathbf{i}_\theta \sin\theta. \tag{4.1}$$

Substituting (4.1) in (1.13) and (1.14) and carrying out the required differentiations show that

$$E_r = \cos\theta\left(k_i^2 + \frac{\partial^2}{\partial r^2}\right)\Pi_{zi}, \tag{4.2}$$

$$E_\theta = -\sin\theta\left(k_i^2 + \frac{1}{r}\frac{\partial}{\partial r}\right)\Pi_{zi}, \tag{4.3}$$

$$H_\phi = i\omega\epsilon_i \sin\theta \frac{\partial}{\partial r}\Pi_{zi}, \tag{4.4}$$

where

$$\Pi_{zi} = \frac{iI\,\mathrm{d}z}{4\pi\omega\epsilon_i}\cdot\frac{\exp(ik_i r)}{r}$$

$$k_i^2 = \omega^2\mu_0\epsilon_i \quad \text{and} \quad i = 1 \text{ or } 2. \tag{4.5}$$

The Hertz vector within the dielectric sphere may be obtained from a linear combination of solutions proportional to e^{ik_1r}/r and to e^{-ik_1r}/r. The considerations are simplified if the secondary fields are finite at the origin. The Hertz vector within the dielectric sphere may be assumed to be of the form

$$\Pi_{z1} = \frac{A}{r}e^{ik_1r} + \frac{B}{r}\sin k_1 r. \tag{4.6}$$

The fields outside the sphere should represent an outgoing wave at large distances from the sphere. This suggests

$$\Pi_{z2} = \frac{C}{r}e^{ik_2r}. \tag{4.7}$$

The constant A characterizes the primary excitation and according to (4.5) is related to the dipole moment $I\mathrm{d}z$ as

$$A = \frac{iI\,\mathrm{d}z}{4\pi\omega\epsilon_1}. \tag{4.8}$$

Continuity of E_θ and H_ϕ across the boundary of the sphere requires that the two conditions

$$\Pi_{z1}+\frac{1}{k_1^2r^2}\left[\frac{\partial}{\partial r}(r\Pi_{z1})-\Pi_{z1}\right]=\Pi_{z2}+\frac{1}{k_2^2r^2}\left[\frac{\partial}{\partial r}(r\Pi_{z2})-\Pi_{z2}\right] \tag{4.9}$$

and:

$$\frac{\partial}{\partial r}(r\Pi_{z1})-\Pi_{z1}=\frac{\partial}{\partial r}(r\Pi_{z2})-\Pi_{z2} \tag{4.10}$$

be satisfied for $r=a$. Substitution of (4.6) and (4.7) in (4.9) and (4.10) results in

$$A=\frac{Ce^{-iy}}{2iz}\left\{e^{iz}\left[\left(1+\frac{1}{iy}-\frac{1}{y^2}\right)(-1+iz)\right.\right.$$

$$\left.+(1+iy)\left(1-\frac{1}{iz}-\frac{1}{z^2}\right)\right]+e^{-iz}\left[\left(1+\frac{1}{iy}-\frac{1}{y^2}\right)(1+iz)\right.$$

$$\left.\left.-(1+iy)\left(1+\frac{1}{iz}-\frac{1}{z^2}\right)\right]\right\}, \tag{4.11}$$

where

$$z=k_1a, \tag{4.12}$$

$$y=k_2a. \tag{4.13}$$

For spheres of a small radius $k_1a,\ k_2a \ll 1$, and a series expansion of $e^{\pm iz}$ results in

$$C=A\frac{3\epsilon_2}{\epsilon_1+2\epsilon_2}. \tag{4.14}$$

With C and A determined, the radiation fields of the dipole are computed by substituting (4.7), (4.8) and (4.14) in (4.3). It follows that

$$E_{\theta 2}=\frac{i\omega\mu_0 I\,dz\sin\theta}{4\pi r}\,\frac{3\epsilon_2}{\epsilon_1+2\epsilon_2}\,e^{ik_2r}. \tag{4.15}$$

The radiated power is to be calculated from

$$P=\frac{1}{2}\left(\frac{\epsilon_2}{\mu_0}\right)^{1/2}\int_0^{2\pi}\int_0^{\pi}|E_{\theta 2}|^2r^2\sin\theta\,d\theta\,d\phi. \tag{4.16}$$

The radiation resistance follows from (4.16) as

$$R=\frac{2P}{I_0^2}=\left(\frac{3\epsilon_2}{\epsilon_1+2\epsilon_2}\right)^2\left(\frac{\mu_0}{\epsilon_2}\right)^{1/2}\frac{k_2^2(dz)^2}{6\pi}. \tag{4.17}$$

where I_0 designates the peak current.

The capacity of the dipole antenna is

$$C = \frac{\epsilon_1 A}{dz}, \tag{4.18}$$

where A is its effective area. The radiation power factor [Wheeler, 1947], which denotes the available product of bandwidth and efficiency, is computed from (4.17) and (4.18) as

$$p = \frac{R}{X} = R\omega C = \frac{9\epsilon_1\epsilon_2}{(\epsilon_1 + 2\epsilon_2)^2} \frac{k_2^3 A\, dz}{6\pi} \tag{4.19}$$

After designating the effective antenna volume $A ds$ by V and letting $\epsilon = \epsilon_1/\epsilon_2$, (4.19) becomes

$$p = \frac{9\epsilon}{(\epsilon+2)^2} \frac{k_2^3 V}{6\pi}. \tag{4.20}$$

The radiation power factor p of (4.20) is increased slightly for $1 < \epsilon < 4$, but decreases for larger values of ϵ.

If the spherical boundary between the two regions at $r = a$ lies in the far fields of the dipole, $k_1 a$ and $k_2 a \gg 1$. Only the terms of the square brackets of (4.11), which are proportional to y or z, are significant, and C is computed as

$$C = A\, e^{iy} \frac{(\epsilon_1)^{1/2}}{(\epsilon_1)^{1/2} \cos z + i(\epsilon_2)^{1/2} \sin z.} \tag{4.21}$$

Substituting (4.7), (4.8), and (4.21) in (4.3), the radiated electric fields become

$$E_{\theta 2} = \frac{i\omega\mu_0 I\, dz \sin\theta}{4\pi r} e^{-ik_2(r-a)} \cdot \left[\cos k_1 a + i\left(\frac{\epsilon_2}{\epsilon_1}\right)^{1/2} \sin k_1 a\right]^{-1}. \tag{4.22}$$

The radiation resistance of the dipole is computed as in (4.16) and (4.17). It follows that

$$R = \left(\frac{\mu_0}{\epsilon_2}\right)^{1/2} \frac{k_2^2\, dz^2}{6\pi} \frac{1}{\cos^2 k_1 a + \frac{\epsilon_2}{\epsilon_1} \sin^2 k_1 a} \tag{4.23}$$

The radiation power factor p is obtained from (4.18) and (4.23) as

$$p = R\omega C = \frac{(\epsilon_1/\epsilon_2)}{\cos^2 k_1 a + \frac{\epsilon_2}{\epsilon_1} \sin^2 k_1 a} \frac{k_2^3 A\, dz}{6\pi}. \tag{4.24}$$

The radiation power factor p of (4.24) is increased with increasing values of $\epsilon = \epsilon_1/\epsilon_2$. This increase is quadratic with ϵ if $k_1 a = (n+\frac{1}{2})\pi$ and linear if $k_1 a = n\pi$.

Similar calculations have also been made for biconical antennas embedded in a dielectric sphere [Galejs, 1962]. The radiation power factor of biconical antenna exhibits characteristics similar to that of the elementary dipole, but the power factor of the wide-angle biconical antenna is decreased monotonically with increasing values of ϵ_1. Such a change of the radiation power factor was also deduced for small capacitor-type antennas by Wheeler [1947]. However, for comparable aperture heights, the wide-angle biconical antenna exhibits approximately $4\frac{1}{2}$ times larger radiation power factor than the capacitor-type antenna. This can be explained by the larger physical height of the wide-angle biconical antenna.

4.2. Lossy Cores for Loops and Capacitor Type Antennas

Small electric and magnetic antennas have been discussed by Wheeler [1947, 1958]. He indicates that for a given antenna volume both antenna types exhibit comparable radiation power factors p (or products of bandwidth and efficiency). In the absence of dielectric losses the efficiencies are determined by antenna and tuner losses. However, significant differences between the two antennas occur in the case of lossy dielectric cores. Such cores may be required by mechanical considerations if core materials of negligible losses are too costly.

In this section a loop antenna wound on a spherical core will be compared with a top-loaded cylindrical dipole.

4.2.1. *Loop Antenna*

The coil of the loop antenna is wound with N turns on a spherical dielectric core of radius a, permittivity ϵ_1, conductivity σ_1, and permeability μ_0. Free space permittivity ϵ_0 and permeability μ_0 are assumed for the outside space. The sphere is excited by a surface current density J_s, the principal component of which is in the ϕ direction. After neglecting the θ component of J_s, the Hertz vector can be shown to have a single scalar component Π_ϕ. Π_ϕ is expressible as a summation of Legendre functions, where the coefficients of the individual terms depend on the J_s distribution, or on the spread and density of the coil winding. The voltage along the coil is proportional to the line integral of Π_ϕ, and the core impedance, $Z = V/I$, where I is the coil current. The calculations of Galejs [1963] show that the coil reactance is given by

$$X = \left(\frac{\mu_0}{\epsilon_0}\right)^{1/2} \pi N^2 (k_2 a) f_1(\theta_1). \tag{4.25}$$

The radiation resistance is

$$R_r = \left(\frac{\mu_0}{\epsilon_0}\right)^{1/2} \frac{\pi}{3} N^2 (k_2 a)^4 f_2(\theta_1), \tag{4.26}$$

and the loss resistance due to dielectric losses is

$$R_L = \left(\frac{\mu_0}{\epsilon_0}\right)^{1/2} \pi \frac{\epsilon_1}{\epsilon_0} N^2 (k_2 a)^3 \tan \delta_1 f_3(\theta_1), \tag{4.27}$$

where

$$\tan \delta_1 = \sigma_1 / (\omega \epsilon_1).$$

In the above expressions $k_2 = \omega(\mu_0 \epsilon_0)^{1/2}$ and $f_i(\theta_1)$ depend on the distribution of the surface current density $J_\phi(\theta)$.

For a coil of a constant pitch with respect to the coil axis $(dz/(a d\phi) = \tan \alpha =$ constant, where ϕ is measured along the coil), the surface current density is of the form

$$J_\phi(\theta) = J_\phi \sin \theta. \tag{4.28}$$

It can be shown that $f_1(\theta_1) = f_2(\theta_1) = 2/9$, $f_3(\theta_1) = 2/135$. The length of wire in such a coil is

$$l = \frac{\pi \alpha}{2 \tan \alpha}. \tag{4.29}$$

The parameters of a short coil can be readily evaluated if the surface current density is assumed to be of the form

$$J_\phi(\theta) = \begin{cases} J_\phi / \sin \theta & \text{for} \quad \theta_1 < \theta < (\pi - \theta_1) \\ 0 & \text{elsewhere.} \end{cases} \tag{4.30}$$

The pitch of this coil

$$\frac{dz}{a d\phi} = -\tan \alpha \sin^2 \theta \tag{4.31}$$

is decreasing toward the poles of the sphere $(\theta = 0, \pi)$. However, for short coils of $\Delta = (\pi/2 - \theta_1) < 20°$, the winding of the coil (4.28) does not differ appreciably from the center portion of the winding (4.30). Calculations show that

$$f_1(\theta_1) = S_1(\theta_1) f(\Delta), \tag{4.32}$$

$$f_2(\theta_1) = \frac{1}{2}\left[\frac{\sin \Delta}{\Delta}\right]^2 f(\Delta), \tag{4.33}$$

$$f_3(\theta_1) = S_3(\theta_1) f(\Delta), \tag{4.34}$$

where

$$S_1(\theta_1) = \frac{1}{\Delta^2} \sum_n \frac{1}{n(n+1)} \left[P_n(\cos\theta_1) \right]^2, \tag{4.35}$$

$$S_3(\theta_1) = \frac{1}{\Delta^2} \sum_n \frac{1}{n(n+1)(2n+1)(2n+3)} \left[P_n(\cos\theta_1) \right]^2, \tag{4.36}$$

$$f(\Delta) = 4\Delta^2 \left[\log\left(\frac{1+\sin\Delta}{1-\sin\Delta} \right) \right]^{-2} \approx 1 - \frac{1}{3}\Delta^2 \quad \text{for } \Delta \leqslant 1, \tag{4.37}$$

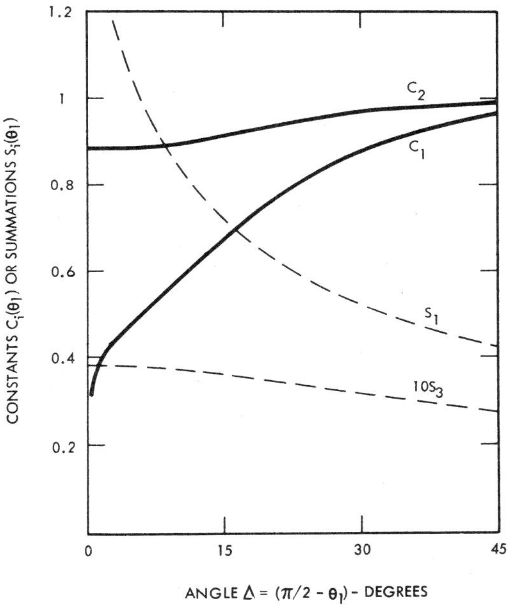

FIG. 4.1. Parameters for determining loop characteristics.

and where n may assume only values of odd positive integers. Plots of $S_1(\theta_1)$ and $S_3(\theta_1)$ are shown in Fig. 4.1. The length of wire in this coil is

$$l = 2\Delta a/\tan\alpha. \tag{4.38}$$

The radiation power factor of the coil (or the product of available bandwidth and efficiency) is computed from (4.25) and (4.26) as

$$p = \frac{R_r}{X} = \frac{1}{3}(k_2 a)^3 C_1(\theta_1). \tag{4.39}$$

The ratio of radiation to loss resistances is obtained from (4.26) and (4.27) as

$$r = \frac{R_r}{R_L} = 5\frac{\epsilon_0}{\epsilon_1} \frac{k_2 a}{\tan\delta_1} C_2(\theta_1). \tag{4.40}$$

This ratio is approximately equal to antenna efficiency, provided $R_r \ll R_L$. The constants $C_1(\theta_1)$ and $C_2(\theta_1)$ are plotted in Fig. 4.1 for $J_\phi(\theta)$ of (4.30), and are equal to unity for $J_\phi(\theta)$ of (4.28).

The above equations for loop antennas are valid only for a uniform current flow through the loop. Phase shifters may be required in larger loops to achieve this condition.

The spherically shaped dielectric cores are encountered infrequently in engineering applications. Hence, it is desirable to derive approximate expressions for antenna parameters that would also apply to other core shapes. The reactance X and the radiation resistance R_r of the coil are in the first approximation unaffected by the presence of the dielectric, and existing free-space formulas may be used for estimating X and R_r. An approximate calculation of the loss resistance R_L may be based on the assumption that the magnetic field inside the core is uniform and equal to the magnetic field that exists in the center of a short coil of radius a.

In a cylindrical coordinate system where the z axis is aligned with the axis of the loop, the magnetic field component H_z is given by (7.10.8) of Smythe [1950] as

$$H_z = \frac{IN}{2a}. \tag{4.41}$$

The associated electric field component E_ϕ is computed from Faraday's law

$$\int_0^{2\pi} E_\phi \rho \, \mathrm{d}\phi = -\frac{\mathrm{d}}{\mathrm{d}t} \int_0^{2\pi} \int_0^{\rho} B_z \rho \, \mathrm{d}\rho \, \mathrm{d}\phi \tag{4.42}$$

as

$$|E_\phi| = \frac{1}{2} \omega \mu_0 \rho H_z. \tag{4.43}$$

The power loss in the dielectric is given by the equation

$$P_L = \sigma_1 \int |E_\phi|^2 \rho \, \mathrm{d}\rho \, \mathrm{d}\phi \, \mathrm{d}z, \tag{4.44}$$

and the loss resistance R_L becomes

$$R_L = \frac{P_L}{I^2} = \sigma_1 \frac{\pi}{8} \left[\frac{\omega \mu_0 N}{a}\right]^2 \cdot \int_0^{\rho_1} \mathrm{d}\rho \, \rho^3 \int_{z_1(\rho)}^{z_2(\rho)} \mathrm{d}z, \tag{4.45}$$

where the limits of integration depend on the geometry of the core, and where I designates the RMS value of the current.

For a spherical dielectric core, (4.45) will be compared with the more accurate expression (4.27). For a sphere where

$$z_{1,2}(\rho) = \pm (a^2 - \rho^2)^{1/2},$$

the integrals of (4.45) can be readily evaluated to give

$$R_L = \frac{\pi\epsilon_1}{30\epsilon_0}\left(\frac{\mu_0}{\epsilon_0}\right)^{1/2} N^2(\beta_2 a)^3 \tan\delta_1. \tag{4.46}$$

The resistance R_L of (4.46) is approximately 10 percent less than R_L for $J_\phi(\theta)$ of (4.30) with short coils ($\Delta < 20°$), but is approximately twice the R_L for $J_\phi(\theta)$ of (4.28) which is achieved by a coil covering all of the sphere. The crude approximation of the magnetic field in (4.41) is therefore useful for obtaining an estimate of R_L for short coils and an upper bound of R_L for wide coils in a spherical geometry. The same general conclusion may be expected to apply to hemispherical, ellipsoidal, or other related core geometries.

4.2.2. *Dipole Antenna*

The dielectric-loaded dipole antenna has been treated in the biconical geometry of Fig. 4.2a by Galejs [1964]. However, for purposes of illustrating the differences with loop antennas, the simple cylindrical capacitor

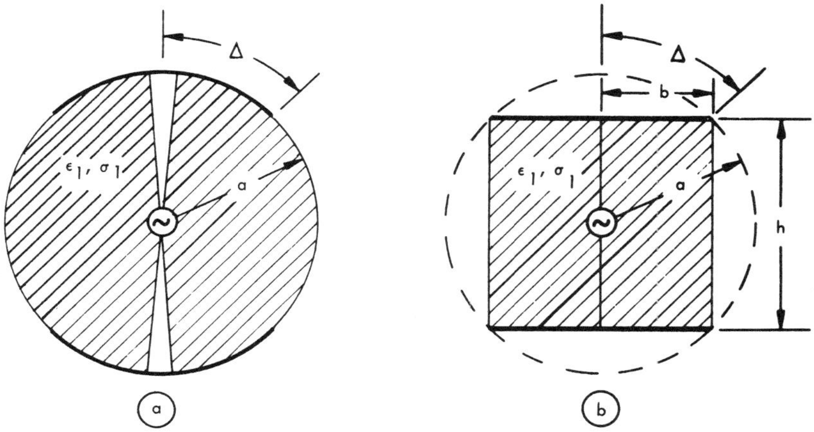

FIG. 4.2. Biconical and cylindrical dipole antennas.

shape of Fig. 4.2b is sufficient. In the absence of dielectric loading ($\epsilon_1 = \epsilon_0$), its capacity is

$$C = k_a \frac{\epsilon_0 \pi b^2}{h}, \tag{4.47}$$

where $b = a \sin\Delta$ and $h = 2a\cos\Delta$. The shape factor of the capacitor k_a is given by Vilbig [1960] as

$$k_a = 1 + \frac{h}{\pi b}\left[\log\frac{16\pi b}{h} + 1\right]. \tag{4.48}$$

With dielectric loading ($\epsilon_1 \neq \epsilon_0$), the capacity is approximated by

$$C = \left(k_a - 1 + \frac{\epsilon_1}{\epsilon_0}\right) \frac{\epsilon_0 \pi b^2}{h}. \tag{4.49}$$

The radiation conductance is independent of dielectric loading [Wheeler, 1947; Galejs, 1962], and is given by

$$G_r = \frac{\pi}{6}\left(\frac{\epsilon_0}{\mu_0}\right)^{1/2} k_a^2 (k_2 b)^4. \tag{4.50}$$

The power loss in the dielectric may be computed as

$$P_L = \sigma_1 \int |E|^2 \, dv = \frac{\sigma_1}{\epsilon_1} C_i V^2, \tag{4.51}$$

where V is the voltage applied to the capacitor, and the capacity C_i of the antenna due to the energy storage inside the dielectric is

$$C_i = \frac{\epsilon_1 \pi b^2}{h}. \tag{4.52}$$

These power losses may be accounted for by a loss conductance

$$G_L = \frac{P_L}{V^2} = \frac{\sigma_1 \pi b^2}{h}. \tag{4.53}$$

The radiation power factor of the dipole antenna is computed from (4.49) and (4.50) as

$$p = \frac{G_r}{\omega C} = \frac{1}{6} \frac{k_a^2}{k_a - 1 + (\epsilon_1/\epsilon_0)} \frac{h}{b} (k_2 b)^3. \tag{4.54}$$

The ratio of radiation conductance G_r to loss conductance G_L is obtained from (4.50) and (4.53) as

$$r = \frac{G_r}{G_L} = \frac{\epsilon_0}{6\epsilon_1} \frac{k_a^2}{\tan \delta_1} \frac{h}{b} (k_2 b)^3. \tag{4.55}$$

This ratio is the same as the ratio of radiation resistance R_r to loss resistance R_L, because $R_i = G_i(\omega C)^{-2}$ for $i = r$ or L.

4.2.3. *Comparison of the Two Antenna Types*

The loop antenna of a spherical dielectric core of radius a will be compared with a dipole antenna that fits in a sphere of same radius as indicated in Fig. 4.2b. The dielectric constant is the same for both antennas, but not the dielectric loss tangents. The parameters of the two antennas are distinguished by subscripts m (magnetic) and e (electric).

The radiation power factors of the two antennas are related from (4.39) and (4.54) as

$$\frac{p_m}{p_e} = \frac{2[k_a - 1 + (\epsilon_1/\epsilon_0)]C_1(\theta_1)a^3}{k_a^2 b^2 h}. \tag{4.56}$$

For $\Delta = 45°$ and $\epsilon_1/\epsilon_0 = 3$, $p_m/p_\epsilon = 1.2$ and 1.16 for $J_\phi(\theta)$ of (4.28) and (4.30); p_m is about the same as p_e. However, the ratio of (p_m/p_e) is decreased by decreasing ϵ_1/ϵ_0.

The ratio of the radiation to loss resistance ratio of the two antennas follows from (4.40) and (4.55) as

$$\frac{r_m}{r_e} = \frac{30C_2(\theta_1)a^3}{k_a^2 h b^2} \frac{\tan \delta_{1e}}{\tan \delta_{1m}(k_2 a)^2} = K_1 \frac{\tan \delta_{1e}}{\tan \delta_{1m}(k_2 a)^2}. \tag{4.57}$$

For $\Delta = 45°$, $K_1 = 3.15$ and 3.1 for $J_\phi(\theta)$ of (4.28) and (4.30). For equal efficiencies ($r_m = r_e$) and $k_2 a' \ll 1$, the magnetic antenna may have much higher losses ($\tan \delta_{1m} \gg \tan \delta_{1e}$). Assuming $k_2 a = 0.1$, $\tan \delta_{1m}$ may exceed 300 $\tan \delta_{1e}$. The requirements imposed on the insulating qualities of the dielectric are considerably less stringent in a loop-type antenna. This can be very significant in antenna design problems where the cost of a high quality dielectric may become prohibitive. This difference between the two antenna types may be explained qualitatively by examining the behavior of the electric field in the vicinity of the antenna. For equal amounts of radiated power, both antennas will exhibit equal radiation resistances R_r and approximately the same far fields. However, the electric field in the near zone of the electric dipole is $(k_2 r)^{-1}$ times higher than with the magnetic dipole ($r =$ distance to the center of the dipole). The losses that are proportional to $|E|^2$ will be $(k_2 a)^{-2}$ higher, which explains the functional form of (4.57).

The previous comparison is clearly in favor of the loop antenna. This is an obvious consequence of differences in the E field dependence between electric and magnetic dipoles.

4.3. Lossy Shells Surrounding Loops and Dipoles

The characteristics of small dipole antennas embedded in dielectric spheres have been determined in Section 4.1 by starting the derivations with dipole fields of homogeneous space, by introducing an additional solution in the finite dielectric region, and by matching the solutions at the dielectric boundary. An analogous development can be also carried out for a magnetic dipole antenna in a dielectric region. The analysis may proceed along similar lines for dipoles placed in the center of a lossy dielectric shell. However, it will be algebraically more involved because of an additional dielectric boundary. Raemer [1962] has presented a numerical method for calculating the relative transmission loss caused by a concentric shell of lossy dielectric (plasma) enclosing an elementary electric or magnetic dipole. More recently, Row [1964, 1965] has considered the same problem; he has derived several simple closed form expressions. His results are summarized in this section.

An infinitesimal electric or magnetic dipole is surrounded by a lossy dielectric shell of conductivity σ_2, outer radius b, and thickness d. Under conditions in which displacement currents in the dielectric shell are negligible ($\sigma_2 \gg \omega\epsilon_2$) and the shell is small in terms of free space wave length ($k_0 b \ll 1$), the radiation efficiencies η_e and η_m of the electric and magnetic dipole are given by simple expressions in several limiting cases.

The complex wave number of the dielectric shell is approximated by

$$k_2 \simeq k_0\left(\frac{i\sigma_2}{\omega\epsilon_0}\right)^{1/2} = (1+i)\left(\frac{\omega\mu_0\sigma_2}{2}\right)^{1/2} = \frac{1+i}{\delta}, \tag{4.58}$$

where δ is the skin depth, and $k_0 = \omega(\mu_0\epsilon_0)^{1/2}$. When the shell is small and thin compared to skin depth ($|k_2 b| \ll 1$, $|k_2 d| \ll 1$), the efficiencies η_j (defined as the ratio between the power radiated into the free space region outside the shell and the power dissipated in the shell itself) are computed as

$$\eta_e \simeq \frac{\omega\epsilon_0}{\sigma_2}\frac{b}{d}(k_0 b)^3, \tag{4.59}$$

$$\eta_m \simeq \frac{\omega\epsilon_0}{\sigma_2}\frac{b}{d}(k_0 b). \tag{4.60}$$

The same expressions of η_e and η_m also apply when the shell is large and thin compared to the skindepth ($|k_2 b| \gg 1$, $|k_2 d| \ll 1$). When the shell is large and thick compared to the skin depth ($k_2 b \gg 1$, $|k_2 d| \gg 1$),

$$\eta_e \simeq \left(\frac{2\omega\epsilon_0}{\sigma_2}\right)^{1/2} 4(k_0 b)^4 \exp(-2\mathrm{Re}\,k_2 d), \tag{4.61}$$

$$\eta_m \simeq \left(\frac{2\omega\epsilon_0}{\sigma_2}\right)^{1/2} 4(k_0 b)^2 \exp(-2\mathrm{Re}\,k_2 d). \tag{4.62}$$

In all of the above cases

$$\frac{\eta_m}{\eta_e} = (k_0 b)^{-2}. \tag{4.63}$$

Since $k_0 b \ll 1$, this formula shows that for fixed radiated power, the power dissipated in the shell about an electric dipole is much larger than for the magnetic dipole. Furthermore, this result is essentially independent of the shell thickness, and for a given radiation efficiency, it is possible to tolerate a much higher conductivity dielectric in the near zone of a small magnetic-type antenna than with an electric type. The relative efficiencies of dipoles surrounded by a lossy dielectric shell (4.63) are the same as the relative efficiencies of small loops and electric dipoles in solid spheres of a lossy dielectric (4.57). In both cases the electric dipole has higher ohmic losses because the electric field near an electric dipole is proportional to r^{-3}, while the field is proportional to r^{-2} for magnetic dipole antennas.

Row [1965] has also considered a finite loop antenna in a conducting spherical shell by extending the method used by Wait [1957] who analyzed an insulated loop immersed in an infinite conducting medium. A loop of radius a is placed inside a spherical shell of lossy dielectric material of inner radius b and thickness d. The spherical shell is surrounded by free space on the outside. The fields of the above dielectric regions are expressed in terms of spherical wave functions. Expressions have been derived for the radiation resistance of the loop and for the impedance change of the loop due to placing it in the shell. It is shown that the minimum impedance change for a given shell and loop size is experienced when the loop is placed in the equatorial plane of the shell. The subsequent equations refer to this condition. The displacement currents are considered to be negligible within the conducting shell ($\sigma_2 \gg \omega\epsilon_2$), and as a consequence the complex wave number in the dielectric shell is given by (4.58).

For a cavity that is small relative to skindepth ($|k_2 b| \gg 1$), and for a thick shell ($d \gg b$), the change of loop impedance is

$$\Delta Z \simeq \frac{\sigma_2(\omega\mu_0)^2(\pi a^2)^2}{6\pi b}\cdot\left[1+\frac{9}{280}\left(\frac{a}{b}\right)^4+\frac{5}{704}\left(\frac{a}{b}\right)^8+\dots\right] \tag{4.64}$$

This result does not depend on the thickness of the shell d, and is the same as for spherical cavities in an infinite conducting medium [Wait and Spies, 1964]. For thick shells, the radiation resistance is negligible and the antenna efficiency approaches zero.

For a thin shell, $|k_2 d| \ll 1$, the radiation resistance is in the first approximation unaffected by the presence of the shell and is given by

$$R_{\text{rad}} \simeq \frac{\pi}{6}\left(\frac{\mu_0}{\epsilon_0}\right)^{1/2}(k_0 a)^4\left[1+\frac{(k_0 a)^4}{1400}+\dots\right]. \tag{4.65}$$

The impedance change is computed as

$$\Delta Z \simeq -i\frac{\pi}{6}\left(\frac{\mu_0}{\epsilon_0}\right)^{1/2}(k_2 a)^2(k_0 a)\frac{ad}{b^2}\cdot\left[1+\frac{9}{56}\left(\frac{a}{b}\right)^4+\frac{45}{704}\left(\frac{a}{b}\right)^8+\dots\right]. \tag{4.66}$$

For small loops, the power radiated by the loop will be small relative to the power dissipated in the dielectric shell. Therefore the radiation efficiency can be computed as a ratio between the radiation resistance and the real part of the incremental impedance ΔZ. The radiation efficiency is computed from equation (4.65) and (4.66) as

$$\eta \simeq \frac{\omega\epsilon_0}{\sigma_2}(k_0 b)\frac{b}{d}\left[\frac{1+(k_0 a)^4/1400+\ldots}{1+\frac{9}{56}\left(\frac{a}{b}\right)^4+\ldots}\right]. \tag{4.67}$$

The leading term of (4.67) can be seen to be the same as in the radiation efficiency of an elementary electric dipole (4.60).

These calculations are based on an assumed uniform loop current. The first order effects of a non-uniform current are considered in Chapter 10 for loops in a stratified dielectric. Loops in finite layers of lossy dielectric are discussed in Section 10.3.

4.4. Exercises

4.1. Derive the reactance X, the radiation resistance R_r and the loss resistance R_r of the spherical loop antenna of (4.25) to (4.27).

4.2. Derive the efficiency expressions (4.59) and (4.60) for dipoles surrounded by lossy shells.

4.3. Derive the impedance changes (4.64) and (4.66) for loops surrounded by lossy shells.

4.5. References

GALEJS, J. (1962) Dielectric loading of electric dipole antennas, *NBS J. Res. (Radio Prop.)* **66D** (5), 557–62.

GALEJS, J. (1963) Small electric and magnetic antennas with cores of a lossy dielectric, *NBS J. Res. (Radio Prop.)* **67D** (4), 445–51.

GALEJS, J. (1964) Capacitor type biconical antenna, *NBS J. Res. (Radio Science)* **68D** (2), 165–72.

KELLER, J. P. and KELLER, M. B. (1950) A point dipole in spherically symmetric media, *Math. Rev.* **12** (3), 224. Also NYU Mathematics Research Group, Research Rep., No. EM-16.

KRAICHMAN, M. B. (1962) Impedance of a circular loop in an infinite conducting medium, *NBS J. Res. (Radio Prop.)* **66D** (4), 499–503.

POLK, C. (1959) Resonance and supergain effects in small ferromagnetically or dielectrically loaded biconical antennas, *Proc. Symposium on Electromagnetic Theory,* the University of Toronto, Ontario, Canada, June 15–20 (published as a special supplement to the *IRE Trans. on Ant. and Prop.* **AP-7**, S4-14-423.

RAEMER, H. R. (1962) Radiation form linear electric and magnetic antennas surrounded by a spherical plasma shell, *IRE Trans. on Ant. and Prop.* **AP-10** (1), 69–78.

ROW, R. V. (1964) Radiation efficiency of electric and magnetic dipole antennas surrounded by a small spherical shell of lossy dielectric, *IEEE Trans. on Ant. and Prop.* **AP-12** (5), 646–7.

ROW, R. V. (1965) Insulated loop antenna in a conducting spherical shell, *IEEE Trans. on Ant. and Prop.* **AP-13** (2), 216–18.

SMYTHE, W. R. (1950) *Static Dynamic Electricity,* McGraw-Hill, New York, NY.

SOMMERFIELD, A. (1949) *Partial Differential Equations in Physics,* Academic Press, New York, NY.

TAI, C. T. (1948) On the theory of biconical antennas, *J. Appl. Phys.* **19** (12) 1155–60.

TAI, C. T. (1949) Application of a variational principle to biconical antennas, *J. Appl. Phys.* **20** (11), 1076–84.

VILBIG, F. (1960) *Lehrbuch der Hockfrequenztechnik,* 1, Akademische Verlagsges, Frankfurt am Main, 147.

WAIT, J. R.(1952) The magnetic dipole antenna immersed in a conducting medium, *Proc. IRE* **40** (10), 1244–5.

WAIT, J. R. (1957) Insulated loop antenna immersed in a conducting medium, NBS *J. Res.* **59**, 133–7.

WAIT, J. R. and SPIES, K. P. (1964) A note on the insulated loop antenna immersed in a conducting medium, *NBS J. Res., Radio Science* **68D**, 1249–50.

WAIT, J. R. (1967) Electromagnetic radiation from conical structures, in *Antenna Theory*, ed. R. E. Collin and F. J. Zucker, McGraw-Hill, New York, NY.

WHEELER, H. A. (1947) Fundamental limitations of small antennas, *Proc. IRE* **35** (12), 1479–84.

WHEELER, H. A. (1958) The spherical coil as an inductor shield or antenna, *Proc. IRE* **46** (9), 1595–1602.

CHAPTER 5

TRANSMISSION LINE THEORY OF BURIED INSULATED ANTENNAS

ANTENNA impedance can be calculated only after determining the current distribution along the antenna, which requires the solution of an integral equation following Chapter 2. This procedure is not necessary for insulated antennas near a highly conducting medium where transmission line theory can be used to obtain estimates of current distribution and of antenna impedance. The early travelling wave antennas were constructed by stringing wire between telephone poles [Bevarage *et al.*, 1932; Bailey *et al.*, 1928], and it has been quite natural to apply transmission line theory to such antennas. Also, buried antennas can be treated using transmission line theory: such calculations have been reported among others by Moore [1951], Ghose [1960], Guy and Hasserjian [1963], and King [1964]. To a large extent, this work is based on the theory of the coaxial line with imperfectly conducting walls [Stratton, 1941]. A number of further topics are considered by Wait [1967]. He discusses elementary sources and long wires in the presence of stratified lossy media and lists a very comprehensive bibliography.

In this chapter the transmission line theory is applied to the calculation of single-antenna self-impedance and of mutual impedances between two antennas. The surface excitation by a buried vertical line is considered as a further example. Such transmission line considerations become inaccurate for short antennas and for media of low losses when the conduction currents become comparable in magnitude to the displacement currents. These limitations of the transmission line theory are discussed in more detail in Chapter 9.

5.1. Impedance of the Buried Insulated Wire

The characteristics of an insulated wire will be considered first for large burial depth. The analysis will be carried out for a coaxial line whose outer conductor is infinite in extent. The radius of the inner conductor is designated by 'a' and the inner radius of the outer conductor is b. Region 1 is the inner conductor of high, but finite conductivity, Region 2 is a low loss dielectric, and Region 3 is finitely conducting. The current I of the coaxial lines center conductor does not exhibit any variations in the ϕ direction and its fields are similarly symmetric. Such fields consist of symmetric TM modes, which can be expressed in terms of cylindrical wave functions [Stratton, 1941]. The total current of the inner conductor is assumed to have the form $I = I_0 \exp(\gamma z)$, and the magnetic field components $H_{\phi j}$ of the three dielectric regions ($j = 1, 2$ or 3) are expressed as

$$H_{\phi 1} = \frac{I}{2\pi a} \frac{J_1(\lambda_1 \rho)}{J_1(\lambda_1 a)}, \tag{5.1}$$

$$H_{\phi 2} = -\frac{ik_2^2}{\omega \mu_2 \lambda_2}[AJ_1(\lambda_2 \rho) + BY_1(\lambda_2 \rho)], \tag{5.2}$$

$$H_{\phi 3} = -\frac{ik_3^2}{\omega \mu_3 \lambda_3} CH_1^{(1)}(\lambda_3 \rho), \tag{5.3}$$

where $J_n(x)$ and $Y_n(x)$ are the Bessel functions of the first and second kind of order n; $H_n^{(1)}(x)$ is the Hankel function of the first kind of order n; $k_j^2 = \omega^2 \mu_j \epsilon_j + i\omega \mu_j \sigma_j$; and $\lambda_j^2 = k_j^2 + \gamma^2$. The constants A, B, and C are determined by requiring $H_{\phi j}$ and

$$E_{zj} = \frac{1}{(-i\omega\epsilon_j + \sigma_j)} \frac{1}{\rho} \frac{\partial}{\partial \rho} (\rho H_{\phi j}) \tag{5.4}$$

to be continuous across the dielectric boundaries at $\rho = a$ and b. It follows that

$$-\frac{A}{B} = \frac{Y_0(\lambda_2 a) - \dfrac{\lambda_1 k_2^2}{\lambda_2 k_1^2} \dfrac{J_0(\lambda_1 a)}{J_1(\lambda_1 a)} Y_1(\lambda_2 a)}{J_0(\lambda_2 a) - \dfrac{\lambda_1 k_2^2}{\lambda_2 k_1^2} \dfrac{J_0(\lambda_1 a)}{J_1(\lambda_1 a)} J_1(\lambda_2 a)}$$

$$= \frac{Y_0(\lambda_2 b) - \dfrac{\lambda_3 k_2^2}{\lambda_2 k_3^2} \dfrac{H_0^{(1)}(\lambda_3 b)}{H_1^{(1)}(\lambda_3 b)} Y_1(\lambda_2 b)}{J_0(\lambda_2 b) - \dfrac{\lambda_3 k_2^2}{\lambda_2 k_3^2} \dfrac{H_0^{(1)}(\lambda_3 b)}{H_1^{(1)}(\lambda_3 b)} J_1(\lambda_2 b)} \tag{5.5}$$

The parameters λ_1 and λ_3 are approximated by k_1 and k_3 respectively, because γ is expected to be of the same magnitude as k_2, which is much less than either k_1 or k_3. The Bessel functions of (5.5) may be approximated noting that $\lambda_1 a > 1$, $\lambda_2 \rho \ll 1$, and $\lambda_3 b \ll 1$. It follows that the denominators of (5.5) are equal to unity. Equating both numerators results in

$$\log a + \frac{k_2^2}{\lambda_2^2} \frac{J_0(k_1 a)}{k_1 a J_1(k_1 a)} = \log b + \frac{k_2^2}{\lambda_2^2} \frac{H_0^{(1)}(k_3 b)}{k_3 b H_1^{(1)}(k_3 b)}. \tag{5.6}$$

Solving (5.6) for $\lambda^2_2 = k^2_2+\gamma^2$ gives

$$\gamma = ik_2\left\{1-\frac{1}{\log(b/a)}\left[\frac{J_0(k_1a)}{k_1aJ_1(k_1a)}-\frac{H_0^{(1)}(k_3b)}{k_3bH_1^{(1)}(k_3b)}\right]\right\}^{1/2} \tag{5.7}$$

which is approximated for $|k_1a| \gg 1$ and $|k_3b| \ll 1$ by

$$\gamma = ik_2\left\{1+\frac{1}{\log(b/a)}\left[\frac{(i)^{1/2}}{a(\omega\mu\sigma_1)^{1/2}}+i\frac{\pi}{4}+0.1159-\log(b(\omega\mu\sigma_3)^{1/2})\right]\right\}^{1/2}. \tag{5.8}$$

The shunt admittance per unit length of the wire is computed from its capacity as

$$y_s = -2\pi i\omega\epsilon_2/\log(b/a) \tag{5.9}$$

for $|i\omega\epsilon_2| \gg \sigma_2$. For transmission lines the shunt admittance y_s and the series impedance per unit length z_s are related to the propagation constant γ

$$\gamma = (y_sz_s)^{1/2}. \tag{5.10}$$

The series impedance z_s is computed in the limit of $\sigma_1 \to \infty$ by inserting (5.8) and (5.9) as

$$z_s = R-i\omega L = \omega\mu_0\left\{\frac{1}{8}-\frac{i}{2\pi}\cdot[0.1159-\log(a(\omega\mu_0\sigma_3)^{1/2})]\right\}. \tag{5.11}$$

The characteristic impedance of the line is given by

$$Z_0 = \left(\frac{z_s}{y_s}\right)^{1/2} = -\frac{z_s}{\gamma} = -\frac{\gamma}{y_s}, \tag{5.12}$$

where (5.8), (5.9), and (5.11) may be used for γ, y_s, and z_s respectively.

The impedance Z_g of a transmission line measured at the generator is related to its terminating impedance Z_t by

$$Z_g = Z_0\frac{Z_t+Z_0\tanh(-\gamma l)}{Z_0+Z_t\tanh(-\gamma l)}, \tag{5.13}$$

where l is the length. The open circuit impedance ($Z_t = \infty$) is

$$Z_g = \frac{Z_0}{\tanh(-\gamma l)}. \tag{5.14}$$

The short circuit impedance ($Z_t = 0$) is

$$Z_g = Z_0\tanh(-\gamma l). \tag{5.15}$$

The currents I and voltages V at the two ends of the transmission line are related as

$$\frac{I_t}{I_g} = \left[\cosh(-\gamma l) + \frac{Z_t}{Z_0}\sinh(-\gamma l)\right]^{-1}, \tag{5.16}$$

$$\frac{V_t}{V_g} = \left[\cosh(-\gamma l) + \frac{Z_0}{Z_t}\sinh(-\gamma l)\right]^{-1}. \tag{5.17}$$

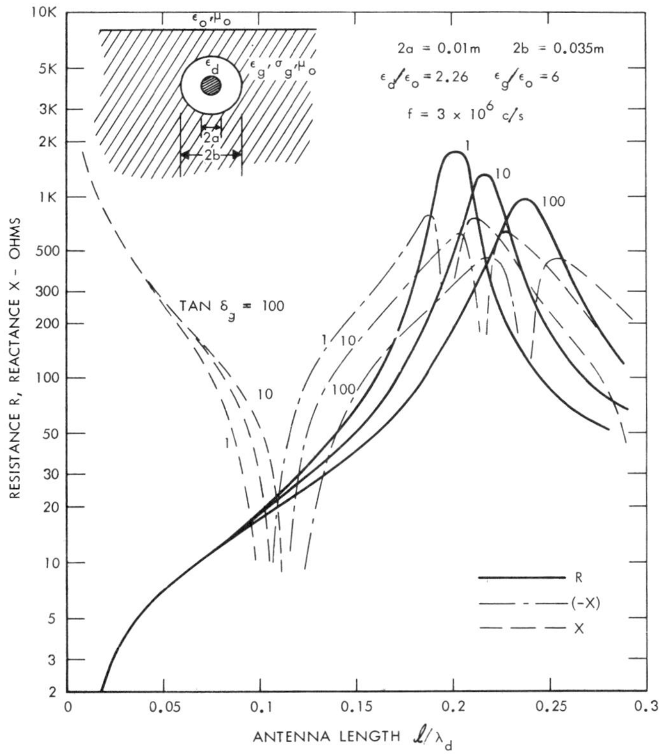

FIG. 5.1. Impedance of a buried antenna $Z = R + iX$. Transmission line theory.

For an open-circuited line $Z_t = \infty$ and $I_t = 0$; for a short-circuited line $Z_t = 0$ and $V_t = 0$.

Guy and Hasserjian [1963] have presented curves for computing the propagation constants γ and the input impedance of an antenna with either shorted or insulated ends. They have also considered the effects of the finite burial depth of the antenna; they point out that the transmission properties for a single buried insulated wire close to the interface are nearly identical to those for a buried wire at large depth. However the attenuation constant varies for intermediate depths and reaches a maximum at a depth approximately equal to 0.3 skin depths.

This similarity between wire characteristics at the surface and at large depths can be established very simply by examining the series impedance per unit length of the wire z_s [Carson, 1929]. An insulated wire that lies on or near the surface of the ground exhibits a series impedance z_s identical to (5.11), except that the term 0.1159 is replaced by 0.6159 [Carson, 1926; Sunde, 1949]. (A similar change in z_s would be produced by decreasing r_3 by a factor of $e = 2.718\ldots$). Even for small burial depths, the shunt admittance y_s remains the same as in a homogeneous lossy medium; $\gamma = (y_s z_s)^{1/2}$ will exhibit only a small change in its imaginary part, and a small change will occur in phase velocity as the burial depth of the wire is changed from infinity to zero.

An example of the impedance characteristics of the buried insulated wire is shown in Fig. 5.1. The antenna is buried near the interface and its ends are insulated from the surrounding medium. The impedance of a similar antenna has been calculated in Chapter 9 using more accurate variational techniques. The limitations of the transmission line approximations will be discussed in Chapter 9.

5.2. Mutual Impedance of Two Buried Antennas

Ths self-impedance Z_{11} of one antenna in the presence of the second antenna and the mutual impedance Z_{12} of the two antennas will be computed after first determining the driving point impedances of the antennas for symmetrical (Z_s) and anti-symmetrical excitation (Z_a) following King and Harrison [King, 1956].

For two identical center-excited antennas, the voltages and currents at the driving points are related by the circuit relations

$$\begin{aligned} V_1 &= Z_{11} I_1(0) + Z_{12} I_2(0), \\ V_2 &= Z_{12} I_1(0) + Z_{11} I_2(0), \end{aligned} \tag{5.18}$$

where the currents (and voltages) are considered at $x = 0$. For symmetrical antenna excitation $I_1(0) = I_2(0)$, the driving point impedance of the antenna is defined as

$$Z_s = \frac{V_j}{I_j(0)} = Z_{11} + Z_{12}. \tag{5.19}$$

For antisymmetrical excitation $I_1(0) = -I_2(0)$, the antenna impedance is

$$Z_a = \frac{V_j}{I_j(0)} = Z_{11} - Z_{12}. \tag{5.20}$$

After determining Z_s and Z_a, the self-impedance Z_{11} and the mutual impedance Z_{12} may be computed as

$$Z_{11} = \frac{1}{2}(Z_s + Z_a), \tag{5.21}$$

and

$$Z_{12} = \frac{1}{2}(Z_s - Z_a). \tag{5.22}$$

The above impedance relations will be applied to the calculation of mutual impedance between two cylindrical wires of radii a with a cylindrical insulating layer of radius b and permittivity ϵ_2 in an infinite lossy medium of permittivity ϵ_3 and conductivity σ_3. For symmetrical and antisymmetrical antenna excitations, the effective distributed series impedance of one of the antennas can be written by analogy with (5.19) and (5.20) as $z_{11} \pm z_{12}$, where z_{11} and z_{12} are the distributed self and mutual series impedances per unit length of the wire, respectively. The effective distributed shunt admittance y_{11} is in the first approximation the same as for a single antenna in (5.9). The complex propagation constant is computed for the two excitations as

$$\gamma_{s,a} = [(z_{11} \pm z_{12})y_{11}]^{1/2} = \gamma[1 \pm z_{12}/z_{11}]^{1/2}, \tag{5.23}$$

where $\gamma = (z_{11}y_{11})^{1/2}$ is the propagation constant along a single antenna. The upper sign (+) in the right-hand-side of (5.23) is used with a symmetrical excitation (subscript s), the lower sign (−) with an antisymmetrical excitation (subscript a). z_{11} is computed from (5.10) as

$$z_{11} = -k^2/y_{11}. \tag{5.24}$$

The distributed mutual impedance z_{12} is defined as the negative of the axial electric field at wire 2 due to unity current in wire 1, i.e.

$$z_{12} = -\frac{E_{z3}(h)}{I_1}, \tag{5.25}$$

where h denotes the distance between the wires.

$E_{z3}(h)$ is computed by substituting (5.3) in (5.4) and by noting that $2\pi b H_{\phi 3}(b) = I_1$. It follows that

$$C = \frac{i\omega\mu_0 I_1}{2\pi k_3 b H_1^{(1)}(k_3 b)}, \tag{5.26}$$

and z_{12} is obtained as

$$z_{12} = -\frac{i\omega\mu_0 H^{(1)}(k_3 h)}{2\pi k_3 b H_1^{(1)}(k_3 b)}. \tag{5.27}$$

Substituting (5.9), (5.24), and (5.27) in (5.23)

$$\gamma_{s,a} = ik_2\left[1 + \frac{H_0^{(1)}(k_3 b) \pm H_0^{(1)}(k_3 h)}{\log(b/a)k_3 b H_1^{(1)}(k_3 b)}\right]^{1/2}. \tag{5.28}$$

For large h, γ_s and γ_a approach γ of (5.7). For $h \to b$, $|\gamma_s|$ will be increased, while $\gamma_a \to ik_2$.

The input impedance of an insulated antenna of total length $2l$ is

$$Z_{s,a} = 2Z_{os,a} \coth(-\gamma_{s,a} l), \tag{5.29}$$

where the characteristic impedance of the line is

$$Z_{os,a} = -\gamma_{s,a}/y_{11}. \tag{5.30}$$

After computing Z_s and Z_a, Z_{11} and Z_{12} are obtained from (5.21) and (5.22).

The mutual impedances of buried insulated wires with insulated ends of the antennas are illustrated in Fig. 5.2. These impedance data will be compared with more accurate variational calculations in Chapter 9.

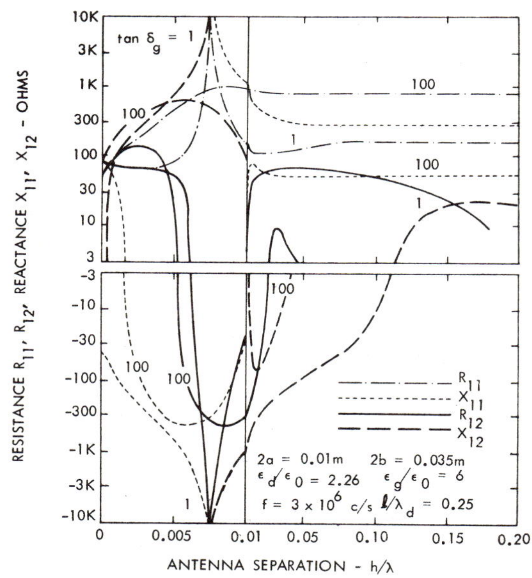

FIG. 5.2. Impedances of insulated buried antennas. Transmission line theory for infinite burial depth.

5.3. Surface Excitation by a Buried Vertical Line

The fields of subsurface sources are usually determined from field expressions of elementary dipoles that are obtained after evaluating Sommerfeld type integrals [Sommerfeld, 1949; Norton, 1936, 1937; Wait,

1961; Baños, 1966]. However, the fields in the lossless medium are also uniquely determined from the distribution of the tangential electric fields on the interface of the two media. These surface fields can be estimated from transmission line theory when the buried wires are long relative to the skin depth $\delta = (2/\omega\mu_0\sigma_3)^{1/2}$ of the lossy medium. Such calculations are given in this Section.

A wire of radius a is located in the center of a vertical cylindrical hole of radius b and of length l. The wire is excited at the bottom of the hole and may be either open-circuited [Fig. 5.3a] or short-circuited [Fig. 5.3b]. The radius of the hole b is small relative to skin depth δ: suitable approximations are introduced in (5.2), (5.3), (5.5), and (5.26). It follows that the field components $H_{\phi 2}$ and $H_{\phi 3}$ of regions $2(a \leqslant r \leqslant b)$ and $3(b \leqslant r)$ are given by

$$H_{\phi 2} = \frac{I(z)}{2\pi\rho}, \tag{5.31}$$

$$H_{\phi 3} = \frac{ik_3 I(z)}{4} H^{(1)}(k_3\rho), \tag{5.32}$$

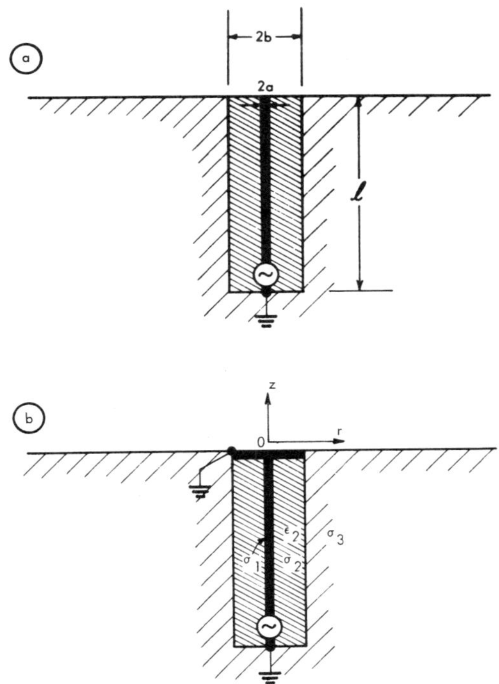

FIG. 5.3. Geometry of the vertical antenna.

where $I(z) = I_0[\exp(\gamma z) + R \exp(-\gamma z)]$, $H_n^{(1)}(x)$ is the Hankel function of first kind of order n, $k_j^2 = \omega^2\mu_0\sigma\epsilon_j + i\omega\mu_0\sigma_j$, and where γ is given by (5.8). The electric field components are determined from the curl relation

$$\mathbf{E}_j = \frac{1}{\sigma_j - i\omega\epsilon_j}\left[-\mathbf{i}_\rho\frac{\partial}{\partial z}H_{\phi j} + \frac{\mathbf{i}_z}{\rho}\frac{\partial}{\partial\rho}(\rho H_{\phi j})\right], \tag{5.33}$$

where $\mathbf{i}_r$ and $\mathbf{i}_z$ are unit vectors. The far-field component E_z for $z = 0$ is related to the surface excitation $E_\rho(\rho', \phi, 0) = E(\rho)$ from (5.33) and (3.28) by applying the condition $\rho \gg \rho'$ as

$$E_z = \frac{k_0 \exp(ik_0\rho)}{\rho}\int_0^\infty E(\rho')J_1(k_0\rho')\rho'\,\mathrm{d}\rho', \tag{5.34}$$

where $J_n(x)$ is a Bessel function of first kind of order n and $k_0 = \omega(\mu_0\epsilon_0)^{1/2}$. The lower limit of the integral may be replaced by b for $k_0 b \ll 1$.

For an open-circuited transmission line $R = -1$ and (5.32) and (5.33) give

$$E(\rho') = \frac{i\omega\mu_0\gamma}{2k_3}I_0 H_1^{(1)}(k_3\rho'). \tag{5.35}$$

The current amplitude I_0 is related to the current distribution along the line $I(z)$ for $|\gamma| \ll |k_3|$ and $|k_3 l|$ or $l/\delta \gg 1$ by

$$I_0 = \frac{k_3^2}{2\gamma}\int I(z)\,\mathrm{e}^{-ik_3 z}\,\mathrm{d}z. \tag{5.36}$$

Substituting (5.35) in (5.34), carrying out the integration indicated in (5.34), and applying (5.36) it follows that

$$E_z = \frac{i}{2\pi}\left(\frac{\mu_0}{\epsilon_0}\right)^{1/2}\frac{k_0}{n^2\rho}\,\mathrm{e}^{ik_0\rho}\int I(z)\,\mathrm{e}^{-ik_3 z}\,\mathrm{d}z, \tag{5.37}$$

where

$$n = k_3/k_0 = (i\sigma_3/\omega\epsilon_0)^{1/2}.$$

The subsurface fields $E_z(z)$ of an elementary vertical dipole located above the surface at $z' = 0$ are given by

$$E_z(z) = \frac{i}{2\pi}\left(\frac{\mu_0}{\epsilon_0}\right)^{1/2}\frac{k_0}{n^2\rho}\,\mathrm{e}^{ik_0\rho}\,I(z')e^{ik_3|z-z'|}\,\mathrm{d}z'. \tag{5.38}$$

Because of reciprocity the same expression applies to fields at z' due to a dipole at z. Interchanging the coordinates z' and z, integration of the current distribution gives (5.37).

For an ideal short circuit of the transmission line (a screen of $\sigma = \infty$ covers the whole plane of $z = 0$) $R = 1$ and $E_\rho(\rho', \phi, 0) = E(\rho') = 0$. For a nonideal short circuit that covers only the vicinity of the cable the currents extending radially from the ground system will generate a nonzero E_ρ component, the magnitude of which can be estimated. The axial current distribution $i_z = \sigma E_z$, which follows from the transmission line equations (5.32) and (5.33) for the $z = 0$ plane with $R = 1$, is assumed to give rise to a radial current I_r within the upper layer of the ground surface. Applying the equation of continuity and (5.33)

$$I_\rho = -2\pi\sigma \int_\rho^\infty E_z \rho \, d\rho = -2\pi \int_\rho^\infty \frac{\partial}{\partial \rho}(\rho H_\phi) \, d\rho = 2\pi\rho H_\phi \tag{5.39}$$

The radial electric field caused by I_ρ near the surface of the ground is computed for an exponential depth attentuation of the linear current density $J_\rho(z) = J_\rho(0) \exp(-ik_3 z)$. Thus,

$$I_\rho = \int_{-\infty}^{0} J_\rho(z) \, dz = -J_\rho(0)/(ik_3).$$

The series impedance of a ground annulus of height Δz, thickness $\Delta\rho$, and an average radius ρ is $Z = \Delta\rho/(2\pi\rho\sigma\Delta z)$. Hence, the electric field near the ground surface can be computed as

$$E(\rho) = \frac{Z}{\Delta\rho} J_\rho(0)\Delta z = -\frac{ik_3 I_\rho}{2\pi\rho\sigma} = -\left(\frac{\omega\mu_0}{i\sigma_3}\right)^{1/2} H_\phi. \tag{5.40}$$

This equation has been used for estimating ground losses in the vicinity of antennas by Wait and Pope [1954]. It is valid for a radial current flow which is exponentially attentuated with the depth, and it becomes inaccurate in the direct vicinity of grounding electrodes. Substituting (5.40) and (5.32) with $R = 1$ in (5.34), an integration gives

$$E_z = -\frac{I_0}{\pi}\left(\frac{\mu_0}{\epsilon_0}\right)^{1/2} \frac{\exp(ik_0\rho)}{\rho n^3} \tag{5.41}$$

For a short-circuited transmission line, $|k| \ll |k_3|$ and $|k_3 l|$ or $l/\delta \gg 1$, I_0 is related to $I(z)$ as

$$I_0 = -\frac{ik_3}{2} \int I(z) \, e^{-ik_3 z} \, dz. \tag{5.42}$$

After substituting (5.42), (5.41) is seen to be identical with the buried dipole expression (5.37).

The previous calculation has shown that fields of buried antenna elements can be determined after computing the surface excitation E_ρ near the antenna. This is a consequence of the equivalence principle that states that fields of a source-free region are uniquely determined by the tangential electric field distribution on its boundary surface [Harrington, 1961]. However, the present calculations show that relatively crude field estimates like the ones used in (5.35) or (5.40) are quite adequate for practical calculations. The radial electric field E_ρ and the magnetic field component H_ϕ exhibit the same radial variation $[\sim H_1^{(1)}(k_3\rho)]$ at the surface as in a great depth. There is no need to consider a perturbation of this radial variation by the lossless half space, at least for purposes of computing the radiated fields. This is contrary to propagation along thin wires (Section 6.1.) or narrow slots (Section 6.2.1) located in the interface of two media, and where the presence of a lossless half space decreases the effective propagation constant of the lossy medium by a factor of $(2)^{-1/2}$.

5.4. References

BAILEY, A., DEAN, S. W. and WINTRINGHAM, W. T. (1928) Receiving system for long wave trans-Atlantic radio telephony, *Proc. IRE* **16**, 1645–1705.

BAÑOS, ALFREDO (1966) *Dipole Radiation in the Presence of a Conducting Half-Space*, Pergamon Press, Oxford.

BEVARAGE, H. H., RICE, C. W. and KELLOGG, E. W., (1923) The wave antenna, *AIEE Trans.* **42**, 215–66.

CARSON, J. R. (1926) Wave propagation in overhead wires with ground return, *BSTJ* **5**, 539–54.

CARSON, J. R. (1929) Ground return impedance of underground wire with earth return, *Bell Sys. Tech.* **8**, 94–98.

GALEJS, J. (1965) Surface excitation by a buried vertical line, *IEEE Trans. on Ant. and Prop.* **AP–13** (4), 645–6.

GHOSE, R. N. (1960) The radiator-to-medium coupling in an underground communication system, *Proc. Nat. Elec. Conf.* **16**, 279–89.

GUY, A. W. and HASSERJIAN, G. (1963), Impedance properties of large subsurface antenna arrays, *IEEE Trans. on Ant. and Prop.* **AP–11**, (3), 232–40.

HARRINGTON, R. F. (1961) *Time-harmonic Electromagnetic Fields*, McGraw-Hill, New York, NY.

KING, R. W. P. (1956) *Theory of Linear Antennas*, Harvard University Press, Cambridge, Mass.

KING, R. W. P. (1964) Theory of the terminated insulated antenna in a conducting medium, *IEEE Trans. on Ant. and Prop.* **AP–12**, 305–18.

MOORE, R. K. (1951) Theory of radio communication between submerged submarines, Ph.D. Thesis, Cornell University, Ithaca, New York.

NORTON, K. A. (1936; 1937) The propagation of radio waves over the surface of the earth and in the upper atmosphere, Part I, *Proc. IRE* **24**, 1367–87 (October 1936); Part II, *Proc. IRE* **25**, 1203–36 (September 1937).

SOMMERFELD, A. (1949) *Partial Differential Equations in Physics*, Academic Press, New York, NY, ch. VI.

STRATTON, J. A. (1941) *Electromagnetic Theory*, McGraw-Hill, New York, NY, sections 9.19 and 9.21.

SUNDE, E. D. (1949) *Earth Conduction Effects in Transmission Systems*, D. VAN NOSTRAND, New York, NY.

WAIT, J. R. (1961) The electromagnetic fields of a horizontal dipole in the presence of a conducting half-space, *Can. J. Phys.* **39**, 1017–28.

WAIT, J. R. and POPE, W. A. (1954) The characteristics of a vertical antenna with a radial conductor ground system, *Appl. Sci. Res.* **B4**, 177–95.

WAIT, J. R. (1967) Electromagnetic fields of sources in lossy media, *Antenna Theory*, ed. by R. E. Collin and F. J. Zucker, McGraw-Hill, New York, NY.

CHAPTER 6

ANTENNAS IN THE INTERFACE BETWEEN TWO MEDIA

THIN-WIRE antennas and long slots can be located in the interface of two media. A wire at the interface between air and lossy ground is considered as one such example; the propagation constant of its current can be readily estimated using the transmission line relations of Section 5.1. Long slots in the interface between air and ground are considered as a second example. For very narrow slots, an iterative procedure, similar to Hallen's method of determining current distribution along linear antenna, is used for obtaining an approximate field distribution along the slots. For wide slots, the integral equation for the electric field distribution is solved by Fourier transforms.

6.1. Thin Insulated Wires

A thin wire is assumed to be half buried in the earth with its axis in the plane of the ground surface. The series impedance per unit length of wire z_s remains nearly the same as at an infinite burial depth following Section 5.1, but its finite shunt admittance per unit length y_s, which is proportional to the wire capacity, is decreased by a factor of two, if the conductivity of the ground $\sigma_g \gg \omega\epsilon_0$. Therefore the propagation constant $\gamma = (y_s z_s)^{1/2}$ for a wire in the air-to-ground interface, is smaller by a factor of $\sqrt{2}$ than the propagation constant in the homogeneous ground

$$\gamma_g = i(\omega^2\mu_0\epsilon_g + i\omega\mu_0\sigma_g)^{1/2}.$$

Coleman [1950] has investigated the modes of propagation for wires buried in a stratified medium: he shows that for a thin wire there is an exponential mode of propagation, of which the propagation constant is that of the medium. The distant layers have no effects on this propagation constant, but for wires lying in the interface of the two media, the propagation constant is given by

$$\gamma = \left(\frac{\gamma_1^2 + \gamma_2^2}{2}\right)^{1/2}, \tag{6.1}$$

where γ_1 and γ_2 are propagation constants of the two media. For a wire of finite radius, a mode of propagation exists closely approximating this exponential mode of very thin wires.

This result has been further discussed by Sunde [1949], who also gives an iterative procedure for computing corrections to the exponential mode of propagation.

6.2. Long Slots

An infinitely long slot is assumed in a perfectly conducting screen separating a lossy and a lossless dielectric. This problem idealizes a slot antenna, which is constructed by placing a wire mesh on the ground surface. As a consequence of the rapid field decay along the slot, the model of an infinitely long slot is adequate for representing a slot of finite length L, if $L \gg$ skindepth in dielectric δ. The same geometry is also a crude approximation to an island antenna as proposed by Morgan [1960] and Gould [1961]. In this approximation the sea is represented as a perfectly conducting screen; the island and the bottom of the sea are represented as a semi-infinite lossy dielectric.

The slot of Fig. 6.1 is excited by a current source that lies in the interface between the two media. Then the tangential magnetic fields of the two media

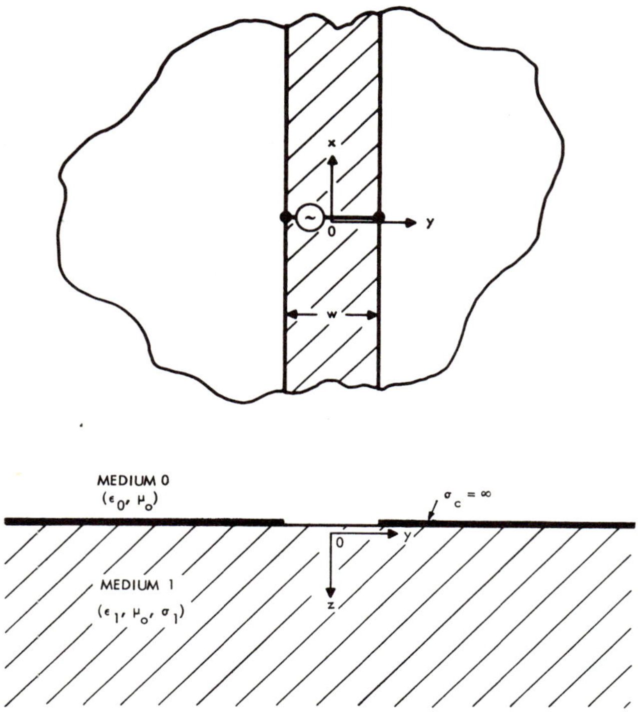

FIG. 6.1. Geometry of the slot.

in the slot plane differ by this source current density. The magnetic fields in the two media are related to the integral of the tangential electric field over the slot. In connection with the previous boundary condition, this gives an integral equation for determining the distribution of the electric field along the slot. The techniques for solving this integral equation are different for narrow slots (slot width $w \ll$ skindepth in dielectric δ) and for wide slots ($w \gg \delta$).

6.2.1. *The Narrow Slot*

The magnetic field component H_x^n of the medium $n(n = 0$ or $1)$ is obtained from (3.7) for $z = 0$ as

$$H_x^n = \frac{(-1)^n}{2\pi i\omega\mu_0} \iint\limits_{(\text{slot})} \mathrm{d}x'\, \mathrm{d}y'\, E_y(x', y') \left[k_n^2 + \frac{\partial^2}{\partial x^2}\right] G_n, \tag{6.2}$$

where

$$G_n = \frac{\exp(ik_n r)}{r}, \tag{6.3}$$

$$k_n = (\epsilon_n\mu_0\omega^2 + i\mu_0\sigma_n\omega)^{1/2}, \tag{6.4}$$

$$r = [(x-x')^2 + (y-y')^2]^{1/2}. \tag{6.5}$$

The source current I_y flows across the slot at $x = 0$. Therefore H_x^0 and H_x^1 are related by

$$H_x^1 - H_x^0 = I_y\delta(x). \tag{6.6}$$

Substituting (6.2) in (6.6) the integral equation may be written as

$$\left[\frac{\partial^2}{\partial x^2} + k^2\right] \iint\limits_{(\text{slot})} \mathrm{d}x'\, \mathrm{d}y'\, E_y(x', y')(G_1 + G_0)$$

$$= -2\pi i\omega\mu_0 I_y \delta(x) - \frac{k_1^2 - k_0^2}{2} \iint\limits_{(\text{slot})} \mathrm{d}x'\, \mathrm{d}y'\, E_y(x', y')(G_1 - G_0), \tag{6.7}$$

where

$$k^2 = \frac{k_0^2 + k_1^2}{2}. \tag{6.8}$$

The significance of this particular rearrangement of the integral equation is discussed following (6.28). Equation (6.7) may be considered as a differential equation for the double integral of its left-hand side. Noting that the solution of

$$v'' + k^2 v = h(x) \tag{6.9}$$

is

$$v = \frac{1}{k}\int_a^x h(\xi)\sin k(x-\xi)\,d\xi + C_1 \sin kx + C_2 \cos kx, \tag{6.10}$$

the integral equation (6.7) becomes

$$\iint_{(\text{slot})} dx'\,dy' E_y(x',y')\Bigg\{(G_1+G_0) + \frac{k_1^2-k_0^2}{2}$$

$$\cdot\int_0^x [G_1(\xi,y,x',y') - G_0(\xi,y,x',y')]\frac{\sin k(x-\xi)}{k}\,d\xi\Bigg\}$$

$$= A(x)\sin kx + B\cos kx, \tag{6.11}$$

where

$$A(x) = \frac{2\pi i\omega\mu_0 I_y}{k}u(x) + A_1 \tag{6.12}$$

and $u(x)$ is a unit step function. Equation (6.11) may be rearranged by applying the identity

$$E_y(x',y') = E_y(x,y') - E_y(x,y') + E_y(x',y'). \tag{6.13}$$

This results in

$$\iint_{(\text{slot})} dx'\,dy'\,E_y(x,y')(G_1+G_0) = A(x)\sin kx + B\cos kx - \Lambda(x), \tag{6.14}$$

where

$$\Lambda(x) = \iint_{(\text{slot})} dx'\,dy'\Bigg\{[E_y(x',y') - E_y(x,y')](G_1+G_0)$$

$$+\frac{k_1^2-k_0^2}{2}E_y(x',y')\int_0^x [G_1(\xi,y,x',y')$$

$$-G_0(\xi,y,x',y')]\frac{\sin k(x-\xi)}{k}\,d\xi\Bigg\}. \tag{6.15}$$

This rearrangement of the integral equation is the essence of Hallen's method [1938], which makes it possible in our case to obtain $E_y(x)$ on the left-hand side of (6.14) by substituting less accurate expressions of E_y into $\Lambda(x)$.

In the x' integrals of (6.14) the variable of integration is changed from x'

to r. Then the integral may be evaluated with the aid of (4.3.11) or (1.3.8) and (2.3.8) of Erdelyi [1954] to give

$$\int_{-\infty}^{\infty} \frac{e^{ik_n r}}{r}\,dx' = 2\int_{|y-y'|}^{\infty} \frac{e^{ik_n r}}{\sqrt{[r^2-(y-y')^2]}}\,dr$$
$$= \pi[-Y_0(k_n|y-y'|)+iJ_0(k_n|y-y'|)]$$
$$\approx 2\left[\log\frac{2}{|y-y'|k_n}+i\frac{\pi}{2}-C\right], \tag{6.16}$$

where

$$r=[(x')^2+(y-y')^2]^{1/2},$$

$C=0.5772\ldots$ is Euler's constant, and it is assumed that $|k_n(y-y')|\ll 1$. (This inequality is satisfied for slot widths $w\ll\delta$, where $\delta=[2/(\omega\mu_0\sigma_1)]^{1/2}$ is the skindepth in the lower medium.) The y' integrals of (6.14) and (6.15) may be evaluated using a static approximation for $E_y(y')$ in the slot. The electric field across a slot of width w in an infinite conducting screen in free space is given by Smythe [1950] as

$$E_y(y)=-\frac{V}{\pi[(w/2)^2-y^2]^{1/2}}, \tag{6.17}$$

where V is the voltage across the slot and $|2y|<w$. Using this value of E_y, it follows that [Stevenson, 1948; Watson, 1947]

$$\int_{-w/2}^{w/2} dy'\,E_y(x,y')\log\frac{A}{|y-y'|}=-V(x)\log\frac{4A}{w}. \tag{6.18}$$

The cases $y=0$ and $(w/2)$ of this integral may be readily verified by using the integrals (324.52c) and (325.20) of Groebner and Hofreiter [1958]. Applying (6.16) and (6.18), (6.14) becomes

$$-C_1V(x)=A(x)\sin kx+B\cos kx-\Lambda(x), \tag{6.19}$$

where

$$C_1=4\log\frac{8}{w[k_0k_1]^{1/2}}-4C+2i\pi$$
$$\approx 4\log\frac{8}{wk_0}+\log\frac{\omega\epsilon_0}{\sigma_1}-4C+i\frac{3\pi}{2}. \tag{6.20}$$

It may be noted that the numerical values of the integral (6.18) and of the constant C_1 of (6.19) and (6.20) are only slightly affected by the choice of

$E_y(y')$. An assumption of $E_y = -V/w$ leads to a modified constant C', where $(C_1 + 1.2) > C' > (C_1 - 1.5)$, if y of (6.18) is considered to vary from $y = 0$ to $y = w/2$. There is a minor change if $C_1 \gg 1$. The integrals of (6.15) remain finite as $w \to 0$, which makes the y integrations trivial.

Equation (6.19) represents the reduced integral equation. Approximate solutions of this equation may be found if $\Lambda(x)$ is small compared to the terms proportional to $A(x)$ and B.

The zero order solution of (6.19) is obtained by neglecting $\Lambda(x)$. This results in

$$-C_1 V(x) = A(x) \sin kx + B \cos kx. \tag{6.21}$$

$A(x)$ and B may be determined from the requirement that

$$V(x) = 0 \tag{6.22}$$

for $x = \pm\infty$. Substituting (6.12) in (6.21) gives

$$iB = -A_1 = -\frac{i\pi\omega\mu_0 I_y}{k}. \tag{6.23}$$

Thus the zero order solution of the integral equation is simply

$$V_0(x) = \frac{\pi\omega\mu_0 I_y}{kC_1} e^{ik|x|} \tag{6.24}$$

where k and C_1 are given by (6.8) and (6.20), respectively. The voltage is exponentially attenuated with the same propagation constant as that for a thin wire in (6.1). $V_0(x)$ represents a good approximation to $V(x)$ of (6.19) as long as $\Lambda(x)$ is negligible.

It is possible to obtain different zero order solutions of the integral equation by a suitable rearrangement of (6.7), and by making k of (6.8) and (6.27) equal to

$$k = \left(\frac{mk_1^2 + nk_0^2}{m+n}\right)^{1/2} \tag{6.25}$$

where m and n are arbitrary constants. However, the appropriate change in the right-hand side of (6.7) makes the triple integral of (6.15) divergent for $w \to 0$, unless $m = n$. The only solution that gives a finite value of $\Lambda(x)$ for $w \to 0$ is with $m = n$ in (6.25), which makes k of (6.25) equal to k of (6.8).

The first order solution $V_1(x)$ is obtained from (6.19) and (6.15) after substituting $V_0(x)$ of (6.24) as $[-E_y(x, y')\Delta y']$ and applying the condition (6.22).

The resulting magnitudes of $V_0(x)$ of (6.24) and of $V_1(x)$ are plotted in Fig. 6.2 for $\sigma_1 = 10^{-3}$ mho/m, $f = 100$ kc/s, and slot widths $w = 10$ and 1 m where $V_1(x) = V_0(x)$ for $w = 0$. It is seen that $|V_1(x)|$ exhibits a more rapid initial decay than $|V_0(x)|$, but approximates $|V_0(x)|$ closely for $|kx| > 3$.

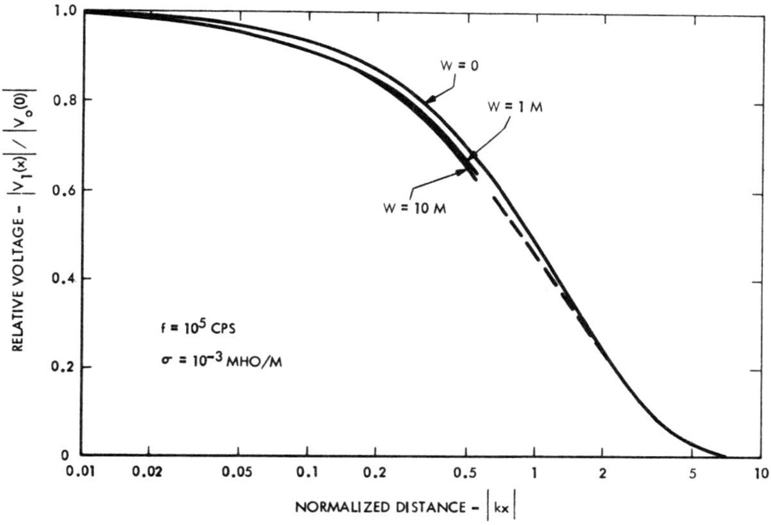

FIG. 6.2. Distribution of the voltage along the slot.

However, there is a more gradual decrease of $V_1(x)$ for $|kx| \gg 1$ because of the term of $V_1(x)$, which is proportional to e^{ik_0x}.

The difference between $V_0(x)$ and $V_1(x)$ curves is decreased with decreased slot widths or with increasing values of C_1. Hence, $V_1(x)$ will provide a small correction to $V_0(x)$ if $|C_1| \gg 1$. To further illustrate the frequencies f and slot widths w where this inequality is satisfied, the real part of C_1 of (6.20) is plotted in Fig. 6.3 for $\sigma_1 = 10^{-3}$ mho/m. The slot widths w, where the skin depth

$$\delta = [2/(\omega\mu_0\sigma_1]^{1/2} > w,$$

are indicated by dashes in Fig. 6.3. Equation (6.16), and hence the subsequent development, is not valid for these combinations of w and f.

The zero order and first order admittances of the slot

$$Y_0 = \frac{I_y}{V_0(0)} \tag{6.26}$$

and

$$Y_1 = \frac{I_y}{V_1(0)} \tag{6.27}$$

are computed from (6.24) and from (32) Galejs [1962].

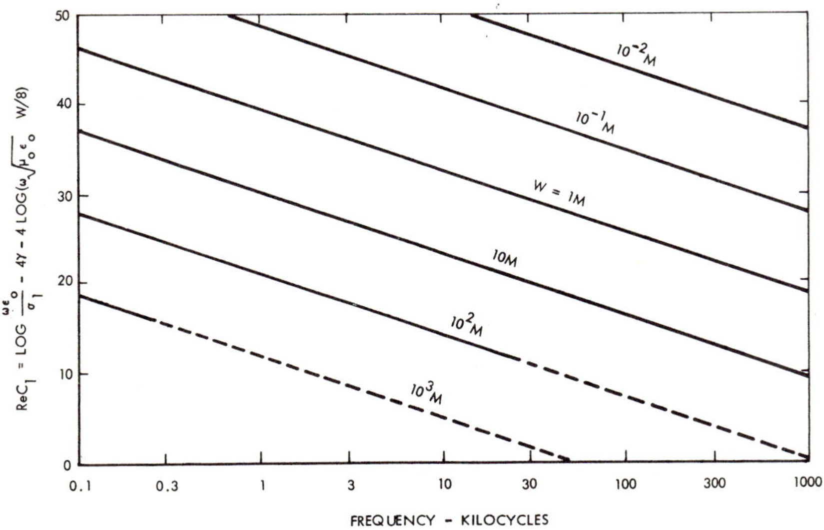

FIG. 6.3. Real part of the parameter C_1 for $\sigma_1 = 10^{-3}$ mho/m and $\epsilon_1/\epsilon_0 = 5$.

The computation of the voltage distribution $V_1(x)$ is fairly laborious. However, the magnetic dipole moment Kdl of the slot may be computed more readily. For $|k_1| \gg k_0$ it can be shown that

$$K_1 dl = 2\int_0^\infty V_1(x)\,\mathrm{d}x \approx -\frac{2V_0(0)}{ik}\,\frac{\mathrm{Re}C_1 + 0.88}{\mathrm{Re}C_1 + (i3\pi/2)}. \tag{6.28}$$

If $|C_1| \gg 1$, the magnetic dipole moment (6.28) is somewhat larger than

$$K_0 dl = 2\int_0^\infty V_0(x)\,\mathrm{d}x = -\frac{2V_0(0)}{ik}. \tag{6.29}$$

6.2.2. *The Wide Slot*

The excitation current I_y of the slot shown in Fig. 6.1 will exhibit no y variation if the width of the slot w is much less than the wavelength along the wire λ_d. This requirement may be met with a dielectric wire insulation of suitable thickness if w is small relative to free-space wavelength λ. Equations (6.2)–(6.6) still apply to the wide slot. Ignoring the transverse variation of the fields across the slot,

$$E_y(x, y) \approx E_y(x). \tag{6.30}$$

This will yield too low values of the electric field E_y near the edge of a zero thickness screen. However, the error is decreased with increasing thickness of the screen and may be small for the geometries suggested by

Morgan [1960] and Gould [1961]. Applying (6.30), the y' integral of (6.2) is evaluated with the aid of the relation

$$\int_{-w/2}^{w/2} dy' \frac{e^{ik_n r}}{r} = 2K_0(-ik_n|x-x'|) - \left[\int_{w/2-y}^{\infty} + \int_{w/2+y}^{\infty}\right] du \frac{e^{ik_n r}}{r}, \tag{6.31}$$

where $u = y' - y$ and where $K_n(z)$ is the modified Bessel function of the second kind of order n. If the displacement currents of the lossless medium are neglected $(k_0 = 0)$,

$$\int_{-w/2}^{w/2} dy' \frac{e^{ik_0 r}}{r} \approx \sinh^{-1}\frac{w/2-y}{|x-x'|} + \sinh^{-1}\frac{w/2+y}{|x-x'|}. \tag{6.32}$$

Substituting (6.30) and (6.31) in (6.2), (6.6) results in

$$\int_{-\infty}^{\infty} dx'\, E_y(x')\left[\sum_{n=0}^{1} \frac{ik_n K_1(-ik_n|x-x'|)}{|x-x'|}\right] + \pi i\omega\mu_0 I_y \delta(x) = \Lambda(x), \tag{6.33}$$

where

$$\Lambda(x) = \frac{1}{2}\int_{-\infty}^{\infty} dx'\, E_y(x')\left\{\sum_{n=0}^{1}\left[\frac{\partial^2}{\partial x^2} + k_n^2\right]\cdot\left[\int_{w/2-y}^{\infty} + \int_{w/2+y}^{\infty}\right] du \frac{e^{ik_n r}}{r}\right\} \tag{6.34}$$

is a correction due to the finite width of the slot w. If $\Lambda(x)$ of (6.34) is neglected, $E_y(x')$ is readily obtained with the aid of Fourier transforms [Galejs, 1962]. It follows that

$$E_y(x) = \frac{\omega\mu_0 I_y}{\pi x(k_1^2 - k_0^2)}[k_0 K_1(-ik_0 x) - k_1 K_1(-ik_1 x)], \tag{6.35}$$

which is identical to the fields produced by a long horizontal wire [Sunde, 1949]. For $k_0 x \ll 1$, $E_y(x)$ is approximated by

$$E_y(x) = \frac{i\omega\mu_0 I_y}{\pi}\frac{k_1^2}{k_1^2 - k_0^2}\cdot\left[\frac{1 + ik_1 x K_1(-ik_1 x)}{k_1^2 x^2} - D_1(x)\right], \tag{6.36}$$

where

$$D_1(x) = \frac{k_0^2}{2k_1^2}\left[\log\frac{2}{(-ik_0 x)} + \frac{1}{2} - C + \ldots\right] \tag{6.37}$$

and $C = 0.5772$ is Euler's constant. It is seen that $D_1(x)$ becomes negligible for $k_0 \ll k_1$. A plot of $|E_y(x)|$ is shown in Fig. 6.4 for $k_0 \ll |k_1|$. For $|k_1 x| \gg 1$, $|E_y(x)/(\omega\mu_0 I_y/\pi)| \sim |k_1 x|^{-2}$. For $|k_1 x| \ll 1$, $|E_y(x)/(\omega\mu_0 I_y/\pi)| \sim |-\log k_1 x|$, but it remains finite if the excitation current I_y flows in a wire of finite radius a, since (6.36) applies only to the fields outside the wire $(x > a)$.

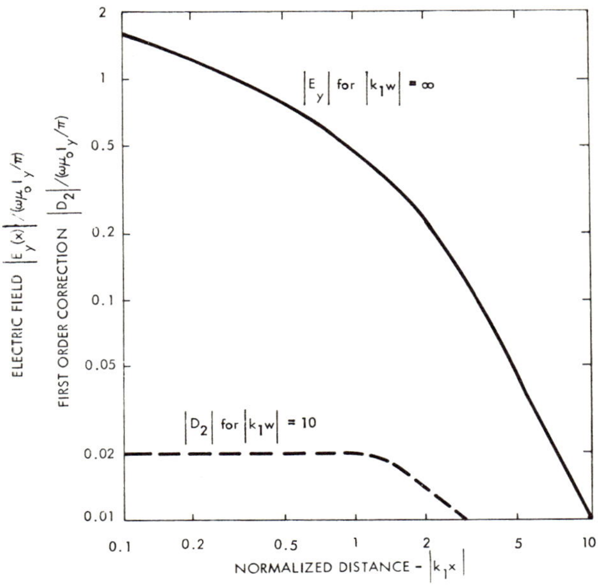

FIG. 6.4. Longitudinal variation of the transverse electric field in a wide slot and its first order correction.

The effects of a finite slot width w are considered with neglected displacement currents of the lossless medium ($k_0 = 0$). Calculation shows that the fields in a slot of a finite width w are approximated by (6.36) if $D_1(x)$ is replaced by a function $D_2(x)$. The magnitude of D_2 has been indicated for $|k_1 w| = 10$ by the dashed curve in Fig. 6.4.

The magnetic dipole moment of the slot is computed by integrating the electric field over the slot. Neglecting displacement currents of the lossless medium ($k_0 = 0$), the magnetic dipole moment is

$$Kdl = w \int_{-\infty}^{\infty} E_y(x)\, \mathrm{d}x = -\frac{\omega\mu_0 I_y w}{k_1}. \tag{6.38}$$

The fields excited by the magnetic dipole moment Kdl of (6.38) are the same as the fields excited by a horizontal electric dipole of dipole moment $I_y w$ [Norton, 1937].

6.2.3. *The Radiation Efficiency of the Slot*

The radiation efficiency of the slot *Eff* is given by the ratio of the radiated power P_r to the applied power P_i.

The power radiated by a magnetic dipole of an rms dipole moment Kds above an infinite, perfectly conducting plane is given by

$$P_r = 40(\omega\epsilon_0)^2 |Kdl|^2, \tag{6.39}$$

where Kdl is given by (6.36) for the narrow slot and by (6.61) for the wide slot.

The power applied to the narrow slot is simply

$$P_i = \mathrm{Re}\,Y[V_0(0)]^2 \tag{6.40}$$

if $V_0(0)$ is the rms voltage across the slot. The radiation efficiency of the narrow slot is obtained as

$$\eta = \frac{P_r}{P_i} = \frac{40(\omega\epsilon_0)^2}{\mathrm{Re}\,Y}\left|\frac{2}{k}\left\{1-\frac{k_1^2-k_0^2}{2k^2C_1}\left[i\pi+\frac{1}{2}\sum_{n=0}^{1}(-1)^n\log\frac{k_n+k}{k_n-k}\right]\right\}\right|^2. \tag{6.41}$$

For the wide slot, the fields outside the exciting wire are identical to those of a long horizontal wire. Hence the input impedance of the two wires will be the same. Ignoring the power dissipation of the line termination, the power applied to the wide slot is

$$P_i = R\int_0^w I_y^2\,dl \approx RI_y^2 w, \tag{6.42}$$

where I_y is the rms source current and where R is the series resistance per unit length of the feed wire. For an infinite wire, R is given by [Carson, 1926]

$$R = R_i + \frac{\omega\mu_0}{8}, \tag{6.43}$$

where R_i is the internal wire resistance, $(\omega\mu_0\sigma_1)^{1/2}h \ll 1$, and h is the height of the wire above ground surface. The above equation is applicable to wires of finite length w if $(\omega\mu_0\sigma_1)^{1/2}w \gg 1$ [Wolf, 1943]. The validity of (6.43) can be also established by computing the mutual impedance between two current elements from (6.36) and by averaging the real part of this mutual impedance over the cross section of the wire. The radiation efficiency of the slot is obtained from (6.39), (6.42), and (6.43) as

$$\eta = \frac{P_r}{P_i} = \frac{40[\omega(\mu_0\epsilon_0)^{1/2}]^4}{|\epsilon_1\mu_0\omega^2 + i\mu_0\sigma_1\omega|}\,\frac{w}{R}. \tag{6.44}$$

For frequencies above 1 kc/s, the internal wire resistance R_i may be much less than $\omega\mu_0/8$ and

$$\eta = \frac{320\omega\epsilon_0 w}{(\epsilon_1/\epsilon_0)|1+i(\sigma_1/\omega\epsilon_1)|}. \tag{6.45}$$

The approximations (6.38), (6.42), and (6.43) restrict the validity of the efficiency figures (6.44) and (6.45) to cases where $k_0 \ll |k_1|$ and $w \ll \lambda = 2\pi/[\omega(\mu_0\epsilon_0)^{1/2}]$.

The radiation efficiencies computed for dry ground from (6.45) have been plotted as solid curves in Fig. 6.5. The dashed portion of the curves refers to narrow slot widths ($w < \delta$). The dotted and dashed curve is indicative of

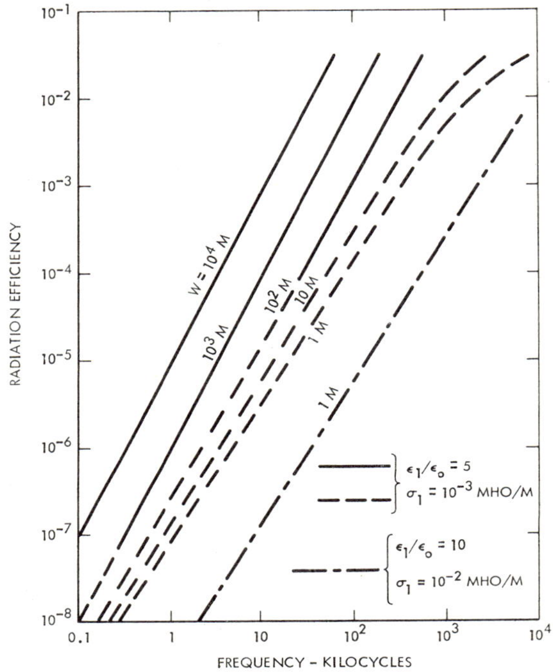

FIG. 6.5. Radiation efficiency of the slot.

the efficiency decrease of narrow slots caused by an increasing conductivity σ_1.

In the geometry of Fig. 6.1 the lossy and the lossless medium can be bounded by a rectangular and a parallel-plane waveguides. Such a geometry has been analyzed by Staras [1963], who obtains similar results regarding the field distribution along the slot.

6.3. Exercises

6.1. Obtain the first order solution $V_1(x)$ of the narrow slot in Section 6.2.1. Obtain simplified forms of $V_1(x)$ for $|kx| \ll 1$ and $|kx| \gg 1$.

6.2. Using the results of Exercise 6.1, derive the magnetic dipole moment

$$K_1 dl = \int_{-\infty}^{\infty} V_1(x)\, dx.$$

6.3. Derive the fields (6.35) of the wide slot.

6.4. References

CAMPBELL, G. A. and FOSTER, R. M. (1948) *Fourier Integrals for Practical Applications*, D. Van Nostrand, New York, NY.

CARSON, J. R. (1926) Wave propagation in overhead wires with ground return, *Bell Sys. Tech. J.* **5**, 539–54.

COLEMAN, B. L. (1950) Propagation of electromagnetic disturbance along a thin wire in a horizontally stratified medium, *Phil. Mag.* **41**, 276–88.

ERDELYI, A. (1954) *Tables of Integral Transforms*, McGraw-Hill, New York, NY, vol. 1.

GALEJS, J. (1962) Excitation of slots in a conducting screen above a lossy dielectric half space, *IRE Trans. Ant. and Prop.* **AP-10**, (2), 436–43.

GOULD, R. N. (1961) Some preliminary experimental tests of a novel method of radiating at very low frequencies, *Nature* **190**, 332–3.

GROEBNER, W. and HOFREITER, N. (1958) *Integraltafel*, Springer-Verlag, Vienna, Austria, Part 2.

HALLEN, E. (1938) Theoretical investigations into the transmitting and receiving qualities of antennae, *Nova Acta Regiae Soc. Sci. Upsaliensis* ser. IV, **11**, 1–44.

MORGAN, M. G. (1960) An island as a natural very-low-frequency transmitting antenna, *IRE Trans. on Ant. and Prop.* **AP-8**, 528–30.

NORTON, K. A. (1937) The physical reality of space and surface waves in the radiation field of radio antennas, *Proc. IRE* **25** 1192–1202.

OBERHETTINGER, F. (1957) *Tabellen Zur Fourier Transformation*, Springer-Verlag, Berlin, Germany.

SMYTHE, W. R. (1950) *Static and Dynamic Electricity*, McGraw-Hill, New York, NY.

STARAS, H. (1963) *Analysis of a Natural ELF Slot Antenna in Electromagnetic Theory and Antennas*, ed. by E. C. Jordan, Pergamon Press, pp. 1183–1200.

STEVENSON, A. F. (1948) Theory of slots in rectangular waveguides, *J. Appl. Phys.* **19**, 24–38.

SUNDE, E. D. (1949) *Earth Conduction Effects in Transmission Systems*, D. Van Nostrand, New York, NY.

WATSON, W. H. (1947) *The Physical Principles of Waveguide Transmission and Antennas Systems* (see Sec. 10.), Clarendon Press, Oxford.

WOLF, A. (1943) The impedance of a grounded wire, *Geophysics*, **7**, 414–18.

CHAPTER 7

CAVITY-BACKED SLOT ANTENNAS

THE rectangular waveguide with an infinite flange was discussed in Section 3.2; radiation from slots cut in rectangular waveguides have been investigated by Stevenson [1948] and Watson [1947]; and rectangular cavity-backed slots have been investigated by Cohen [1954] and Vasiliadis and Tanner [1960] without considering the effects of the cavity and higher order modes on the field distribution in the aperture. Such considerations are not applicable to very small cavities where the cavity will affect the fields in the aperture plane.

In this chapter slot performance will be investigated in conjunction with small cavities. Also discussed will be conditions where the usually assumed sinusoidal (or single mode) voltage variation along the slot is inadequate. Such problems arise in missile-borne antennas and ground based dielectric filled antennas where antenna volume is limited by space or cost consideration. For rectangular slots, the slot admittance will be formulated variationally [Galejs, 1963a] and will be also computed using the Hallen's iterative method [Galejs, 1963b]. The annular cavity-backed slot is considered only variationally [Galejs and Thompson, 1962]. These problems can be formulated by extending the work reported in Section 3.2 and 3.3.

7.1. Rectangular slots

7.1.1. *Integral Equation for the Field Variation in the Slot*

Figure 7.1 depicts the slot that is backed by a rectangular cavity of volume $x_i y_i z_i$. The slot is symmetrical with respect to the x-dimension of the cavity. The y coordinate of the slot center y_c is arbitrary. The slot is excited by a current source connected across its center, and the tangential magnetic fields inside and outside of the slot plane will differ by this source current density. The magnetic fields inside and outside the slot are related to the integral of the tangential electric fields over the slot plane. This relationship, in connection with the previous boundary condition gives an integral equation for determining the distribution of the electric field in the slot plane.

The fields in the rectangular cavity can be derived from scalar functions, Ψ_i and Φ_i, similar to (3.10) through (3.15). On the cavity bottom the

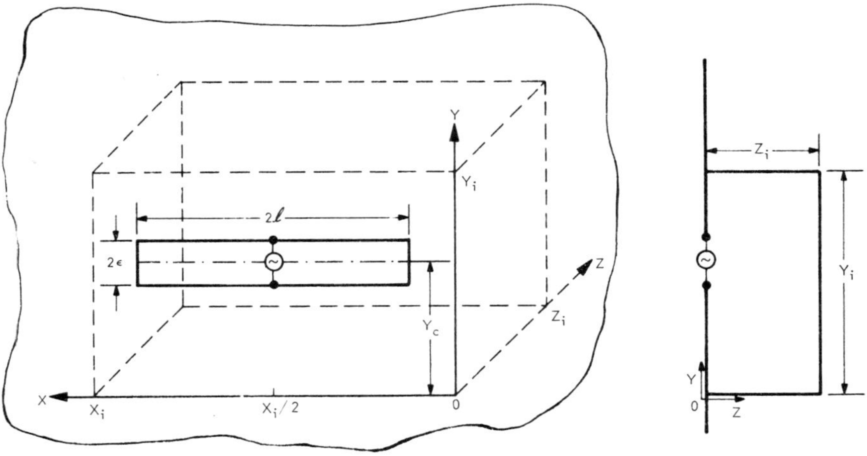

FIG. 7.1. Geometry of the slot and cavity.

tangential electric fields vanish. Following (3.10) and (3.11) this derivation requires that $\Psi_i = \partial\Phi_i/\partial z = 0$ at $z = z_i$. These conditions are satisfied if Ψ_i and Φ_i of (3.8) and (3.9) are replaced by

$$\Psi_i = \sum_n \sum_m a_{nm} \cos \beta_{xi}x \cos \beta_{yi}y\,(e^{\gamma_i z_i} - e^{-\gamma_i(z-2z_i)}), \tag{7.1}$$

$$\Phi_i = \sum_n \sum_m b_{nm} \sin \beta_{xi}x \sin \beta_{yi}y\,(e^{\gamma_i z_i} + e^{-\gamma_i(z-2z_i)}), \tag{7.2}$$

where $\beta_{xi} = n\pi/x_i$, $\beta_{yi} = m\pi/y_i$, $\gamma_i = i\beta_c = i\,[k_i^2 - \beta_{xi}^2 - \beta_{yi}^2]^{1/2}$. The electric component E_x is neglected in the aperture, and the relationship of the coefficients a_{nm} and b_{nm} to the aperture field E_y is similar to that of the coefficients and aperture field in Section 3.2. The resulting values of a_{nm} and b_{nm} are substituted in (7.1) and (7.2). Then H_{xi} is computed from (3.13) as

$$H_{xi} = -\frac{1}{4\pi i\omega\mu_0} \iint\limits_{\text{(slot)}} ds' \left\{ E_y(x', y') \left[k_i^2 + \frac{\partial^2}{\partial x'^2} \right] G_i \right\}, \tag{7.3}$$

where

$$G_i = -\frac{4\pi}{x_i y_i} \sum_{m=0}^{\infty} \sum_{n=1}^{\infty} \frac{\epsilon_m \epsilon_n}{\beta_c} \cot \beta_c z_i \sin \beta_x x \cos \beta_y y \sin \beta_x x' \cos \beta_y y'$$

$$\epsilon_m = 1, \quad \text{if} \quad m = 0 \tag{7.4}$$

$$\epsilon_m = 2, \quad \text{if} \quad m \neq 0.$$

The H_{x0} component of the magnetic field in the $z = 0$ plane outside the cavity is obtained from (3.7) or (6.2) as

$$H_{x0} = \frac{1}{2\pi i\omega\mu_0} \iint\limits_{\text{(slot)}} ds'\, E_y(x', y') \left[k^2 + \frac{\partial^2}{\partial x'^2} \right] G_0, \tag{7.5}$$

where

$$G_0 = \frac{\exp(ikr)}{r} \tag{7.6}$$

with $r = [(x-x')^2 + (y-y')^2]^{1/2}$ and with $k = \omega(\mu_0\epsilon_0.)^{1/2}$

The source current I_y flows across the slot at $x = x_i/2$. Therefore, H_{x0} and H_{xi} are related by

$$H_{xi} - H_{x0} = I_y \delta\left[x - \frac{x_i}{2}\right]. \tag{7.7}$$

Substituting (7.3) and (7.5) in (7.7) one obtains the integral equation

$$\iint_{(\text{slot})} ds'\, E_y(x', y') \left\{\left[k_i^2 + \frac{\partial^2}{\partial x^2}\right. G_i + 2\left[k^2 + \frac{\partial^2}{\partial x^2}\right] G_0\right\}$$

$$= -4\pi i \omega \mu_0 I_y \delta\left[x - \frac{x_i}{2}\right]. \tag{7.8}$$

7.1.2. *Variational Solution*

Multiplying both sides of (7.8) by $E_y(x, y)$ and integrating over the slot gives the slot admittance as

$$Y = \frac{I_y}{V_0} = \frac{1}{V_0^2}\left[\iint_{(\text{slot})} ds\, E_y(x, y) . \iint_{(\text{slot})} ds'\, E_y(x', y') G\right], \tag{7.9}$$

where

$$G = \frac{1}{4\pi i \omega \mu_0}\left\{\left[k_i^2 + \frac{\partial^2}{\partial x^2}\right] G_i + 2\left[k^2 + \frac{\partial^2}{\partial x^2}\right] G_0\right\} \tag{7.10}$$

$$V_0 = V\left(\frac{x_i}{2}\right) = -\int_{y_c-\epsilon}^{y_c+\epsilon} dy\, E_y\left[\frac{x_i}{2}, y\right]. \tag{7.11}$$

Y of (7.9) is functionally similar to Z of (1.35); the stationarity of (7.9) can be demonstrated using the method shown in Section 1.5.

The fields in the slot are assumed to be of the form

$$E_y(x, y) = [A f_A(x) + B f_B(x)] f(y). \tag{7.12}$$

where A and B are the yet undetermined amplitudes of the trial functions $f_A(x)$ and $f_B(x)$, and where $f(y)$ is the static approximation [Smythe, 1950]

$$f(y) = -\frac{1}{\pi[\epsilon^2 - (y - y_c)^2]^{1/2}}. \tag{7.13}$$

Substitution of (7.10) in (7.7) results in

$$Y = \frac{I_y}{V_0} = \frac{A^2 \gamma_{AA} + 2AB\gamma_{AB} + B^2 \gamma_{BB}}{[AF_A + BF_B]^2}, \tag{7.14}$$

where $F_n = f_n(x_i/2)$,

$$\gamma_{NM} = \iint\limits_{\text{(slot)}} \mathrm{d}s\, f_N(x) f(y) \iint\limits_{\text{(slot)}} \mathrm{d}s'\, f_M(x') f(y') G. \tag{7.15}$$

Considering the ratio of (A/B) as a variable in (7.14), the requirement that Y be stationary leads to

$$\frac{\mathrm{d}Y}{\mathrm{d}(A/B)} = 0. \tag{7.16}$$

The (A/B) ratio is given by (2.38), and it follows that

$$Y = \frac{\gamma_{AA}\gamma_{BB} - \gamma_{AB}^{\;2}}{\Delta} \tag{7.17}$$

with

$$\Delta = F_B^2 \gamma_{AA} - 2F_A F_B \gamma_{AB} + F_A^2 \gamma_{BB}. \tag{7.18}$$

The slot admittance (7.17) has been computed for the field distribution $E_y(x, y)$ of (7.12) with "optimized" coefficients A and B. The slot admittance can also be computed for the simpler field distribution

$$E_s(x, y) = A f_A(x) f(y). \tag{7.19}$$

It follows from (7.14) that

$$Y_s = \frac{\gamma_{AA}}{F_A^{\;2}}. \tag{7.20}$$

The trial functions $f_m(x)$ of (7.12) should satisfy the condition

$$f_M\left(\frac{x_i}{2} \pm l\right) = 0. \tag{7.21}$$

Following Section 2.3, suitable trial functions are

$$f_A(x) = \sin\left\{k\left[l - \left|x - \frac{x_i}{2}\right|\right]\right\}. \tag{7.22}$$

and

$$f_B(x) = 1 - \cos\left\{k\left[l - \left|x - \frac{x_i}{2}\right|\right]\right\}. \tag{7.23}$$

The assumed form of $f_M(x)$ provides a smooth transition between a sinusoidal distribution for deeper cavities to rapidly attenuated fields for shallow cavities, as will be seen from subsequent numerical calculations. It is stipulated that the form of the trial function is general enough and that it will closely approximate the actual distribution.

A trial function in the form of a power series

$$V(x) = \sum a_n x^n$$

may have no advantages if only a few terms of the series are considered. The integrals γ_{NM} of (7.15) are split in two parts as

$$\gamma_{NM} = \gamma^i_{NM} + \gamma^0_{NM}, \tag{7.24}$$

where the first part has G_i of (7.4) and the second part has G_0 of (7.6) in the integrand.

The integrals γ^i_{NM} are considered first. The y integrations result in Bessel functions.

The x integrals involve products of trigonometric functions, and are simple to evaluate. The numerical evaluation of the resulting double summations is facilitated by approximating Σ_m for $m \geqslant M \geqslant 2y_i/(\pi z_i)$ by an integral [Galejs, 1963a].

The x integrations of γ^0_{NM} have been carried out as indicated in Section 2.3. The y' integrals can be evaluated with the aid of the relation

$$\frac{1}{\pi^2}\int_{-\epsilon}^{\epsilon} \frac{dy}{(\epsilon^2 - y^2)^{1/2}} \int_{-\epsilon}^{\epsilon} \frac{dy'}{(\epsilon^2 - (y')^2)^{1/2}} \log \frac{b}{|y-y'|}$$
$$= \frac{1}{\pi}\int_{-\epsilon}^{\epsilon} \frac{dy}{(\epsilon^2 - y^2)^{1/2}} \log \frac{2b}{\epsilon} = \log \frac{2b}{\epsilon} \tag{7.25}$$

and with

$$\frac{1}{\pi^2}\int_{-\epsilon}^{\epsilon} \frac{dy}{(\epsilon^2 - y^2)^{1/2}} \int_{-\epsilon}^{\epsilon} \frac{dy' |y'-y|}{(\epsilon^2 - (y')^2)^{1/2}}$$
$$= \frac{2}{\pi^2}\left[\int_{-\epsilon}^{\epsilon} dy + \int_{-\epsilon}^{\epsilon} \frac{dy\, y \sin^{-1}(y/\epsilon)}{(\epsilon^2 + y^2)^{1/2}}\right] = \frac{8}{\pi^2}\epsilon. \tag{7. 26}$$

The resulting expressions of γ^0_{NM} are given by (2.42) to (2.47), except that $\log(4x/\epsilon)$ is substituted for $\log(2x/a)$, and $(8k\epsilon/\pi^2)$ for ka. The multiplicative constant of (2.42) to (2.44) is now defined as $K_\gamma = (\epsilon_0/\mu_0)^{1/2}/(\pi i)$.

The expression for the slot admittance (7.17) is examined first for an idealized cavity that reflects zero admittance (or an open circuit) into the slot plane. With $\gamma^i{}_{NM} = 0$ in (7.24), the slot admittance depends only on $\gamma^0{}_{NM}$. Assuming that the half-width of the slot ϵ is equal to the diameter $2a$ of a cylindrical wire, the admittance Y (7.17) for a narrow slot of $k\epsilon \ll 1$ is related to the impedance (2.39) of a complementary dipole antenna in the same way as the respective constants K_γ. This relation is in agreement with (3.6).

The cavity effects are considered in numerical computations, and the conductance G and the susceptance B of the cavity-backed slot are shown

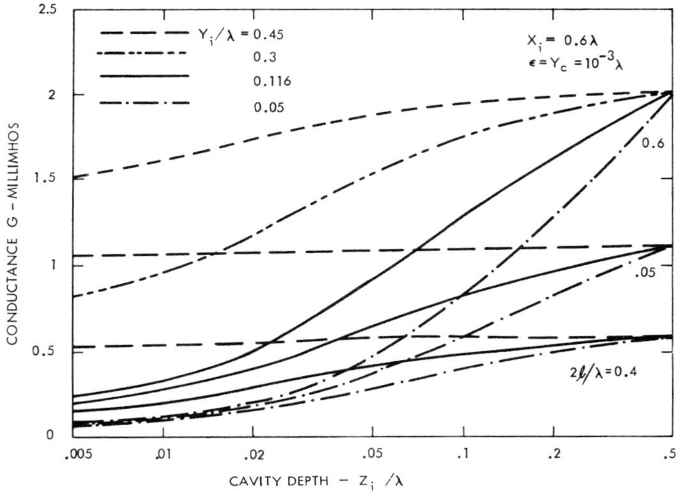

FIG. 7.2. Conductance of the cavity-backed slot.

in Figs. 7.2 and 7.3 for several cavity widths y_i and slot lengths $2l$. The conductance G decreases with decreasing cavity depth z_i, cavity width y_i, and slot length $2l$. The susceptance B may be capacitive for shallow

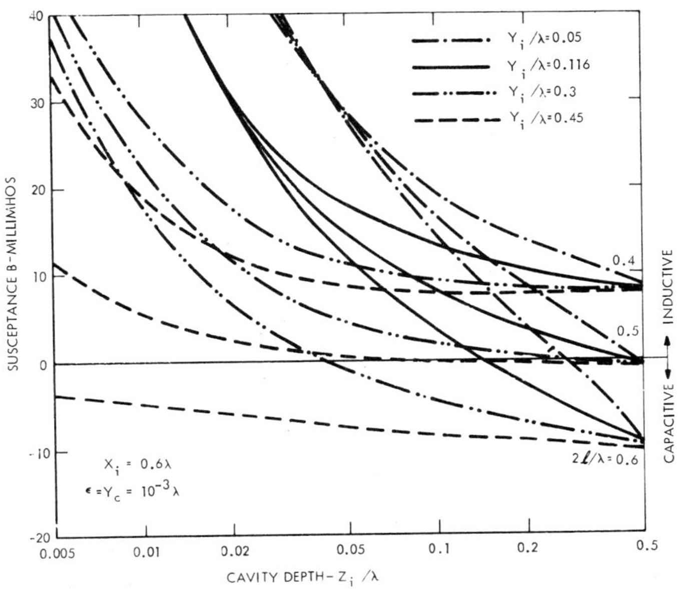

FIG. 7.3. Susceptance of the cavity-backed slot.

cavities if $2l > \lambda/2$. A decreasing cavity depth makes B approach an inductive short circuit ($B \to +\infty$). The resonance, which is defined by $B = 0$, occurs at smaller depths z_i for the wider cavities.

The voltage distribution along the slot computed from (7.12) and (2.38) is shown in Fig. 7.4. For deep cavities ($z_i/\lambda \approx 0.5$), the voltage is proportional to $\sin k(l - |x - x_i/2|)$, which would approximate the voltage distribution along the slot in the absence of the cavity. For resonant cavities

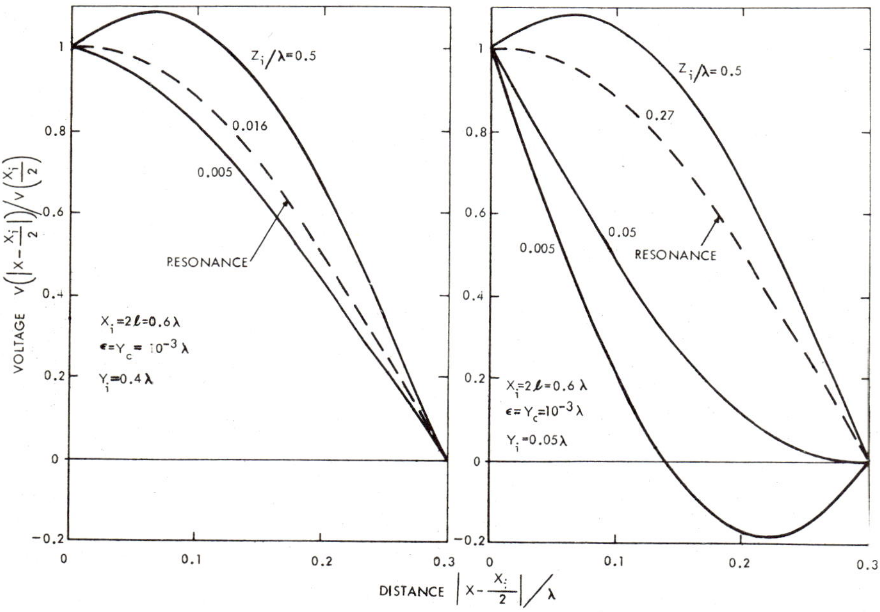

FIG. 7.4. Voltage distribution along the slot.

the relative voltage is approximated by a half-cycle of a sine function. The dropoff of the voltage near the center of the slot becomes more rapid for shallow and narrow cavities. For such cavities more of the current flows around the bottom of the cavity ($x \approx x_i/2$, $z = z_i$) than around the edges of the slot ($x = x_i/2 \pm l, z = 0$).

The sinusoidal field distribution of (7.19) approximates (7.12) only for deeper cavities. Numerical calculations show that $Y_s = G_s + iB_s$ of (7.20) is related to $Y = G + ib$ of (7.17) as

$$G(z_i) < G(z_i = \max) \approx G_s(z_i) = \text{constant} \tag{7. 27}$$

$$B(z_i) < B_s(z_i). \tag{7. 28}$$

An assumption of a sinusoidal voltage distribution $V_s(x)$ results in a slot conductance G_s, which is independent of z_i. G_s is larger than the conductance G of shallow cavities and approximates G of a deeper cavity. The susceptance B is more capacitive than B_s at a given cavity depth z_i.

The calculated slot admittance is compared with measurements in Fig. 7.5. The measured susceptance is in good agreement with calculations when allowing for the uncertain terminal reactance. Measurements provide a

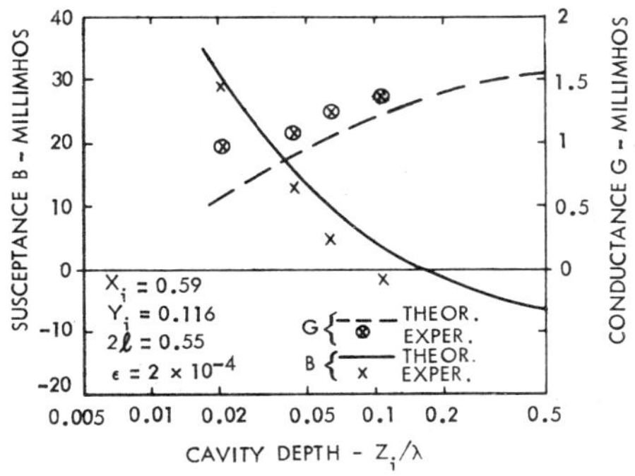

FIG. 7.5. Admittance of the cavity-backed slot.

somewhat higher conductance. This may be due to losses in the cavity walls and in the ground plane, which have not been accounted for in the calculations.

7.1.3. *Hallen's Method*

Considering (8) as a differential equation for the double integral, its solution is

$$\iint_{(\text{slot})} dx'\, dy'\, E_y(x', y')\,(G_i+2G_0) = -\frac{4\pi i\omega\mu_0}{k} I_y u\left[x-\frac{x_i}{2}\right] \sin\left[k\left(x-\frac{x_i}{2}\right)\right]$$

$$+C_1 \sin\left[k\left(x-\frac{x_i}{2}\right)\right] + C_2 \cos\left[k\left(x-\frac{x_i}{2}\right)\right], \quad (7.29)$$

where $u(x)$ is a unit step function. Equation (7.29) may be rearranged by applying the identity (6.13) into

$$2\iint_{(\text{slot})} dx'\, dy'\, E_y(x, y')\,(G_i+2G_0)$$

$$= A(x) \sin\left[k\left(x-\frac{x_i}{2}\right)\right] + B \cos\left[k\left(x-\frac{x_i}{2}\right)\right] - \Lambda(x), \qquad (7.30)$$

where

$$A(x) = -\frac{4\pi i\omega\mu_0}{k} I_y u\left[x-\frac{x_i}{2}\right] + C_1. \tag{7.31}$$

$$B = C_2, \tag{7.32}$$

and

$$\Lambda(x) = \iint_{(\text{slot})} (G_i+2G_0)\,[E_y(x',y')-E_y(x,y')\}\,\mathrm{d}x'\,\mathrm{d}y'. \tag{7.33}$$

This rearrangement of the integral equation makes it possible to obtain $E_y(x)$ on the left-hand side of (7.30) by substituting less accurate E_y expressions in $\Lambda(x)$. For narrow slots ($\epsilon \ll l$), it is permissible to use the static approximation for the y variation of E_y. It follows from (6.16) to (6.18) that

$$2\iint_{(\text{slot})} \mathrm{d}x'\,\mathrm{d}y'\,E_y(x,y')\,G_0 \approx -4V(x)\log\frac{4l}{\epsilon}. \tag{7.34}$$

After applying (7.4) and (6.17), the integral involving G_i is evaluated as

$$\iint_{(\text{slot})} \mathrm{d}x'\,\mathrm{d}y'\,E_y(x,y')\,G_i = \frac{8V(x)}{Y_0}\sum_{m=0}^{\infty}\sum_{n=1}^{\infty}\frac{\epsilon_m\epsilon_n}{n\beta_c}\cot\beta_c z_i \sin\beta_x x\cos\beta_y y$$
$$\cdot J_0(\beta_y\epsilon)\cos\beta_y y_c \sin\left(\beta_x\frac{x_i}{2}\right)\sin\beta_x l. \tag{7.35}$$

The G_0 integral (7.34) as well as the G_i integral (7.35) become logarithmically infinite as $\epsilon \to 0$. In the case of G_i this is seen most readily after approximating G_i by its small r value

$$G_i \simeq \frac{2}{r}, \tag{7.36}$$

which makes (7.35) equal to the approximate value of the G_0 integral in (7.34). This approximation, however, ignores any cavity effects on the G_i integral, which is undesirable in the present investigation. $\Lambda(x)$ of (7.33) remains finite for $\epsilon \to 0$, and the subsequent analysis simplifies after replacing y and y' of the G_0 expression in (7.33) by y_c.

The integral (7.30) now becomes

$$V(x)c(x) = A(x)\sin\left[k\left(x-\frac{x_i}{2}\right)\right] + B\cos\left[k\left(x-\frac{x_i}{2}\right)\right] - \Lambda(x), \tag{7.37}$$

where

$$c(x) = -4\log\frac{4l}{\epsilon} + 2\sum_{m=0}^{\infty}\sum_{n=1}^{\infty} F_{nm}\frac{\sin\beta_x x}{\beta_x}\sin\left(\beta_x\frac{x_i}{2}\right)\sin\beta_x l, \tag{7.38}$$

$$\Lambda(x) = \int_{x_i/2-l}^{x_i/2+l} \left[-2\frac{e^{ik|x-x'|}}{|x-x'|} + \sum_{m=0}^{\infty} \sum_{n=1}^{\infty} F_{nm} \sin\beta_x x \sin\beta_x x' \right]$$

$$\cdot \left[V_0(x') - V_0(x) \right] \mathrm{d}x', \qquad (7.39)$$

$$F_{nm} = 4\pi \frac{\epsilon_n \epsilon_m}{x_i y_i \beta_c} \cot\beta_c z_i \cos^2\beta_y y_c J_0(\beta_y \epsilon). \qquad (7.40)$$

The simplest solution $V_0(x)$ of (7.37) is obtained by neglecting $\Lambda(x)$. $A(x)$ and B are determined from the condition

$$V\left[\frac{x_i}{2} \pm l\right] = 0. \qquad (7.41)$$

This gives

$$V_0(x) = \frac{-2\pi i l_y (\mu_0/\epsilon_0)^{1/2}}{c(x)} \left\{ \sin\left[k\left|x - \frac{x_i}{2}\right|\right] - \tan kl \cos\left[k\left|x - \frac{x_i}{2}\right|\right] \right\}. \qquad (7.42)$$

The slot admittance may be computed from (7.42) as

$$Y_0 = \frac{I_y}{V(x_i/2)} = \frac{1}{2\pi i}\left(\frac{\epsilon_0}{\mu_0}\right)^{1/2} c\left(\frac{x_i}{2}\right) \cot kl. \qquad (7.43)$$

This slot admittance is purely susceptive. Numerical calculations show that the zero order solution of the integral equation (7.42), which ignores $\Lambda(x)$ of (7.39), is accurate only for deep cavities. It provides an incorrect susceptance for short slots that are backed by shallow cavities. Furthermore, the zero order solution does not provide the slot conductance.

The first order solution $V_1(x)$ may be obtained by using $V_0(x)$ of (7.42) for computing $\Lambda(x)$ of (7.39). Because of the difficulties in evaluating the integrals, $V_1(x)$ will be computed with a simpler representation of $V_0(x)$ than (7.42). For a slot approximately $\lambda/2$ long, the voltage distribution along the slot is expected to be nearly cosinusoidal. Assuming a cosinusoidal distribution, the trial function $V_0^t(x)$ becomes

$$V_0^t(x) = B \cos\left[k\left(x - \frac{x_i}{2}\right)\right] \Big/ c(x_1), \qquad (7.44)$$

where x_1 may be selected to be equal to $x_i/2$. Details of these computations are discussed by Galejs [1963b].

The resulting slot susceptance $B_1 = \mathrm{Im}Y_1$ is shown by the solid curves of Fig. 7.6. The dashed curves are obtained for $2l = x_1$ by using closed form approximations to Y_1[Galejs, 1963b]. The dotted and dashed curves represent the variational solution [Galejs, 1963a]. There is a reasonable agreement between the variational solution and the Hallen's first order solution for $2l > 0.5\lambda$ and $z_i > 0.05\lambda$. The first order solution for slots of

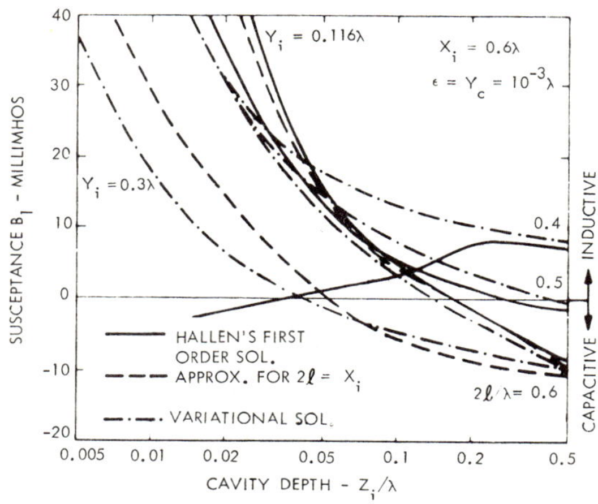

FIG. 7.6. First order slot susceptance.

$2l > 0.5\lambda$ is accurate only with deep cavities ($z_i > 0.25\lambda$), because of the shortcomings of the corresponding zero order solution.

Representative figures of the slot conductance $G_1 = \text{Re}Y_1$ have been listed in Table 7.1 for $z_i = 0.5\lambda$. G_1 is compared with the conductance G_v of the variational solution of Section 7.1.2 and with an admittance estimate G_c based on a complementary dipole configuration. It can be seen from (3.6) that

$$G_c = 2\frac{\epsilon_0}{\mu_0}R, \tag{7.45}$$

where R is the free space input resistance of a complementary dipole antenna. As pointed out before, $V_0^t(x)$ of (7.44) may approximate $V_1(x)$ if $2l \approx \lambda/2$. Therefore, G_1 approximates G_v and G_c in Table 7.1 for $2l = \lambda/2$. $V_0^t(x)$ of (7.44) is too inaccurate for conductance computation at different slot lengths.

The first order slot admittance computed by the Hallen's method appears to corroborate the slot admittance of the variational method in the cases where the voltage distributions of (7.42) and (7.44) are valid. Obvious disagreements occur for very shallow cavities, where the voltage is rapidly attenuated along the slot.

More accurate solutions of the Hallen's method may be obtained by trial functions $V_0^t(x)$ that resemble $V_0(x)$ of (7.42) more closely than $V_0^t(x)$ of

TABLE 7.1. SLOT CONDUCTANCE COMPUTED BY THE HALLEN'S METHOD (G_1), VARIATIONAL METHOD (G_v) AND FROM A COMPLEMENTARY DIPOLE (G_c). $x_i = 0.6\lambda$, $z_i = 0.5\lambda$, $\epsilon = y_c = 10^{-3}\lambda$.

γ_i/λ	$2l/\lambda$	G_1	G_v	G_c
		mmho	mmho	mmho
0.116	0.4	0.83	0.58	0.59
	0.5	1.01	1.11	1.15
	0.6	0.93	2.01	2.34
0.3	0.6	0.93	2.03	2.34

(7.44) or by going to higher order solutions $V_2(x), V_3(x), \ldots$. This involves a considerable analytical effort and is not attempted in this text.

7.2 Annular Slot

Radiation characteristics of annular slot antennas have been considered by Pistolkors [1948], Levine and Papas [1951], and Marcuvitz [1951]. The radiation conductance of an annular slot antenna backed by a hemispherical cavity has been calculated by Wait [1958]. This latter analysis emphasizes the finite conductivity of the cavity wall and does not consider slot susceptance, which determines the bandwidth limitations of the antenna.

The complex admittance of a cavity-backed annular slot antenna is determined in this section. For purposes of analysis, the antenna of Fig. 7.7 is excited by a current sheath in the slot. The slot admittance is determined by a method similar to that of Section 3.3.

7.2.1. *Slot Admittance*

The ϕ-symmetrical TM fields inside the cavity are similar to those of the coaxial line of (3.30), but there are also reflected waves of the higher modes with a z-dependence $\exp(-\gamma_n z)$. E_ρ is related to H_ϕ by (3.34) and the boundary condition of $E_\rho(\rho, -z_i) = 0$ is satisfied if $H_{\phi i}$ is of the form

$$H_{\phi i}(\rho,z) = \frac{A_0}{\rho}\left[e^{ik_i z} + e^{-ik_i(z+2z_i)}\right] + \sum_{n=1}^{\infty} A_n R_n(\rho)\left[e^{\gamma_n z} + e^{-\gamma_n(z+2z_i)}\right], \quad (7.46)$$

where $\gamma_n = i(k_i^2 - \lambda_n^2)^{1/2}$ and $\mathrm{Re}\gamma_n \leq 0$. For a coaxial cavity, $R_n(\rho)$ and λ_n are defined by (3.31) and (3.34). For cylindrical cavities

$$A_0 = 0, \quad (7.47)$$

$$R_n(\rho) = N_n J_1(\lambda_n \rho), \quad (7.48)$$

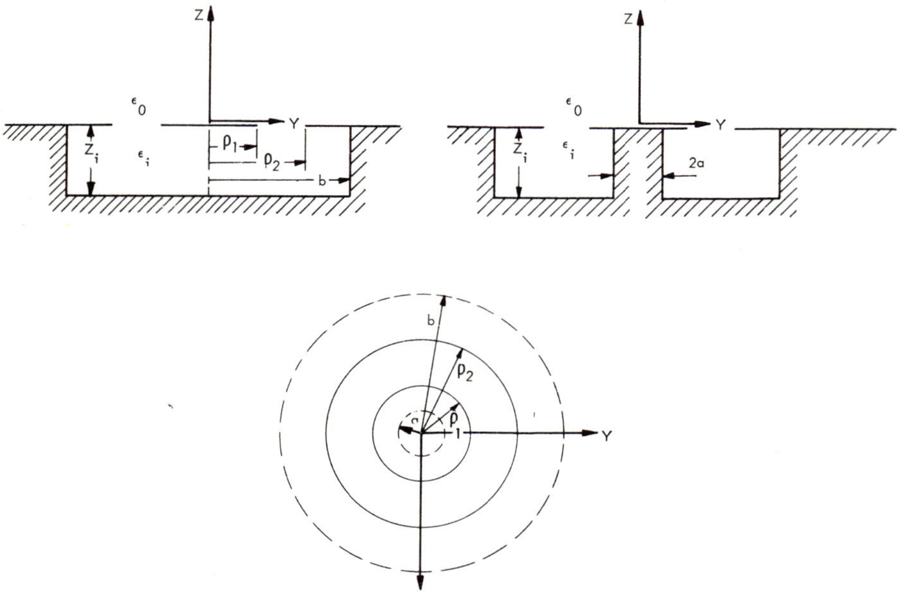

FIG. 7.7. Geometry of cylindrical and coaxial cavities.

where $J_m(x)$ is the Bessel function of the first kind of order m, λ_n is defined by

$$J_0(\lambda_n b) = 0 \tag{7.49}$$

and

$$N_n^{-1} = (b^2/2) J_1^2(\lambda_n b). \tag{7.50}$$

The amplitudes A_n are computed similarly as in (3.35) and (3.36). The magnetic field $H_{\phi 0}(\rho,z)$ outside the slot plane ($z \geqq 0$) is given by (3.29), except that the limits of the first integral should be changed to ρ_1 and ρ_2.

The tangential fields $H_{\phi i}(\rho, 0)$ differ by linear source current density $J_s(\rho)$ across the slot. This results in the integral equation

$$\frac{\cot k_i z_i}{k_i \rho \log (b/a)} \int_{\rho_1}^{\rho_2} E_\rho(\rho')\, d\rho' - \sum_{n=1}^{\infty} \frac{\coth (\gamma_n z_i)}{\gamma_n} R_n(\rho) \int_{\rho_1}^{\rho_2} E_\rho(\rho') R_n(\rho')\, \rho'\, d\rho'$$
$$- \frac{\epsilon_0}{\epsilon_i} \int_{\rho_1}^{\rho_2} E_\rho(\rho')\, \rho'\, d\rho' \int_0^\infty \frac{J_1(\zeta\rho) J_1(\zeta\rho')\zeta\, d\zeta}{(\zeta^2 - k^2)^{1/2}} = \frac{J_s(\rho)}{i\omega\epsilon_i}, \tag{7.51}$$

where the first term of the left-hand side is equal to zero for cylindrical cavities according to (7.47), and where

$$J_s(\rho) \approx \frac{I}{2\pi\rho}. \tag{7.52}$$

Equation (7.51) is brought into stationary form by multiplying (7.51) with $\rho E_\rho(\rho)$ and then integrating the resultant equation from $\rho = \rho_1$ to $\rho = \rho_2$. The voltage across the slot V is defined as

$$V = -\int_{\rho_1}^{\rho_2} E_\rho(\rho)\,\mathrm{d}\rho, \tag{7.53}$$

The aperture fields $E_\rho(\rho)$ are assumed to be of the form (3.38), and the slot admittance is computed as

$$Y = \frac{IV}{V^2} = Y_i + Y_0, \tag{7.54}$$

with

$$Y_i = 2\pi i\omega\epsilon_i \frac{\cot k_i z_i}{k_i \log(b/a)} - \frac{i\omega\epsilon_i}{[\log(\rho_2/\rho_1)]^2} \sum_{n=1}^{\infty} \frac{\coth(\gamma_n z_i)}{\gamma_n} \left[\int_{\rho_1}^{\rho_2} R_n(\rho)\,\mathrm{d}\rho\right]^2 \tag{7.55}$$

$$Y_0 = -\frac{2\pi\omega\epsilon_0}{[\log(\rho_2/\rho_1)]^2} \int_0^{\infty} \frac{\mathrm{d}\zeta}{\zeta(\zeta^2 - k^2)^{1/2}} \left[J_0(\zeta\rho_1) - (\zeta\rho_2)\right]^2. \tag{7.56}$$

Y_i is deduced from the $H_{\phi i}$ fields inside the cavity and represents the admittance reflected by the cavity into the slot plane. Similarly, Y_0 represents the admittance reflected by the outside space ($z \geqq 0$). For comparable aperture dimensions, Y_0 is the same as the admittance $Y_s = y_s Y_c$ of a coaxial line, which can be verified by comparing (7.56) with (3.40) and (3.42). The approximation (3.43) applies also to Y_0 of the cavity-backed annular slot, if a and b of (3,43) are replaced by ρ_1 and ρ_2 respectively. The admittance Y_i depends on the cavity structure, and will be considered separately for cylindrical and coaxial back-up cavities.

7.2.2. *Cylindrical Cavity*

For a cylindrical cavity the first term on the right-hand side of (7.55) is zero because of (7.47), and $R_n(\rho)$ of (7.48) should be used. A slot of $\rho = (\rho_1 + \rho_2)/2 = \lambda/5$ will be considered as an example. For $(\rho_2 - \rho_1)/\rho_1 = 0.05$, the slot admittance is $Y_0 = (4.8 - i37.5)$ mmhos. The susceptance $B = \mathrm{Im}(Y_0 + Y_i)$ is calculated for $n_{max} = 100$, and is plotted in Fig. 7.8. For $(b/\lambda) < 0.384$, all the cavity modes are below cutoff and the cavity reflects a capacitive susceptance. By increasing the cavity radius to $(b/\lambda) < 0.384$, one of the

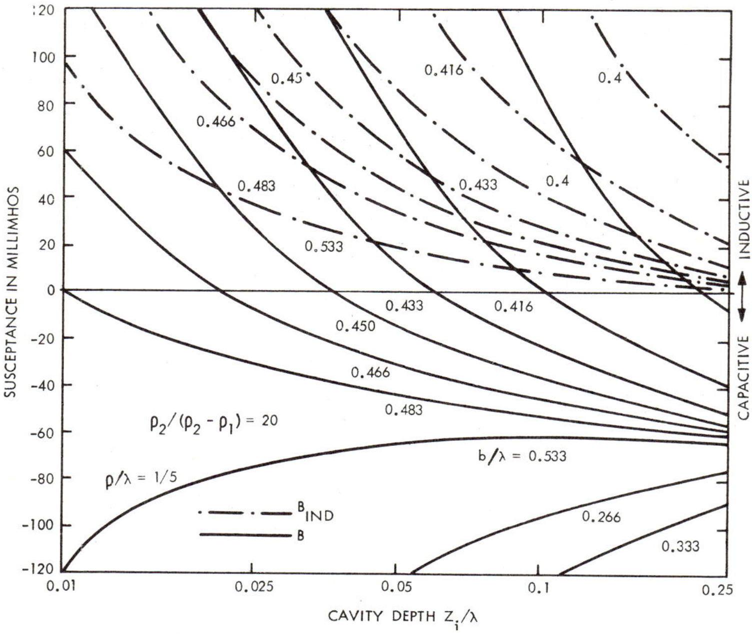

FIG. 7.8. Slot susceptance with a cylindrical cavity.

cavity modes will propagate. This mode reflects inductive susceptance, which is indicated by dots and dashes in Fig. 7.8. The inductive susceptance B_{ind} is sufficient to tune out the capacitive parts B_{cap} of Y_i and Y_0 if the propagating mode is near its cutoff. It is seen that B_{ind} at resonance ($B = 0$) is only slightly less than $|B|$ for $z_i = \lambda/4$, unless resonance occurs for $z_i < 0.03\lambda$. Hence shallow self-resonant cavities exhibit somewhat lower G/B_{cap} ratios than cavities the depth of which approaches $\lambda/4$.

The admittance ratio of $G = \text{Re}Y$ to $-B_{cap} = B_{ind} = |B|_{res}$ at resonance is plotted in Fig. 7.9. The effect of dielectric cavity loading ($\epsilon_i > \epsilon_0$) is indicated by the dotted and dashed curves. For $z_i/\lambda = \text{const}$, the slots of larger radii ρ require a slight increase in cavity radius b, but provide a significant increase of the $G/|B|_{res}$ ratio. $G/|B|_{res}$ ratios of more than 5 percent are achieved with $\rho > \lambda/5$ and $b > 2\lambda/5$. Dielectric cavity loading decreases the cutoff radius of the first propagating cavity mode indicated by vertical dashed lines in Fig. 7.9. The cavity radius b is decreased, but the $G/|B|_{res}$ ratio is also decreased by dielectric loading of the cavity.

For cavities of radius $b < 0.384\lambda$, all the cavity modes are below cutoff and reflect a capacity into the slot plane. Such cavities will be examined in conjunction with slots that have a circumference $2\pi\rho_2$ comparable to a wavelength λ and with slots of $\rho_2 \ll \lambda$.

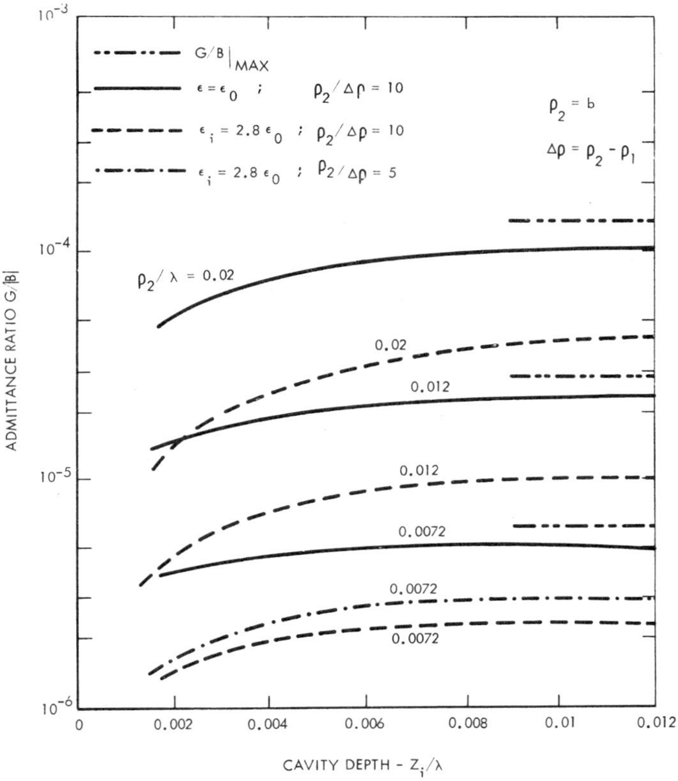

FIG. 7.9. Admittance ratio $G/|B|$ of small annular slots backed by shallow cylindrical cavities.

The admittance ratio $G/|B|$ is decreased for the small slots and is less than 10^{-4} for $b = \rho_2 \leqslant 0.02\lambda$, as seen from Fig. 7.10. For small cavities, $k_i \ll \lambda_q \approx (q-\frac{1}{4})\pi/b$. $G/|B|$ will become approximately proportional to z_i if $z_i \ll \rho_2/\pi$, while B is only slightly dependent on z_i if $z_i > \rho_2/\pi$. An upper bound to the possible increase of $G/|B|$ of a small slot ($\rho_2 \ll \lambda$) with an increasing cavity size can be obtained from the admittance of the same slot in an infinite conducting screen $G_s + iB_s$, which may be readily computed from the impedance of a complementary loop antenna. It follows that

$$\frac{G}{|B|} < \left|\frac{G}{B}\right|_{\max} = \left|\frac{G_s}{2B_s}\right| = \frac{\pi}{12} \frac{(k\rho_2)^3}{\log[(32\rho_2)/(\rho_2-\rho_1)] - 1.75}, \qquad (7.57)$$

The $G/|B|$ ratio of the small slots is compared in Fig. 7.10 with the $G/|B|$

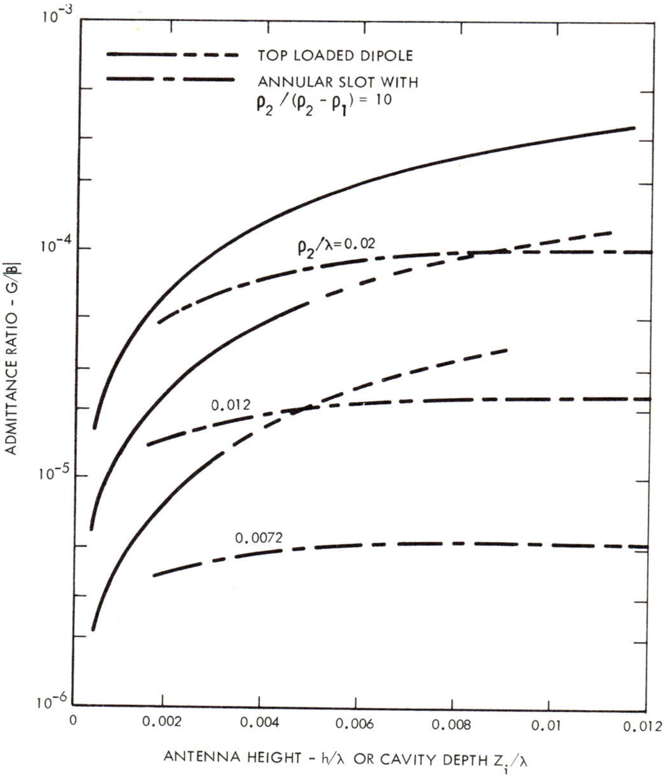

FIG. 7.10. Admittance ratio $G/|B|$ of a top-loaded dipole and of an annular slot backed by a cylindrical cavity.

ratio of a circular top loaded dipole. For a dipole above a conducting ground plane, $G/|B|$ has twice the value indicated in (4.54), which gives

$$\frac{G}{|B|} = \frac{1}{3\epsilon_r} a^2 h k^3, \tag{7.58}$$

where ϵ_r is the effective relative dielectric constant of the antenna, h is the antenna height, and a is its effective radius. The effective radius a becomes equal to the actual radius of the antenna disk ρ_2 if $a \ll h$. The plot of $G/|B|$ of (7.58) for $\epsilon_r = 1$ in Fig. 7.10 indicates that the toploaded dipole exhibits higher values of $G/|B|$ than the slot with a cylindrical cavity of comparable size.

7.2.3. *Coaxial Cavity*

For a coaxial cavity $R_n(\rho)$ (3.31) should be used in (7.55). The susceptance $B = \text{Im}(Y_0 + Y_i)$, as well as the inductive susceptance reflected by the propagating cavity mode B_{ind}, are plotted in Fig. 7.11. Again, the cavity resonance is defined by the condition $B = 0$ or $B_{ind} = -B_{cap}$. It is seen that B_{ind} at resonance is nearly the same as $|B|$ for $z_i = 0.25\lambda$, except when resonance occurs for $z_i < 0.03\lambda$. This implies that a relatively shallow self-resonant cavity ($B = 0$) will exhibit nearly the same $G/|B_{cap}|$ ratio as a $\lambda/4$ deep

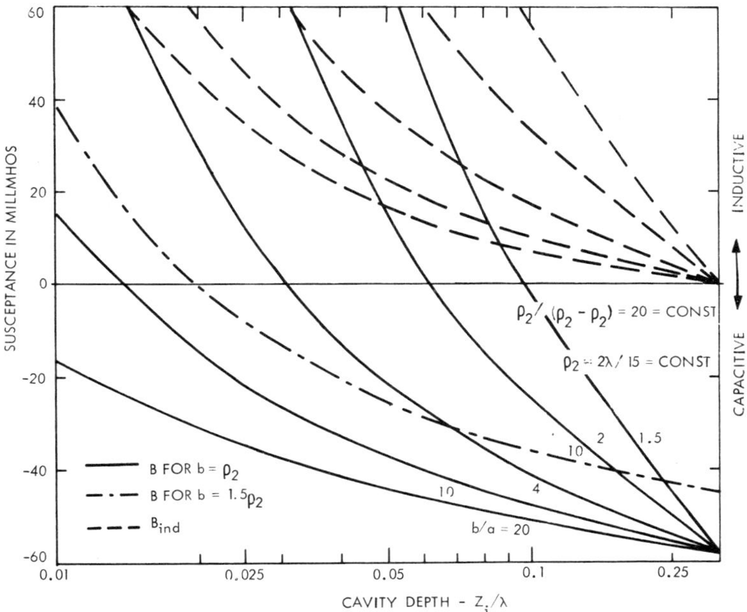

FIG. 7.11. Slot susceptance with a coaxial cavity.

cavity requiring additional inductive tuning. When the cavity depth is less than the resonant depth z_i, the slot susceptance becomes inductive, the slot requires additional capacitive tuning, and the G/B_{ind} ratio is decreased relative to its value at the self-resonance. A self-resonant antenna will exhibit nearly the same bandwidth as a slot with a $\lambda/4$ deep cavity. The bandwidth of the antenna is significantly decreased by decreasing the cavity depth below its value at self-resonance.

The admittance ratio of $G = \text{Re}Y$ to $-B_{cap} = B_{ind} = |B|_{res}$ at resonance is plotted in Fig. 7.12, $G/|B|_{res}$ ratios of more than 5 percent are achieved with $b > \lambda/5$ and $(b/a) < 2$. G/B_{res} of nearly 10 percent is reached with $b \approx \lambda/4$ and $(b/a) < 1.5$. The corresponding cavity depth $z_i > 0.04\lambda$, where the smaller depth figures are associated with the larger (b/a) ratios. For a given $G/|B|_{res}$ ratio, the slots are of approximately the same size for coaxial and cylindrical cavities. The radius of the resonant coaxial cavity, however, is almost one half of the radius of the cylindrical cavity. The effect of a dielectric cavity loading ($\epsilon_i > \epsilon_0$) is indicated by the dotted and dashed curves. Dielectric loading decreases the resonant cavity depth; it also decreases the $G/|B_{res}|$ ratio.

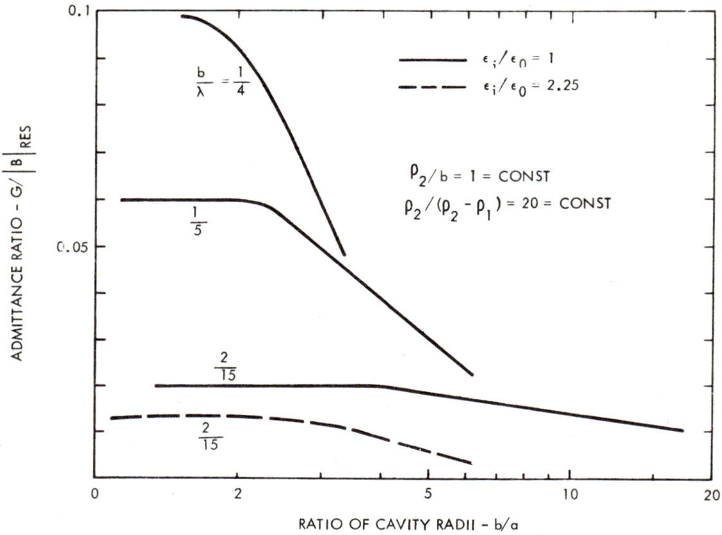

FIG. 7.12. Admittance ratio $G/|B|$ at resonance. Coaxial cavity.

For very small slots of $(\rho_2/\lambda) \ll 1$, the slot susceptance is approximated by the first term of (7.55). Numerical calculations show that this approximates B within a few percent if $(\rho_2/\lambda) < 0.02$. Now B is inductive and independent of the dielectric loading. The resulting G/B is given by

$$\frac{G}{B} = \frac{G}{B_{ind}} = \frac{(k\rho)^4(kz_i) \log(b/a)}{6}, \tag{7.59}$$

which is seen to be much less than $|G/B|_{max}$ of (7.57) of the cylindrical cavity.

7.3. Exercises

7.1 Compute the coefficients γ^i_{NM} of (7.24) for the cavity-backed rectangular slot.

7.2 Derive the expressions γ^0_{NM} indicated following (7.26) for a rectangular slot radiating into free space.

7.3 Derive the solution $V_1(x)$ for the cavity-backed rectangular slot using the zero order approximation $V^t_0(x)$ of (7.44).

7.4 Evaluate the admittance Y_i of (7.55) which is reflected into the plane of the annular slot by (a) a cylindrical cavity, (b) a coaxial cavity.

7.4. References

COHEN, M. H. (1954) On the bandwidth of cavity antennas, *J. Appl. Phys.* **25**, 582–7.

GALEJS, J. (1963a) Admittance of a rectangular slot which is backed by a rectangular cavity, *IEEE Trans. on Ant. and Prop.* **AP-11** (2), 119–26.

GALEJS, J. (1963b) Hallen's method in the problem of a cavity-backed rectangular slot antenna, *NBS J. Res.* **67D** (2), 237–44.

GALEJS, J. and THOMPSON, T. W. (1962) Admittance of a cavity-backed annular slot antenna, *IRE Trans. on Ant. and Prop.* **AP-10** (6), 671–8.

KING, R. W. P. (1956) *Theory of Linear Antennas*, Harvard University Press, Cambridge, Mass.

LEVINE, H. and PAPAS, C. H. (1951) Theory of circular diffraction antenna, *J. Appl. Phys.* **22**, 29–43. (Eq. (6.12) should be multiplied with a factor $2[(r-1)/(r+1)]^2$).

MARCUVITZ, N. (1951) *Waveguide Handbook*, McGraw-Hill, New York, NY, section 4.16.

PISTOLKORS, A. A. (1948) Theory of the circular diffraction antenna, *Proc. IRE* **36**, 54–60.

SMYTHE, W. R. (1950) *Static and Dynamic Electricity*, McGraw-Hill, New York, NY, (see Problem 4.21).

STEVENSON, A. F. (1948) Theory of slots in rectangular waveguides, *J. Appl. Phys.* **19**, 24–38.

VASILIADIS, A. and TANNER, R. L. (1960) Theoretical limitations on the broadbanding potential of antennas, with application to cavity-backed slots and other antennas, Stanford Research Institute, Menloe Park, Calif., Tech. Rep. 71.

WAIT, J. R. (1958) A low-frequency annular-slot antenna, *NBS J. Res.* **60**, 59–64. (Eq. (12) should be multiplied by 2.)

WATSON, W. H. (1947) *The Physical Principles of Waveguide Transmission and Antenna Systems*, Oxford, Clarendon Press (see Section 10).

WHEELER, H. A. (1947) Fundamental limitations of small antennas, *Proc. IRE* **35**, 1479–84.

CHAPTER 8

SLOT ANTENNAS WITH A STRATIFIED DIELECTRIC OR ISOTROPIC PLASMA LAYERS

THE problem of the plasma or dielectric covered antenna has been treated in numerous papers, and the radiation patterns of a variety of sources have received particular attention. Radiation from line source excited dielectric or plasma layers has been considered by Tai [1951] Tamir and Oliner [1962], and Oliner and Tamir [1964]. Slotted cylindrical antennas with a dielectric coating are discussed by Wait [1959]. He considered also the fields of an electric dipole in the presence of a thin plasma sheath [Wait, 1961]. Axial slots on metallic cylinder are discussed by Rush [1963], Swift [1964], and Hasserjian [1965]. Circumferential slots have been investigated by Knop [1961] and radiation from a finite aperture in a coated plane is discussed by Knop and Cohn [1964]. Radiation through multi-layer cylindrical plasma sheath is discussed by Harris [1964] and by Harris *et al.* [1965]. The antenna impedance can be determined from the antenna near fields, which are in general more difficult to compute than the radiation fields. This problem is considered in few papers. Marini [1960] formulates the radiation patterns and also the input impedance of an insulated slotted spherical antenna surrounded by a strongly ionized plasma sheath. But he discusses in detail only the electrically small sphere. This work is extended to weak plasma sheaths by Katzin and Koo [1964]. The mutual admittance between two magnetic line sources in the presence of a thick plasma layer is computed by Yee [1961]. All of the above work assumes the field distribitions along the antenna, and it does not consider the perturbing effects of the plasma layer. An iteration technique is used by Raemer [1962] in considering the effects of a plasma shell on the current distribution along the antenna, but there are no satisfactory results for the antenna impedance. A variational formulation of a slot impedance has been worked out by Cutler [1959]. The fields in the plasma layer and in the outside space are considered as a superposition of waveguide modes. Since there appear to be computational difficulties, this formulation has not yielded workable equations or any numerical results.

Another variational formulation for the slot impedance has been given by Galejs [1946b]. The dyadic Green's function for a plasma layer of finite thickness is obtained from the free space Green's function by the method of images. The fields outside the slot depend on a surface integral of the fields over the slot plane and over the surface of the plasma layer. For plasma layers of a thickness greater than a wavelength, the surface fields are related to the fields in the slot plane with the aid of plane wave reflection coefficients, which leads to an integral equation for the fields in the slot plane. This formulation [Galejs, 1964b] is rather involved and does not apply to thin or stratified plasma layers.

A more straightforward method has been developed by Galejs [1965a, b]. A rectangular wavelength is excited in its principal mode, and radiates into a stratified plasma or dielectric. The fields are expressed as a superposition of TE and TM modes and the slot admittance has been calculated variationally for aperture fields that are approximated by several trial functions. Villeneuve [1965] has used a variational formulation for a waveguide radiating into a plasma layer but he restricts the aperture fields to the principal mode of the waveguide. A similar approach is used by Compton [1964] for computing the aperture admittance of a parallel plate or rectangular waveguide in the presence of a lossy dielectric slab. Galejs [1965b] has calculated the mutual impedance between two waveguides; and Galejs and Mentzoni [1965] have experimentally verified the admittance calculations.

The admittance of an annular slot antenna radiating into a plasma layer has also been calculated [Galejs, 1964a]. The same technique is applied to impedance and efficiency calculations of buried vertical dipole antennas that are approximated for purposes of analysis by a cavity-backed slot antenna covered by layers of lossy dielectric [Galejs, 1966].

8.1. Admittance of a Rectangular Waveguide

8.1.1. *The General Formulation*

The waveguide geometry under consideration is depicted in Fig. 8.1. A rectangular waveguide of cross-sectional area $x_i y_i$ is joined to an infinite flange by a symmetrical aperture of width 2ϵ and length $2l$. Layered isotropic plasma (or dielectric) fills the space $0 \leqslant z \leqslant z_1$. However, there is free space for $z > z_1$. The waveguide is excited in its principal TE_{10} mode by waves that propagate in the positive z direction. There are also reflected waves in this guide. Waves propagating in the positive and negative z directions are excited in the outer space ($z \geqslant 0$). The fields consist of a superposition of TE and TM modes that can be derived from scalar functions Ψ_j and Φ_j respectively. The subscripts $j = i$ or p designate the guide portion or the outside space. The functions Ψ_i and Φ_i are given by (3.8) and (3.9) as

$$\Psi_i = a_{10} \cos \beta_{xi1} x' (e^{\gamma_{i1} z} + Re^{-\gamma_{i1} z}) + \sum_n \sum_m{}' a_{nm} \cos \beta_{xi} x' \cos \beta_{yi} y' \, e^{-\gamma_i z}, \quad (8.1)$$

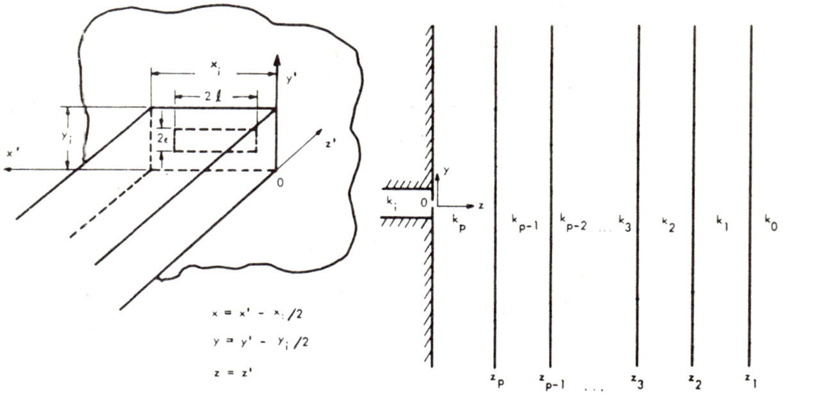

FIG. 8.1. Geometry for admittance calculation.

$$\Phi_i = \sum_n \sum_m b_{nm} \sin \beta_{xi} x' \sin \beta_{yi} y' \, e^{-\gamma_i z}. \tag{8.2}$$

For a laterally unbounded space ($z > 0$), the discrete summations of (8.1) and (8.2) are replaced by integrals and Ψ_p and Φ_p are defined by

$$\Psi_p = \left(\frac{1}{2\pi}\right)^2 \iint_{-\infty}^{\infty} A(u, v) \, e^{iux} \, e^{ivy} (e^{\gamma_p z} + R_a \, e^{-\gamma_p z}) \, du \, dv, \tag{8.3}$$

$$\Phi_p = \left(\frac{1}{2\pi}\right)^2 \iint_{-\infty}^{\infty} B(u, v) \, e^{iux} \, e^{ivy} (e^{\gamma_p z} + R_b \, e^{-\gamma_p z}) \, du \, dv. \tag{8.4}$$

The symbols of (8.1) to (8.4) are defined as

$$\beta_{xi} = \frac{n\pi}{x_i}, \beta_{xil} = \frac{\pi}{x_i}, \beta_{yi} = \frac{m\pi}{y_i}, k_j = (\omega^2 \mu_0 \epsilon_j + i\omega\mu_0\sigma_j)^{1/2}$$

with $j = i$ or p, $\gamma_i = i(k_i^2 - \beta_{xi}^2 - \beta_{yi}^2)^{1/2}$, $\gamma_{il} = i(k_i^2 - \beta_{xil}^2)^{1/2}$, $\gamma_p = i(k_p^2 - u^2 - v^2)^{1/2}$ and $\epsilon_m = 1$ for $m = 0$, $\epsilon_m = 2$ for $m \neq 0$. The prime on the double summation (8.1) designates the omission of its $n = 1$, $m = 0$ term. The field components are related to Ψ_j and Φ_j by (3.10) to (3.15). After (8.1) and (8.2) are substituted in (3.10) and (3.11), the resulting expression can be solved for a_{10}, a_{nm}, and b_{nm}, giving

$$a_{10} = \frac{2/(i\omega\mu_0)}{\beta_{xil} x_i y_i (1+R)} \iint E_y \sin \beta_{xil} x' \, dx' \, dy', \tag{8.5}$$

$$a_{nm} = -\frac{2\epsilon_m/(i\omega\mu_0)}{x_i y_i (\beta_{xi}^2 + \beta_{yi}^2)} \left[\beta_{yi} \iint E_{xi} - \beta_{xi} \iint E_{yi}\right], \tag{8.6}$$

$$b_{nm} = -\frac{2\epsilon_m}{x_i y_i (\beta_{xi}^2 + \beta_{yi}^2)\gamma_i} \left[\beta_{xi} \iint E_{xi} + \beta_{yi} \iint E_{yi}\right] \tag{8.7}$$

where the integrals $\iint E_{xi}$ and $\iint E_{yi}$ are defined by

$$\iint E_{xi} = \iint E_x \cos \beta_{xi} x' \sin \beta_{yi} y' \, dx' \, dy', \tag{8.8}$$

$$\iint E_{yi} = \iint E_y \sin \beta_{xi} x' \cos \beta_{yi} y' \, dx' \, dy'. \tag{8.9}$$

Substituting (8.3) and (8.4) in (3.10) and (3.11) and solving for $A(u, v)$ and $B(u, v)$ gives

$$A(u, v) = \frac{-iv \iint E_{xp} + iu \iint E_{yp}}{i\omega\mu_0 \gamma_p (u^2+v^2)(1+R_a)}, \tag{8.10}$$

$$B(u, v) = \frac{-iu \iint E_{xp} - iv \iint E_{yp}}{\gamma_p (u^2+v^2)(1-R_b)}, \tag{8.11}$$

where $\iint E_{jp}$ is defined by

$$\iint E_{jp} = \iint E_j \, e^{-iux} \, e^{-ivy} \, dx \, dy. \tag{8.12}$$

E_x and E_y refer to the aperture fields at $z = 0$ and $j = x$ or y. After a_{nm}, b_{nm}, $A(u, v)$, and $B(u, v)$ are substituted in (8.1) to (8.4), the magnetic field components H_{xj} and H_{yj} of (3.13) and (3.14) are related to the integrals of the electric field components (8.8), (8.9), and (8.12). The tangential magnetic field components, H_x and H_y, are continuous across the aperture. The H_x of (8.1), (8.2), and (3.13) is equal to H_x of (8.3), (8.4), and (3.13). Also H_y of (8.1), (8.2), and (3.14) is equal to H_y of (8.3), (8.4), and (3.14). The H_x equations are multiplied with E_y and integrated over the slot plane; the H_y equations are multiplied with E_x and also integrated over the slot plane. The resulting two equations contain terms proportional to E_x and E_y integrals of (8.8), (8.9), and (8.12).

After E_y is assumed to be an even functional with respect to x and y, it follows from (3.10) and (3.11) that E_x is an odd function. It is also seen that $E_x(x, y) = E_x(-x, -y)$. Using these properties of E_y and E_x, the two earlier equations can be combined and the normalized slot admittance of the waveguide $y_s = (1-R)/(1+R)$ is computed as

$$y_s = \frac{x_i y_i}{\gamma_{i1} D^2} \left\{ \sum_n \sum_m{}' \frac{\epsilon_m}{x_i y_i \gamma_i} \left[(\beta_{xi}^2 - k_i^2) \left(\iint E_{yi} \right)^2 + (k_i^2 - \beta_{yi}^2) \left(\iint E_{xi} \right)^2 \right] + I \right\}, \tag{8.13}$$

where

$$I = \frac{1}{8\pi^2} \int_{-\infty}^{\infty} \int_{-\infty}^{\infty} \left[F(R) \left(\iint E_{yp} \right)^2 + G(R) \left(\iint E_{xp} \right)^2 \right] du \, dv, \tag{8.14}$$

$$F(R) = \frac{1}{(u^2+v^2)\gamma_p} \left[\gamma_p^2 u^2 \frac{1-R_a}{1+R_a} - k_p^2 v^2 \frac{1+R_b}{1-R_b} \right], \tag{8.15}$$

$$G(R) = \frac{1}{(u^2+v^2)\gamma_p}\left[-\gamma_p^2 v^2 \frac{1-R_a}{1+R_a} + k_p^2 u^2 \frac{1+R_b}{1-R_b}\right] \tag{8.16}$$

$$D = \iint E_y(x', y') \sin \beta_{xil} x' \mathrm{d}x' \mathrm{d}y'. \tag{8.17}$$

The reflection coefficients R_a and R_b of the TE and TM modes, respectively, depend on the dielectric structure for $z > 0$. For stratified dielectric layers shown in Fig. 8.1, the reflection coefficient R_{dj} ($d = a$ or b) in the region of $z_{j+1} \leqslant z \leqslant z_j$ is related to the reflection coefficient $R_{d(j-1)}$ of the region $z_j \leqslant z \leqslant z_{j-1}$ by the expressions

$$R_{aj} = \mathrm{e}^{2\gamma_j z_j} \frac{[\mathrm{e}^{2\gamma_{j-1} z_j} + R_{a(j-1)}] - [(\gamma_{j-1})/\gamma_j][\mathrm{e}^{2\gamma_{j-1} z_j} - R_{a(j-1)}]}{[\mathrm{e}^{2\gamma_{j-1} z_j} + R_{a(j-1)}] + [(\gamma_{j-1})/\gamma_j][\mathrm{e}^{2\gamma_{j-1} z_j} - R_{a(j-1)}]}, \tag{8.18}$$

$$R_{bj} = \mathrm{e}^{2\gamma_j z_j} \frac{[(k_{j-1})/k_j]^2[\mathrm{e}^{2\gamma_{j-1} z_j} + R_{b(j-1)}] - [(\gamma_{j-1})/\gamma_j][\mathrm{e}^{2\gamma_{j-1} z_j} - R_{b(j-1)}]}{[(k_{j-1})/k_j]^2[\mathrm{e}^{2\gamma_{j-1} z_j} + R_{b(j-1)}] + [(\gamma_{j-1})/\gamma_j][\mathrm{e}^{2\gamma_{j-1} z_j} - R_{b(j-1)}]}. \tag{8.19}$$

R_{dj} is derived by considering the scalar functions (8.3) and (8.4) in two adjacent layers and by requiring the tangential field components (3.10), (3.11), (3.13), and (3.14) to be continuous across the interface. The computations should start with $j = 1$, where $R_{d(j-1)} = R_{d0} = 0$ in (8.18) and (8.19). A series of computations then gives R_{d1}, R_{d2}, R_{d3}, and, finally, R_{dp} as R_a or R_b in (8.15) and (8.16). If E_y is assumed to be an even function with respect to the x and y axis, it follows from (8.1), (8.2), (3.10), and (3.11) that E_x is an odd function. The integral (8.12) is then changed to

$$\iint E_{xp} = -\iint E_x \sin ux \sin vy \,\mathrm{d}x \,\mathrm{d}y, \tag{8.20}$$

$$\iint E_{yp} = \iint E_y \cos ux \cos vy \,\mathrm{d}x \,\mathrm{d}y. \tag{8.21}$$

The stationarity of (8.13) can be demonstrated by examining the first order change δy_s due to small changes δE_x and δE_y about their correct values E_x and E_y. This is done in a similar manner to that of Section 1.5. As a result of these calculations [Galejs, 1963c], δy_s is shown to be zero.

Trial functions for the aperture fields can be selected in such a way that the double integrals (8.8), (8.9), (8.17), (8.20), and (8.21) can be evaluated in a closed form. The slot admittance (8.13) is now expressed in terms of a double summation and double integrals of the variables u and v. It is possible to evaluate these double integrals numerically, as was shown by Villeneuve [1965]. In the present development, however, the double integrals are evaluated by approximating them by discrete summations. For a particular constant mesh approximation, these double integrals are reduced to discrete summations that can be identified as modes of a wide rectangular waveguide enclosing the space $z \geqslant 0$.

After the function within the square bracket of (8.14) is denoted by $H(u, v)$, I is rewritten as

$$I = \frac{1}{8\pi^2} \int_{-\infty}^{\infty} \int_{-\infty}^{\infty} H(u, v)\, du\, dv. \tag{8.22}$$

The integrals $\int\int E_{xp}$ and $\int\int E_{yp}$ are odd and even with respect to u or v, but their squares are both even functions in (8.14). Also $F(R)$ and $G(R)$ are even of u and v. Consequently $H(u, v)$ is even with respect to u and v. Changing the variable of integration in (8.22) to n and m, where $u = n\pi/x_0$ and $v = m\pi/y_0$, gives

$$I = \frac{1}{2\pi^2} \int_0^{\infty} \int_0^{\infty} H(u, v)\, du\, dv$$

$$= \frac{1}{2x_0 y_0} \int_0^{\infty} \int_0^{\infty} H(n, m)\, dn\, dm. \tag{8.23}$$

For sufficiently large values of x_0 and y_0, $H(n, m)$ will change only slightly even when n or m are changed by integers. The integral may be broken down in a summation of terms

$$\int_{n_0-1}^{n_0+1} \int_{m_0-1}^{m_0+1} H(n, m)\, dn\, dm \approx 4H(n_0, m_0) \tag{8.24}$$

for nonnegative lower limits of the integrals. The factor 4 is replaced by 2 for n_0 or $m_0 = 0$. With n restricted to odd integers and m to even integers, I is approximated by the summation

$$I \approx \frac{1}{x_0 y_0} \sum_n \sum_m \epsilon_m H(n, m). \tag{8.25}$$

Using this restriction of permissible n and m values, introducing shifted coordinates $x'' = x' + x_0/2$ and $y'' = y' + y_0/2$, and noting that only the squares of $\int\int E_{xp}$ and $\int\int E_{yp}$ enter into the impedance expression, (8.20) and (8.21) are changed to

$$\int\int E_{xp} = \int\int E_x \cos \beta_{xp} x'' \sin \beta_{yp} y''\, dx''\, dy'', \tag{8.26}$$

$$\int\int E_{yp} = \int\int E_y \sin \beta_{xp} x'' \cos \beta_{yp} y''\, dx''\, dy'', \tag{8.27}$$

where $\beta_{xp} = n\pi/x_0$, $\beta_{yp} = m\pi/y_0$. Equations (8.26) and (8.27) are of the same form as (8.8) and (8.9). It can be seen that the modes characterized by

β_{xp} and β_{yp} for $z \geqslant 0$ are representative of a rectangular waveguide whose inside dimensions are x_0 and y_0 and which is symmetrical with respect to the aperture in the $z = 0$ plane [Galejs, 1965a]. The approximation of (8.22) by (8.25) therefore changes the laterally unbounded geometry for $z \geqslant 0$ into a rectangular waveguide of inner dimensions x_0 and y_0.

The slot admittance y_s of (8.13) is formally specified in terms of the electric field components E_y and E_x in the aperture at $z = 0$. It remains to find a compatible set of E_y and E_x for the aperture.

The boundary conditions used in the derivation of y_s still apply as E_x is made very small, and the amplitude of the TE and TM modes of the waveguide may be selected in such a way that $(\int\int E_{xj})^2$ is negligible relative to $(\int\int E_{yj})^2$ in (8.13). The admittance is now determined by specifying only a single electric field component E_y in the aperture. This is in line with earlier work on radiation from waveguide backed slots without dielectric discontinuities [Lewin, 1951; Stevenson, 1948; Watson, 1947], or with dielectric discontinuities [Cohen *et al.*, 1951].

It is also possible to select reasonable trial functions with $E_x \neq 0$. When assuming that the principal waveguide mode $(H_x = H_0 \sin \beta_{xil} x')$ specifies the aperture field component H_x in a completely open smaller guide, $E_x \neq 0$ if y variations are allowed for the E_y component. In this particular case the E_x integrals of the slot admittance in (8.13) may be expressed in terms of E_y integrals, and a separate trial function is not required for the E_x component.

8.1.2. *Slot Admittance with a Negligible E_x Component*

With $E_x(x, y)$ negligible, the electric fields $E_y(x, y)$, are assumed to be of the form (7.12), (7.22), and (7.23), where the double primes of the x and y coordinates are omitted. The parameter k of the trial functions in (7.22) and (7.23) may be made equal to k_i, k_p, k_0, or kept arbitrary. The y variation of $E_y(x, y)$ is neglected, and $f(y)$ of (7.12) becomes

$$f(y) = -\frac{1}{2\epsilon}. \tag{8.28}$$

Substituting (7.12) in (8.13) and utilizing the stationary character of (8.13) for determining A and B, it follows that the normalized slot admittance y_s is given by (7.17) and (7.18), and the parameters γ_{MN} and F_M can be readily computed.

The slot admittance can be also computed for the sinusoidal field distribution $f_A(x)$ of (7.22) with $B = 0$ in (7.12). This gives y_s as in (7.20).

8.1.3. *Slot Admittance with a Non-negligible E_x Component*

The discussion of this section is restricted to small guides of $y_i = 2\epsilon$ and $x_i = 2l$ with an assumed magnetic field distribution $H_x = H_0 \sin \beta_{xil} x'$ in the aperture at $z = 0$. The absence of higher order modes of H_x in the aperture ($H_x \sim a_{10}$) establishes a relation between the amplitudes a_{nm} of TE and b_{nm} of TM modes following (8.1), (8.2), and (3.13) [Galejs, 1963c]. After applying (8.6) and (8.7), $\int\int E_{xi}$ is related to $\int\int E_{yi}$ by

$$\iint E_{xi} = \frac{\beta_{xi}^2 - k_i^2}{\beta_{xi}\beta_{yi}} \iint E_{yi} \tag{8.29}$$

for $m \neq 0$. Also the amplitudes of the TE_{n0} modes of the smaller guide are equal to zero, and E_y exhibits the same x variation as H_x, but may have a different y variation. With the relative amplitudes of TE and TM modes determined, E_x can be also expressed in terms of $(\int\int E_{yi})$, and the integral $\int\int E_{xp}$ computed as

$$\iint E_{xp} = \frac{2x_i^3 x_0 V_0}{\pi^2 \beta_{yp}} \frac{n(\beta_{xil}^2 - k_i^2)(-1)^{(n-1)/2}}{x_0^2 - n^2 x_i^2} \cos \beta_{xp} l \cdot (-1)^{m/2} \left[\frac{\sin \beta_{yp}\epsilon}{\beta_{yp}\epsilon} - Q(\beta_{yp}\epsilon) \right], \tag{8.30}$$

where V_0 is the potential across the center of the slot and $Q(w)$ is defined as

$$Q(w) = Q(\beta\epsilon) = -\frac{2}{V_0} \int_0^\epsilon E_y(u) \cos \beta u \, du \tag{8.31}$$

with $u = y' - \epsilon$. D of (8.17) is now given by $D = -V_0 x_i/2$, and the integrals of E_{yp} and E_{yi} are expressed as

$$\iint E_{yp} = \frac{2x_i x_0^2 V_0 (-1)^{(n-1)/2}}{\pi (x_0^2 - n^2 x_i^2)} (-1)^{m/2} \cos \beta_{xp} l Q(\beta_{yp}\epsilon), \tag{8.32}$$

$$\iint E_{yi} = \begin{cases} \dfrac{V_0 x_i}{2} (-1)^{(m/2)+1} Q\left(\dfrac{m\pi}{2}\right) & \text{for} \quad n = 1, m \neq 0 \\ 0 & \text{for} \quad n \neq 1. \end{cases} \tag{8.33}$$

The above integrals depend on the function $Q(w)$ or on the E_y variation in the y direction. The static solution for rectangular waveguide with $x_0 \gg x_i$ shows that near the waveguide edge at $u = |\epsilon - \delta|$ $(\delta \to 0)$ the electric field component E_y is proportional to $\delta^{-1/3}$. This field variation will be approximated by

$$E_y(u) = \frac{A + Bu^2/\epsilon^2}{(\epsilon^2 - u^2)^{1/3}} (-V_0/\epsilon^{1/3}) \cdot \tag{8.34}$$

The constants A and B are determined by requiring that E_y of (8.34) is equal to the static field near the edge, and that the integral of $E_y(u)$ across the aperture is equal to $-V_0$[or $Q(w)=$ for $\beta=0$]. The resulting expression of $E_y(u)$ has a 2 percent error at the center of the slot $(u=0)$. Substituting (8.34) in (8.31), $Q(w)$ can be evaluated using the integrals (1.3.8) and (1.1.5) of Erdelyi [1954]. A calculation shows that

$$Q(w)=\frac{1}{2}\left(\frac{3\pi}{2}\right)^{1/6}\Gamma\left(\frac{2}{3}\right)\left[\frac{J_{1/6}(w)}{w^{1/6}}-\frac{7}{3}\frac{J_{7/6}(w)}{w^{7/6}}\right]+\frac{7}{3}2^{1/6}\Gamma\left(\frac{7}{6}\right)\frac{J_{7/6}(w)}{w^{7/6}}, \tag{8.35}$$

where $J_n(w)$ is the Bessel function of the first kind of order n and $\Gamma(z)$ is the gamma function. The slot admittance y_s may be computed after substituting (8.29), (8.30), (8.32), and (8.33) into (8.13). The summations are extended over even integers m and odd integers n, except for the primed summation where $n=1$ because of (8.33).

The static solution for the slot in an infinitely thin screen gives $E_y(u) =-(V_0/\pi)(\epsilon^2-u^2)^{-1/2}$ and $Q(w)=J_0(w)$. This will give higher values of integrals than the more accurate form (8.34). When ignoring the y variation of $E_y(x,y)$ in the aperture, $Q(w)$ becomes $\sin w/w$ and the integrals $\int\int E_{xp}$, $\int\int E_{yi}$ and $\int\int E_{xi}$ are equal to zero according to (8.30), (8.33), and (8.29). The slot admittance computed from (8.13) using this particular form of $Q(w)$ in (8.32) is the same as (y_s) in the limit of $k\to\beta_{xil}$.

8.1.4. *Slot Admittance for Thick Plasma Layers*

For thick plasma layers, an alternate representation of fields outside the slot plane is obtained by modifying the solution for a homogeneous medium. This method does not apply to thin or stratified plasma; the essentials of the method will be illustrated only briefly.

The $H_{xp}(x,y)$ component of the magnetic field outside the slot plane $z=0^+$ of Fig. 8.1 is obtained by starting out with the solution for a homogeneous medium [Levine and Schwinger, 1950] and applying the method of images. It is shown by Galejs [1963b] that

$$H_{xp}=\frac{i}{2\pi\omega\mu_0}\sum_{n=\infty}^{\infty}\left\{\iint E_y(x',y')\left(k_p^2+\frac{\partial^2}{\partial x^2}\right)\frac{e^{ik_pR_{on}}}{R_{on}}\,dx'\,dy'\right.$$

$$\left.-\iint_{(z'=h)} E_y(\xi,\eta)\left(k_p^2+\frac{\partial^2}{\partial x^2}\right)\frac{e^{ik_pR_{hn}}}{R_{hn}}\,d\xi\,d\eta\right\}, \tag{8.36}$$

where

$$R_{on}=[(x-x')^2+(y-y')^2+(2nh)^2]^{1/2}, \tag{8.37}$$

$$R_{hn}=[(x-\xi)^2+(y-\eta)^2+(2n+1)^2h^2]^{1/2}, \tag{8.38}$$

and where the x' and y' coordinates of points on the plasma boundary at $z' = h$ have been designated by ξ and η. The first double integral of (8.36) is similar to the integral (3.7) with the exception of different values of R for $n \neq 0$. The second double integral of (8.36) is extended over the surface of the plasma layer. $E_y(\xi, \eta)$ should be related to $E_y(x', y')|_{z'=0}$ in order to obtain an integral equation for the electric fields in the slot plane. The boundary of $z' = h$ is assumed to be in the far fields of the source, and $E_y(\xi, \eta)$ may be calculated from multiple reflections of fields from a superposition of magnetic dipoles. It follows from Galejs [1963b] that

$$E_y(\xi, \eta) = -\frac{ik_p}{2\pi} \iint\limits_{(\text{slot})} \frac{h}{(\eta - y')^2 + (\xi - x')^2}$$

$$\cdot \sum_{m=0}^{\infty} \frac{E_y(x', y')\, dx'\, dy (2m+1)}{(2m+1)^2 h^2 + (\eta - y')^2 + (\xi - x')^2}$$

$$\cdot \exp\{ik_p[(2m+1)^2 h^2 + (\eta - y')^2 + (\xi - x')^2]^{1/2}\}$$

$$\cdot [(\eta - y')^2 (1 - R_{\parallel}) R_{\parallel}{}^m + (\xi - x')^2 (1 - R_{\perp}) R_{\perp}{}^m], \tag{8.39}$$

where the plane wave reflection coefficients $R_{\parallel}$ and $R_{\perp}$ are given by Stratton [1941] as

$$R_{\parallel} = \frac{k_0 \cos \theta_i - k_p \cos \theta_r}{k_0 \cos \theta_i + k_p \cos \theta_r}, \tag{8.40}$$

$$R_{\perp} = \frac{k_0 \cos \theta_r - k_p \cos \theta_i}{k_0 \cos \theta_r + k_p \cos \theta_i}, \tag{8.41}$$

where

$$\frac{k_p}{k_0} = \frac{\sin \theta_r}{\sin \theta_i}, \tag{8.42}$$

$$\theta_i = \tan^{-1} \frac{[(\xi - x')^2 + (\eta - y')^2]^{1/2}}{(2m+1) h} \tag{8.43}$$

and

$$k_0 = \omega (\mu_0 \epsilon_0)^{1/2}. \tag{8.44}$$

After substituting (8.39) in (8.36), H_{xp} is uniquely related to the electric field E_y in the aperture plane. The magnetic fields inside the slot plane ($z = 0^-$) are determined similarly to the method used in Section 8.1.1. It is also possible to derive a stationary expression for the slot admittance in the same manner. Further details of this development are listed by Galejs [1964b].

8.1.5. *Numerical Results*

The numerical calculations illustrate first the variations of the slot admittance when a slot radiates into an open waveguide of increasing size. Calculations were made for x_0/x_i increasing in increments of 0.1 and $y_0 = x_0$. The slot admittance of Fig. 8.2 exhibits systematic variations for smaller values

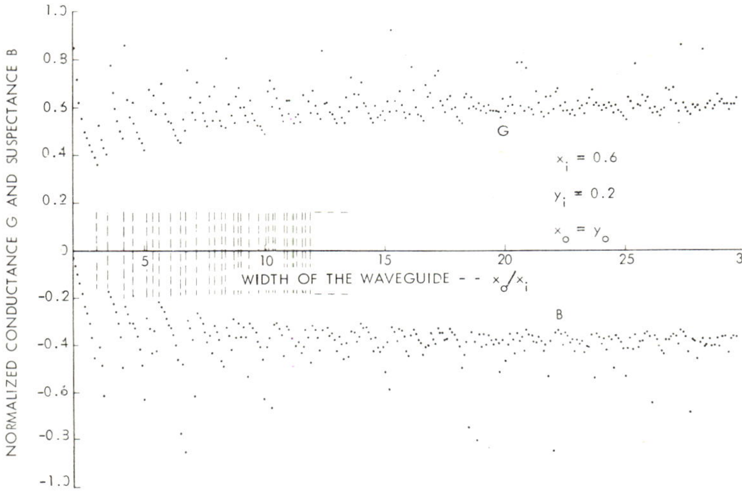

FIG. 8.2. Admittance of the slot radiating into a wide waveguide.

of x_0/x_i, but the admittance points appear to be scattered with smaller deviations for $x_0/x_i > 20$, although there are several widely scattered points. In absence of plasma ($R_a = R_b = 0$) the slot admittance y_s may become infinite at the mode cutoff when $\gamma_p = \gamma_0 \to 0$. Considering the free space admittance, this occurs for

$$x_0 = (\lambda/2)\,[n^2 + m^2]^{1/2},$$

where n are odd and m are even integers. These values of x_0 are indicated by vertical lines in Fig. 8.2 for the smaller values of x_0/x_i. The above values of x_0 correspond to the apparent discontinuity points of the slot admittance. The widely scattered admittance points for larger values of x_0/x_i are very close to these cutoff values of x_0. Although the separation between the cutoff points of the modes is decreased for increasing values of x_0 in Fig. 8.2, one has to approach the cutoff point more closely to obtain a sizeable deviation from the mean of the calculated admittance figures. This accounts for the smaller number of widely scattered admittance points for x_0/x_i large. The free space admittance of the present slot is $y = 0.63 - i0.35$ following Lewin

[1951], or $y = 0.65 - i0.38$ following [Galejs, 1964b]. The waveguide model is, therefore, effective for obtaining an approximation to the free space slot admittance, if the ratio of slot length to the outer waveguide dimension exceeds 10. In order to exclude the widely scattered admittance points from consideration, the slot admittance must be a continuous, gradually changing function of the guide dimensions near the selected outer guide dimension.

Examples of admittance calculations are shown in Fig. 8.3 for a single layer plasma of relative dielectric constant ϵ_p/ϵ_0, loss tangent $\tan \delta_p$, and

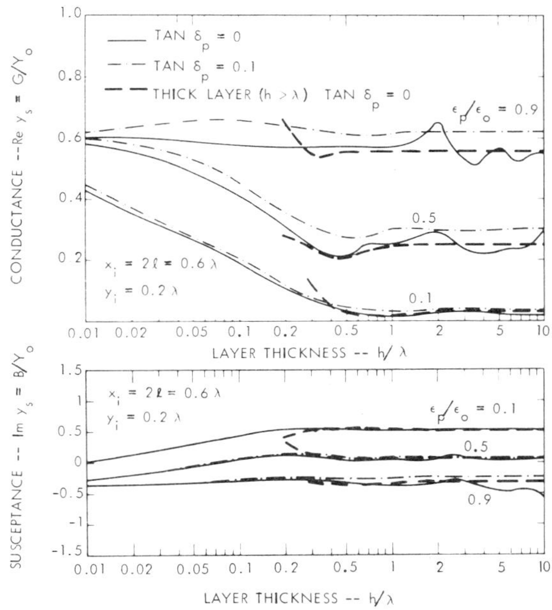

FIG. 8.3. Waveguide admittance; $\epsilon = 0.5 y_0$.

thickness $h = z_p$. The slot admittance starts out with a free space admittance for h/λ much less than 1 and approximates for $h/\lambda \approx 1$ the slot admittance of a thick plasma layer which is obtained using the method of Section 8.1.4. Although the admittance variation is relatively smooth for h/λ much less than 1, it becomes oscillatory for h/λ larger than 1. The slot admittance for thick plasma layers is approximately equal to the average of the oscillatory conductance curve for h large. For relatively thin layers of plasma, the admittance depends less on guide dimensions if the plasma layer provides a sharp discontinuity with respect to free space (e.g. $\epsilon_p/\epsilon_0 \ll 1$). In this case the guide admittance changes are determined principally by the effects of the nearby plasma and free space interface, with guide walls being of only a minor significance. For thick plasma layers, however, the plasma and air interface is at a distance comparable to the distance from the slot to the guide walls, and the admittance depends more on guide dimensions.

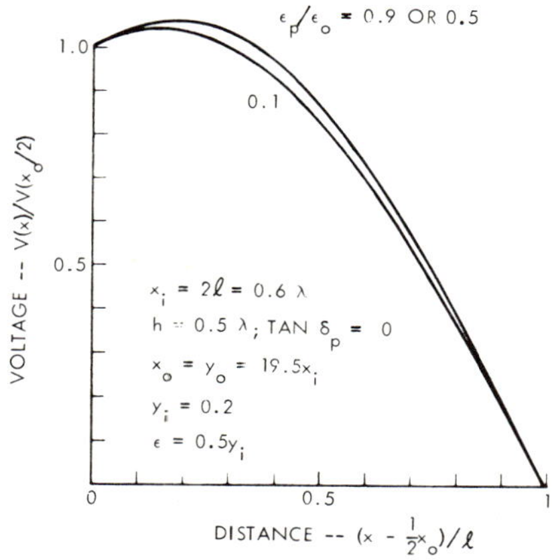

FIG. 8.4. Variational approximation to the voltage distribution along the waveguide excited slot $K = K_0$.

The variational approximation to the voltage distribution along the slots is computed from (7.12), (7.22), and (7.23) with the ratio A/B defined by (2.38) and with the slot and plasma parameters of Fig. 8.3. The voltage distribution, shown in Fig. 8.4, is seen to change only slightly with varying plasma parameters. These calculations are made with $k = k_0$ in the trial functions (7.22)

TABLE 8.1. COMPARISON OF SLOT ADMITTANCES

		$\epsilon_p/\epsilon_0 = 0.1\ h/\lambda = 0.1$	$\epsilon_p/\epsilon_0 = 0.5\ h/\lambda = 0.5$
Two-term trial function $E_y \neq f(y)$	$k = k_0$	$0.1913 + i0.4276$	$0.2160 + i0.05626$
	$k = k_p$	$0.1934 + i0.4153$	$0.2174 + i0.04758$
Principal mode E field, $E_y \neq f(y)$		$0.1881 + i0.4120$	$0.2143 + i0.05226$
Principal mode H_x and $E_y\ (x)$ Static $E_y(y)$	$E_x \neq 0$	$0.1875 + i0.3857$	$0.2128 + i0.04928$
Principal mode $E_y\ (x)$ Static $E_y\ (y)$	$E_x = 0$	$0.1875 + i0.3863$	$0.2128 + i0.04985$

Note: $x_0 = y_0 = 10.5x_i$, $x_i = 2l = 0.6\lambda$, $y_i = 2\epsilon = 0.2\lambda$.

and (7.23). For $k = k_p$ the voltage distribution is hardly changed [Galejs, 1963c]. These distributions are about the same as with the laterally unbounded plasma in fig. 7 of Galejs [1964b]. The calculated admittance is nearly the same for both values of k of the trial functions, as may be seen from the examples of table I. The accuracy of the voltage distribution shown in Fig. 8.4 can be strictly determined only by comparing these variational approximations with a more accurate result, if this were available. However, the assumed form of trial functions is general enough to allow for wide variations of the field distribution along the slot, such as observed in cavity backed slot antennas [see Fig. 7.4]. The close resemblance between the voltage distributions computed with $k = k_0$ and k_p suggests that they may approximate the true distribution. These distributions differ only slightly from the principal waveguide mode ($E_y \sim \sin \pi x'/x_i$) for $x_i = 2l$.

The preceding numerical calculations were based on electric aperture fields with neglected E_x and with a y-independent E_y. These calculations are compared in Table 8.1 with calculations made for the principal mode H_x field in the aperture, where E_y exhibits the static y variation and $E_x \neq 0$. The differences between the two sets of calculations are small. Even in the case where $E_x \neq 0$, the neglect of E_x does not affect the slot conductance and changes the susceptance by less than 1 percent. The change from the static $E_y(y)$ to a y-independent E_y causes a negligible change in G and about 7 percent change in B.

The changes of the susceptance B are larger than the changes of G for the cases listed in table I and further susceptance calculations are shown in Fig. 8.5. There is a change of B caused by a change of $E_y(y)$, but none by the neglect of E_x within the accuracy of the plot.

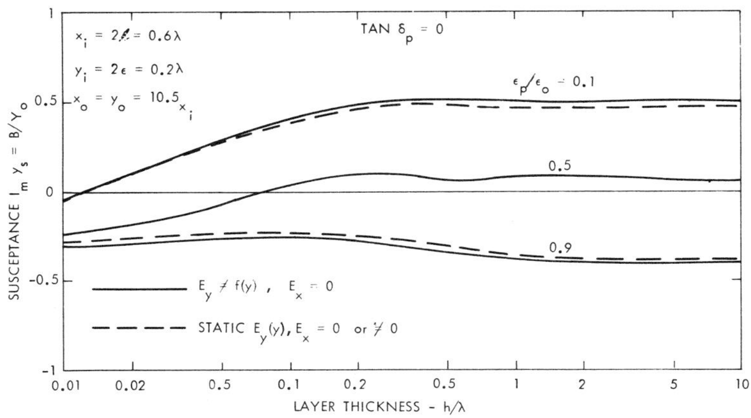

FIG. 8.5. Waveguide susceptance; comparison of solutions for $E_x = 0$ and $E_x \neq 0$.

The slot admittance of an open waveguide ($x_i = 2l$, $y_i = 2\epsilon$) radiating into a laterally unbounded plasma has been recently calculated by Villeneuve [1965]. He assumes the principal waveguide mode for the aperture field E_y with $E_x = 0$, and the slot admittance, which is obtained after numerically evaluating a twofold Fourier transform, should correspond to the admittance computed from (7.20) in the limit of $k \to \beta_{xil} = \pi/x_i$. A comparison of the results by Villeneuve [1965] with computations following Section 8.1.2 [$E_y \neq f(y)$, E_x neglected] and of Section 8.1.3 [H_x of the principal waveguide mode in the aperture, static $E_y(y)$ and $E_x \neq 0$ or $= 0$] is shown in Fig. 8.6. The free space slot admittance of Lewin [1951] is also indicated in

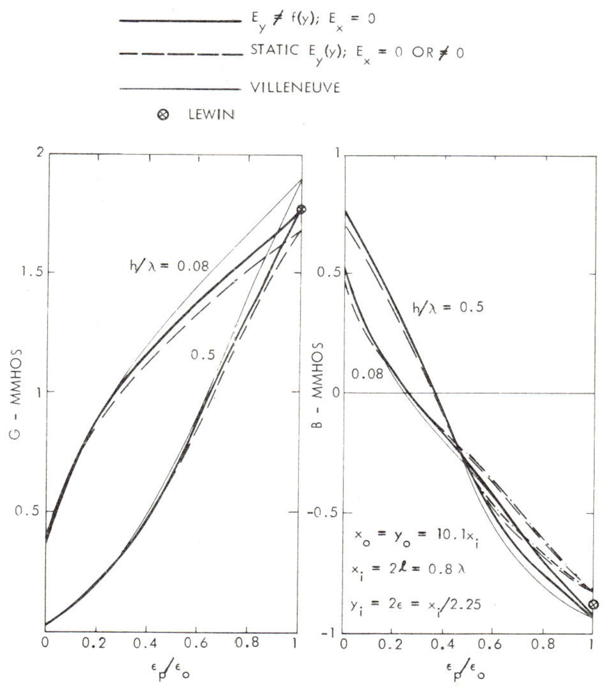

FIG. 8.6. Admittance of the waveguide slot.

Fig. 8.6. The results of Villeneuve [1965] agree with those for $E_y \neq f(y)$ of the smaller values of ϵ_p. Nevertheless, there are differences up to 5 percent for $\epsilon_p/\epsilon_0 > 0.6$. The neglect of E_x in the solution with the static $E_y(y)$ dependence gives a detectable change in the susceptance of Fig. 8.6 because of the increase in x_i/λ. (This is discussed in more detail in Galejs [1963c].)

8.1.6. *Admittance Measurements*

Admittance measurements are carried out first for dielectric layers. This provides a nontrivial check of the theoretical developments. The presence of surface waves for dielectric layers of $\epsilon_p/\epsilon_0 > 1$ causes additional singularities in the integrands of the admittance expression: the numerical evaluation of the integrals using a constant mesh size approximation can be expected to be less accurate than in the plasma case ($\epsilon_p/\epsilon_0 < 1$) where there are no surface waves.

The admittance of an X-band waveguide radiating into a dielectric layer of varying thickness d is shown in Fig. 8.7. The measurements are made with a waveguide of aperture length x_i and width y_i. The calculations for this antenna geometry are made first with the principal waveguide mode for the electric aperture fields and then for a two-term trial function using the same value of $k[k = 0.625\,k_0,\ k_0 = \omega(\mu_0\epsilon_0)^{1/2}]$. The two sets of calculations agree

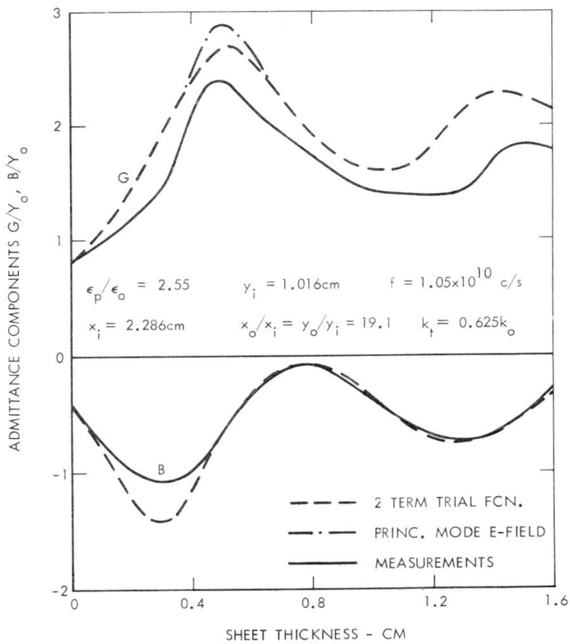

FIG. 8.7. Admittance of a waveguide-covered with a polystyrene sheet.

closely except for sheath thicknesses of approximately 0.5 cm. The aperture field distribution computed with a two-term trial function (Fig. 8.8) exhibits significant deviations from the principal waveguide mode. It is nearly triangular for $d \simeq 0.5$ cm, where the two sets of impedance calculations shown in Fig. 8.7 differ most. The impedance calculations have been repeated also

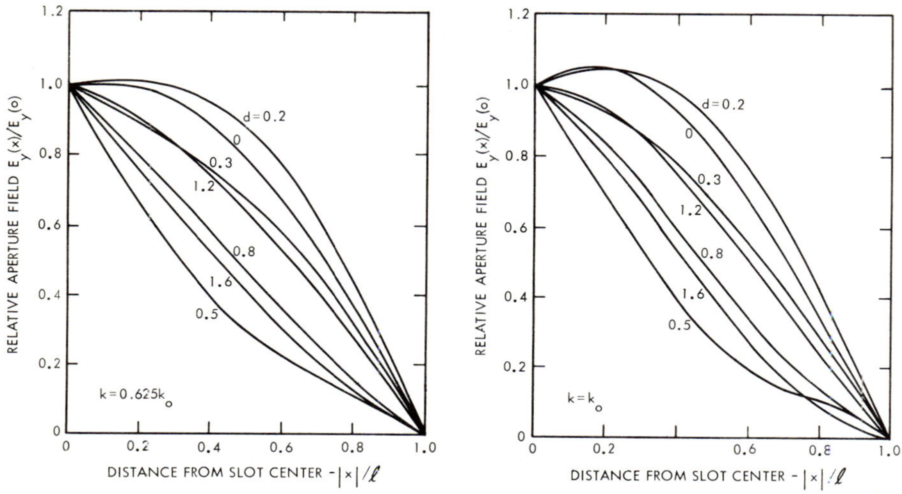

FIG. 8.8. Variational approximation to the aperture fields.

for $k = k_0$ in the trial functions, and do not differ from those with $k = 0.625k_0$. There are minor differences, however, in the computed field distributions, although the same basic trend of the fields are observable. There are substantial deviations of the fields from those of the principal waveguide mode in the aperture. However, because of the stationary character of the admittance expressions the principal waveguide mode can be used as aperture fields for impedance calculations.

The experimental studies of real plasma effects were done with a slot antenna in a gas-filled bell jar. The plasma layer was generated by a pulsed d.c. discharge. The admittance was measured at a frequency of 9.5 Gc/s for a standard X-band waveguide opening into a ground-plane slot as a function of time following the initiation of the discharge [Galejs and Mentzoni, 1965]. Maximum plasma densities occur right after the discharge (time = 0), and the admittance gradually approaches the value for a free-space layer. The calculated admittance data are shown in Fig. 8.9 for the same geometry and for a one-layer approximation of a typical plasma profile. Maximum plasma density corresponds to the lower values of the relative dielectric constant.

The measured results are indicated by solid points. The two values yielding the lowest dielectric constants correspond to post discharge times of 10 and 20 μsec, respectively. The experimental determination of electron densities resulted from double Langmuir probe measurements where it was assumed that the ion temperature is equal to room temperature. If ion temperatures of about 2000°K and 600°K are used, the open circles are obtained. It is seen that in this case there is good agreement with theoretical

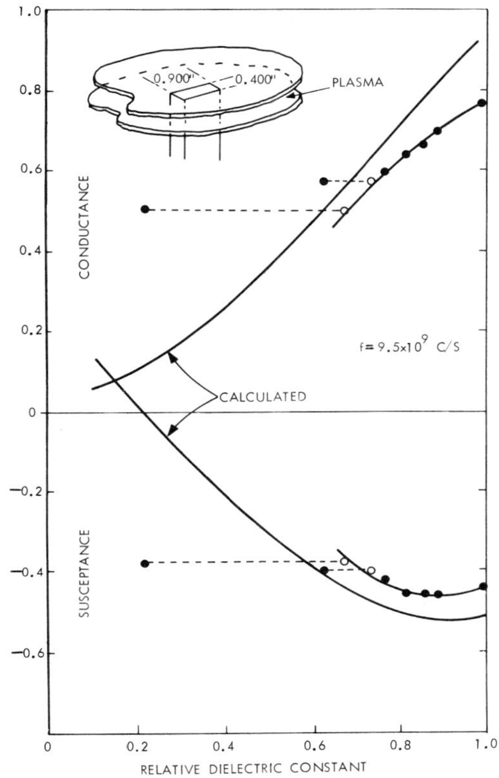

FIG. 8.9. Theoretical and measured admittance of a plasma-covered slot antenna.

results. An elevated ion temperature in the very early afterglow is to be expected.

8.2. Mutual Admittance between Two Rectangular Waveguides

Yee [1961] has computed the mutual admittance for parallel line sources and thick plasma layers, and Galejs [1965b] has considered finite waveguide backed slots. This latter formulation is similar to the one described for self admittances in Section 8.1, and will be further discussed in this section.

The mutual admittance between two rectangular waveguides radiating into a plasma layer is computed in the geometry of Fig. 8.14. The geometry is further specialized by assuming open waveguide apertures ($y_j = 2\epsilon_j$, $x_j = 2l_j$). The plasma layers that are above the $z = 0$ plane are not shown explicitly. In the self-admittance calculations of Section 8.1 the aperture fields of open guides could be approximated by the principal waveguide mode. More exact self-impedance calculations, that allow for different x and

y variations of E_y in the aperture, and that also consider the presence of an E_x component, differ very little from this simplified model. The same conditions are assumed to apply also to mutual admittance calculations for apertures that are parallel and symmetrical. However, these assumptions would require a further justification for slots that are offset in the x direction or that are not parallel. In circuit terminology the mutual admittance between two sets of terminals is defined as the ratio between the short circuit current in the terminals labeled (2) and the impressed voltage at the terminals labeled (1) or

$$Y_{21} = \frac{I_2}{V_1}\bigg|_{V_2=0}. \tag{8.45}$$

In the waveguide problem the currents I_j and the voltages V_j of a principal mode are proportional to the field components H_{xj} and E_{yj}. These aperture field components exhibit a cross-sectional x variation proportional to $\sin \beta_j x^i$ and no y variation. V_j and I_j are defined as [Marcuvitz, 1951]

$$V_j = \left(\frac{2}{x_j y_j}\right)^{1/2} \iint_{(\text{slot } j)} E_{yj} \sin \beta_j x^i \, dx^i \, dy^i = E_j\left(\frac{z_j y_j}{2}\right)^{1/2} \tag{8.46}$$

$$I_j = -\left(\frac{2}{x_j y_j}\right)^{1/2} \iint_{(\text{slot } j)} H_{xj} \sin \beta_j x^i \, dx^i \, dy^i, \tag{8.47}$$

where $\beta_j = \pi/x_j$, $j = 1$ or 2, and $i = {}'\text{or}''$. The magnetic field component, which is measured at terminals (2), and which is due to E_y at terminals (1), will be proportional to E_{y1} or V_1. Thus

$$H_{x2,1} = KV_1 \sin \beta_2 x''. \tag{8.48}$$

Multiplying (8.48) with $\sin \beta_2 x''$ and integrating over the aperture 2 gives

$$-I_2 = KV_1\left(\frac{x_2 y_2}{2}\right)^{1/2} \tag{8.49}$$

A comparison of (8.49) to (8.45) shows that

$$Y_{21} = -K\left(\frac{x_2 y_2}{2}\right)^{1/2} \tag{8.50}$$

The continuity of the magnetic field components across the aperture 2 leads to

$$H_{xp,1} = H_{x2,1}. \tag{8.51}$$

Substituting (8.48) and (8.50) in (8.51) gives

$$H_{xp,1} = -\left(\frac{2}{x_2 y_2}\right)^{1/2} Y_{21} V_1 \sin \beta_2 x''. \tag{8.52}$$

Multiplying both sides of (8.52) with E_{y2} and integrating over the aperture 2 gives

$$Y_{21} = -\frac{1}{V_1 V_2} \iint E_{y2} H_{xp,1}\, dx\, dy, \tag{8.53}$$

where E_{y2} and V_2 are computed after removing the short circuit of aperture 2. The stationary character of (8.53) can be demonstrated as shown in Section 1.5. After computing H_{x1} from (8.3), (8.4), and (3.13), it follows that

$$Y_{21} = \frac{1}{2\pi^2 i\omega\mu_0 (x_1 y_1 x_2 y_2)^{1/2}} \cdot \int_{-\infty}^{\infty} \int_{-\infty}^{\infty} F(R) \left[\frac{\iint E_{yp1}}{E_{y1}(0, y)}\right]\left[\frac{\iint E_{yp2}}{E_{y2}(0, y)}\right] e^{i(ux_c + vy_c)}\, du\, dv, \tag{8.54}$$

where $\iint E_{ypj}$ $(j = 1 \text{ or } 2)$ follows from (8.21) as

$$\iint E_{yp} = -\frac{4\beta_{xil} E_y(0, y) \cos ul \sin v\epsilon}{(u^2 - \beta_{xil}^2)\, v}, \tag{8.55}$$

and where the appropriate values of β_{xil}, l and ϵ of the two waveguides are used in (8.55). The admittance Y_{21} can be normalized by dividing it by the characteristic admittance of one of the waveguides, which is $Y_0 = \gamma_{il}/(i\omega\mu_0)$. For x_c and $y_c \to 0$, and $x_1 = x_2$ and $y_1 = y_2$, the two slots of Fig. 8.10 degenerate into a single slot, and the expression for the mutual admittance (8.54) gives the same self-admittance of a slot as the earlier calculation in

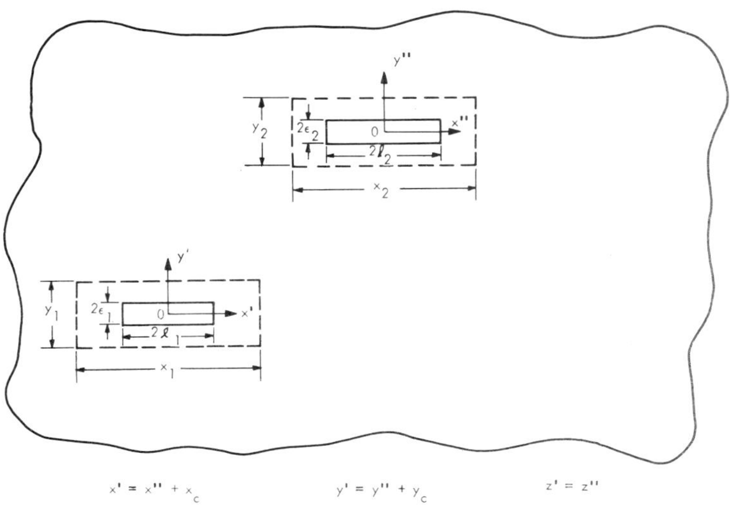

FIG. 8.10. Waveguide geometry for calculating mutual admittance.

(8.13). (The E_x integrals and the double summation of (8.13) are equal to zero when assuming the principal waveguide mode for the aperture fields.)

The double integral (8.54) can be approximated by a double summation as shown in (8.22) to (8.25). When changing to the discrete variables, n and m, where $u = n\pi/x_0$ and $v = m\pi/y_0$, x_0 and y_0 can be interpreted as the dimensions of a large waveguide that encloses the plasma layer laterally.

Numerical calculations are made for two parallel slots ($x_c = 0$) of equal size ($x_1 = x_2$, $y_1 = y_2$) that radiate into free space or in a single plasma layer of thickness z_p. The integrals (8.54) are approximated by summations with $x_0 = y_0 = 10.5x_1$; the accuracy of the procedure is tested by increasing $x_0 = y_0$ to $19.5x_1$. This gives negligible changes in the admittance figures; the mesh size in the integrals corresponding to $x_0 = y_0 = 10.5x_1$ was considered adequately small.

In free space with no plasma layers in front of the two waveguides, the mutual admittance curves of Fig. 8.11 show about the same relative variations as mutual impedance curves of two parallel thin half-wavelength-long dipoles in free space. This applies, in particular, to those values of the slot separation y_c/λ where either the mutual conductance or the susceptance goes through a zero.

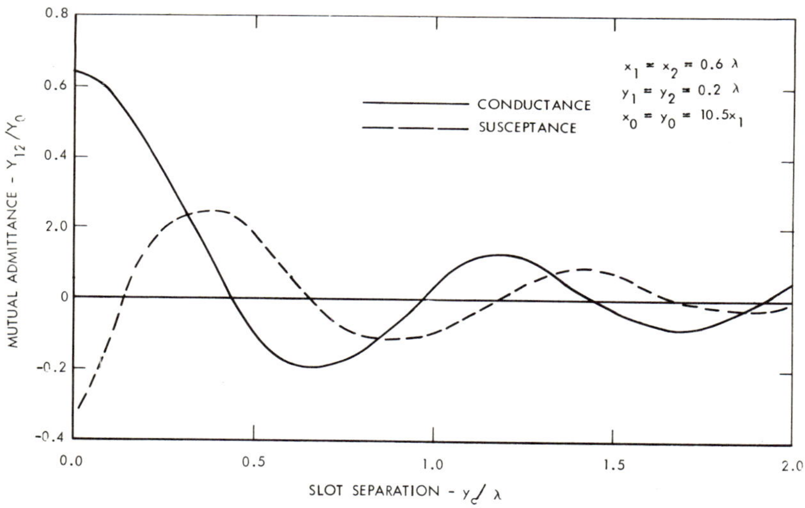

FIG. 8.11. Free space mutual admittance between two waveguide apertures.

The effects of a single lossless plasma layer of thickness $z_p = 0.1\lambda$ and λ are shown in Figs. 8.12 through 8.14 for relative dielectric constants $\epsilon_p/\epsilon_0 = 0.9$, 0.5, and 0.1. The mutual admittance variations of the thin plasma layer of $\epsilon_p/\epsilon_0 = 0.9$ are nearly the same as for the slots in free space. The mutual admittance is decreased by decreasing ϵ_p and by increasing the layer thickness. The plasma layer of $\epsilon_p/\epsilon_0 \ll 1$ tends to decouple two adjacent slots, and there is little resemblance between the mutual admittance curves for free space in Fig. 8.11 and for $\epsilon_p/\epsilon_0 = 0.1$ in Fig. 8.14.

The mutual admittance with a lossy plasma layer is illustrated in Fig. 8.15 for $\epsilon_p/\epsilon_0 = 0.5$ and $\tan \delta_p = 1$. A comparison with Fig. 8.12 shows an

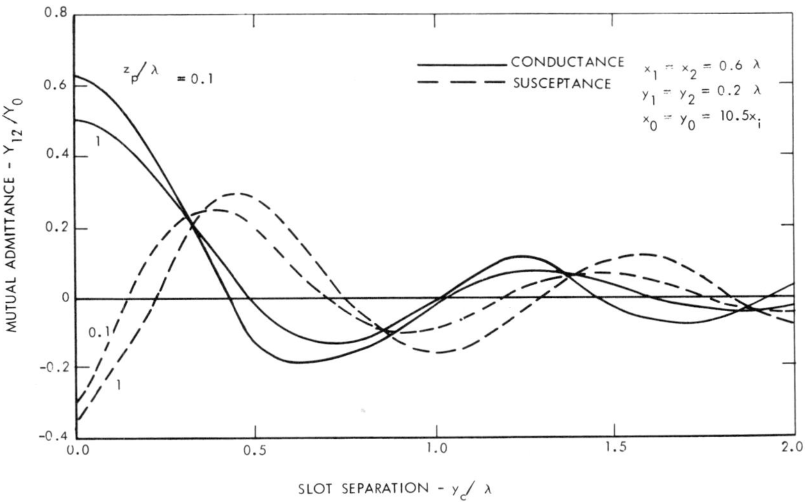

FIG. 8.12. Mutual admittance for a lossless plasma layer of $\epsilon_p/\epsilon_0 = 0.9$.

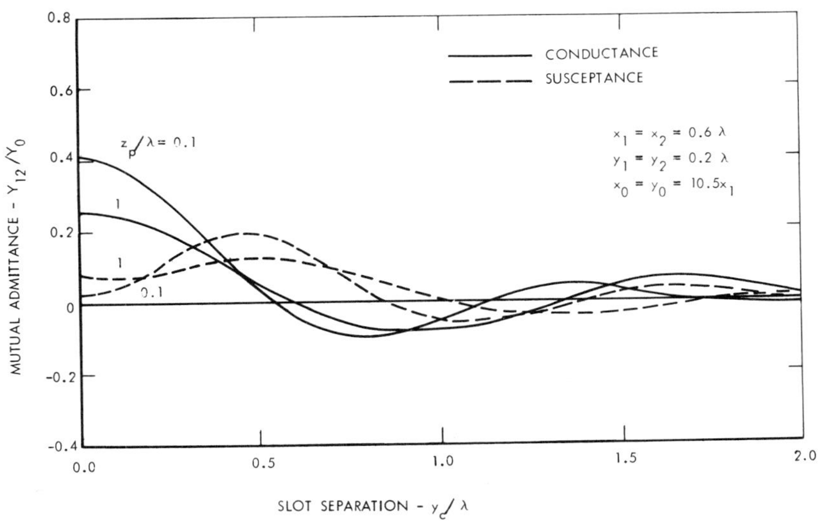

FIG. 8.13. Mutual admittance for a lossless plasma layer of $\epsilon_p/\epsilon_0 = 0.5$.

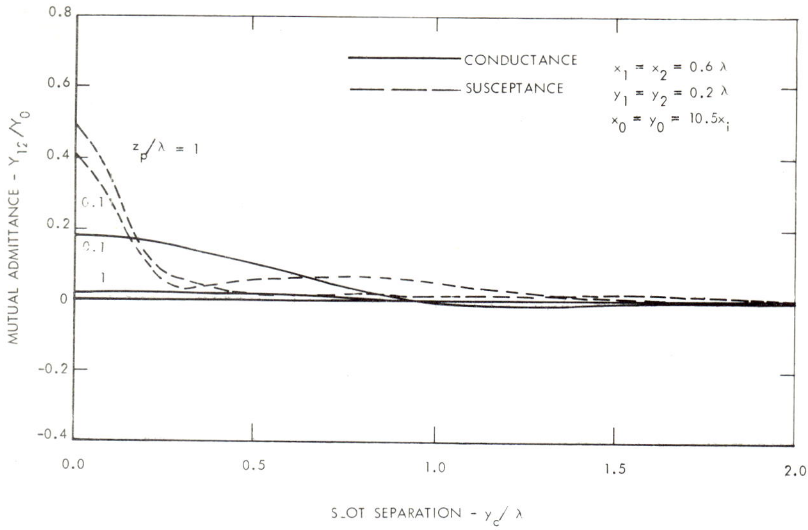

FIG. 8.14. Mutual admittance for a lossless plasma layer of $\epsilon_p/\epsilon_0 = 0.1$.

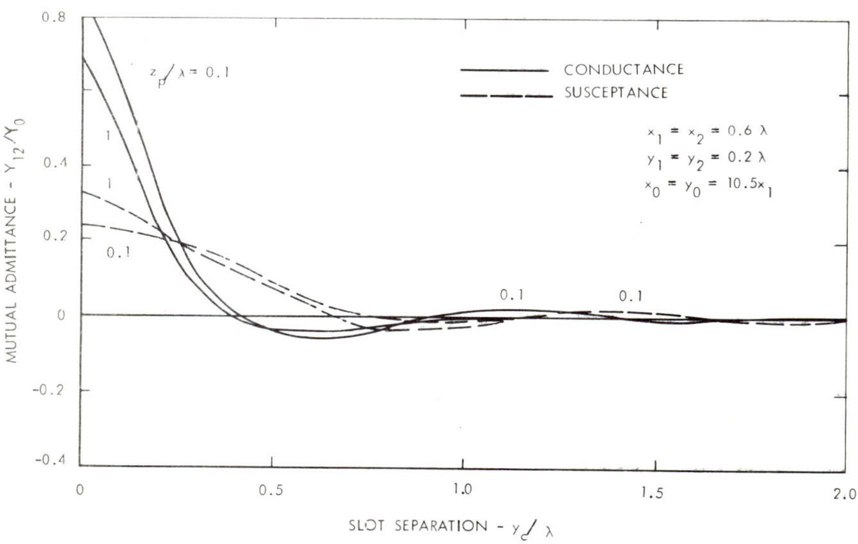

FIG. 8.15. Mutual admittance for a lossy plasma layer of $\epsilon_p/\epsilon_0 = 0.5$, $\tan \delta_p = 1$.

increased self-admittance, but a decreased mutual admittance for slot separations y_c exceeding 0.4λ. There is a further decoupling of the slots by a lossy plasma layer.

The free space mutual admittance between two thin infinitely long slots [Yee, 1961] is shown in Fig. 8.16. The mutual admittance curves of the infinitely long slots of Fig. 8.16 decay more gradually than in Fig. 8.11. Two-dimensional models will therefore indicate too high a mutual admittance figure when applied to finite length slots.

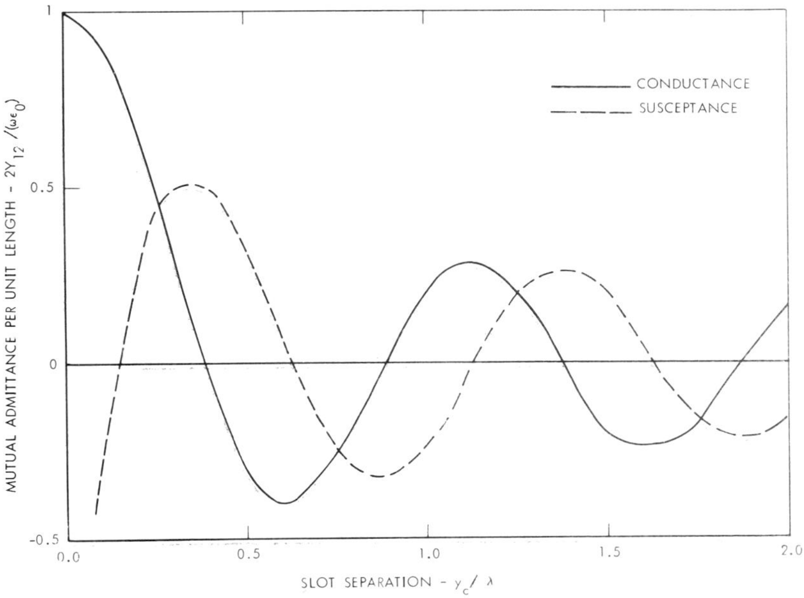

FIG. 8.16. Free-space mutual admittance between two thin infinitely long slots.

8.3. Admittance of an Annular Slot

8.3.1. *General Formulation*

A cavity-backed annular slot is covered with two layers of plasma or dielectric as indicated in Fig. 8.17. The ϕ-symmetrical fields inside the cavity are determined by the method used in Section 7.2.1. $H_{\phi i}$ inside the cavity is given by (7.46) to (7.50). The TM fields of the outside space ($z \geqslant 0$) can be deduced for circular cylindrical coordinates (ρ, ϕ, z) from a scalar function Φ as indicated by (1.21), (1.23), and (1.25). This gives

$$H_{\phi} = i\omega\epsilon \frac{\partial}{\partial\rho}\Phi, \tag{8.56}$$

$$E_{\rho} = \frac{\partial^2}{\partial z \partial\rho}\Phi = \frac{1}{i\omega\epsilon}\frac{\partial}{\partial z}H_{\phi}, \tag{8.57}$$

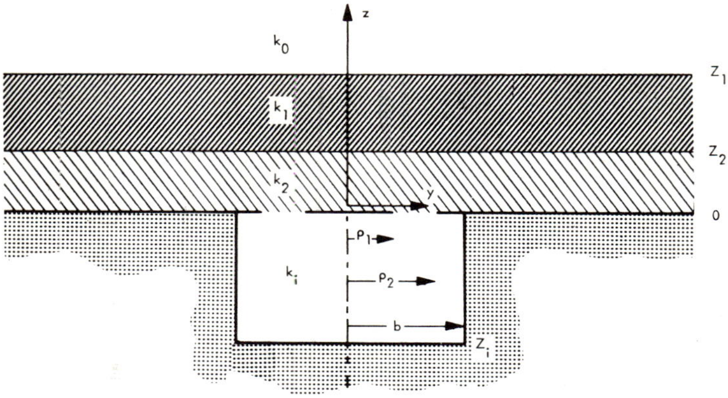

FIG. 8.17. Slot geometry.

$$E_z = \frac{1}{\rho}\frac{\partial}{\partial\rho}\left(\rho\frac{\partial}{\partial\rho}\Phi\right) = -\frac{1}{i\omega\epsilon}\frac{1}{\rho}\frac{\partial}{\partial\rho}(\rho H_\phi). \tag{8.58}$$

The function Φ is given in rectangular coordinates by (8.4). It is transformed into circular cylindrical coordinates using the relations $x = \rho \cos \phi$, $y = \rho \sin \phi$, $u = \lambda \cos \psi$, and $v = \lambda \sin \psi$. The function $B(u, v)$ is expressed as

$$B(u, v) = B(\lambda, \psi) = \sum_n B_n(\lambda) \exp(in\psi). \tag{8.59}$$

The ψ integral is readily evaluated, and (8.4) becomes

$$\Phi_j = \frac{1}{2\pi}\sum_n i^n e^{in\phi} \int_0^\infty B_n(\lambda)\,(e^{\gamma_j z} + R_b e^{-\gamma_j z})\,J_n(\lambda\rho)\lambda\,d\lambda, \tag{8.60}$$

where $\gamma_j = i(k_j^2 - \lambda^2)^{1/2}$, and $J_n(x)$ is the Bessel function of first kind or order n. For ϕ-symmetrical fields, only the $n = 0$ term of (8.60) is considered, and $H_{\phi j}$ of regions $j = 0, 1$, or 2 is computed from (8.56) as

$$H_{\phi j} = \int_0^\infty A_j(\lambda)\,(e^{\gamma_j z} + R_{bj}\,e^{-\gamma_j z})\,J_1(\lambda\rho)\,\lambda\,d\lambda. \tag{8.61}$$

In the slot plane $z = 0$, the radial electric field E_ρ is computed from (8.61) and (8.57) as

$$E_{\rho 2} = \frac{1}{i\omega\epsilon_2}\int_0^\infty A_2(\lambda)\gamma_2(1 - R_{b2})\,J_1(\lambda\rho)\,\lambda\,d\lambda, \tag{8.62}$$

where the expression $\epsilon_2 = \epsilon_{2r}(1 + i \tan \delta_2)$ may be complex. Equation (8.62) can be inverted using the Hankel-transform relation (8.1.1) of Erdelyi [1954], which gives

$$g(\lambda) = \int_0^\infty f(\rho)\,J_1(\lambda\rho)\rho\,d\rho \leftrightarrow f(\rho) = \int_0^\infty g(\lambda)\,J_1(\lambda\rho)\lambda\,d\lambda. \tag{8.63}$$

Multiplying the relation on the right-hand side with $J_1(\mu\rho)\rho$ and integrating with respect to ρ establishes $g(\lambda)$ of the left-hand side if

$$\int_0^\infty J_1(\lambda\rho)J_1(\mu\rho)\rho\,d\rho = \delta(\lambda-\mu)/\mu, \tag{8.64}$$

where $\delta(x)$ is the delta function. Using either (8.63) or (8.64), $A_2(\lambda)$ of the region just outside the aperture is related to the aperture fields $E_\rho(\rho)$ by

$$A_2(\lambda) = \frac{i\omega\epsilon_2}{(1-R_{b2})\gamma_2}\int_{\rho_1}^{\rho_2} E_\rho(\rho)J_1(\lambda\rho)\rho\,d\rho. \tag{8.65}$$

The reflection coefficient R_{b2} given by (8.19) depends on the dielectric structure for $z \geqslant 0$.

The tangential magnetic fields $H_{\phi i}(\rho, 0)$ and $H_{\phi 2}(\rho, 0)$ differ by the linear source current density $J_s(\rho)$ across the slot, where $J_s(\rho)$ is assumed to be of the form

$$J_s(\rho) = \frac{I}{2\pi\rho}.$$

Substituting (8.11) in (8.1) and (8.2), one obtains an integral equation

$$H_{\phi i}(\rho, 0) = H_{\phi 2}(\rho, 0) + J_s(\rho)$$

in the unknown field distribution $E_\rho(\rho)$. A variational principle for determining $E_\rho(\rho)$ is constructed by multiplying (8.65) with $\rho E_\rho(\rho)$ and then integrating from $\rho = \rho_1$ to ρ_2. After defining the voltage across the slot V as

$$V = -\int_{\rho_1}^{\rho_2} E_\rho(\rho)\,d\rho, \tag{8.68}$$

it follows that

$$\frac{IV}{2\pi} = \int_0^\infty \frac{i\omega\epsilon_2}{\gamma_{n2}}\frac{1+R_{b2}}{1-R_{b2}}\left[\int_{\rho_1}^{\rho_1} E_\rho(\rho)J_1(\lambda\rho)\rho\,d\rho\right]^2 \lambda\,d\lambda$$

$$-\sum_{m=1}^{\infty}\frac{i\omega\epsilon_i}{\gamma_m}\frac{1+R_{\text{im}}}{1-R_{\text{im}}}\left[\int_{\rho_1}^{\rho_2} E_\rho(\rho)R_m(\rho)\rho\,d\rho\right]^2, \tag{8.69}$$

where

$$R_{\text{im}} = e^{-2\gamma_m z_i}.$$

and where γ_m and $R_m(\rho)$ are defined as in (7.46).

8.3.2. *One-term Approximation of the Aperture Fields*

The integrations can be carried out for a general power series representation of $E_\rho(\rho)$ and it is formally possible to determine all the coefficients in this power series. [Levine and Papas, 1951; Galejs and Thompson, 1962.] The subsequent numerical work is simplest using the trial function

$$E_\rho(\rho) = \frac{\alpha_0}{\rho}. \tag{8.70}$$

The integrations are straightforward for this form of $E_\rho(\rho)$ and the slot admittance $Y = I/V$ follows as

$$Y = \frac{4\pi}{[\log(\rho_2/\rho_1)]^2}\left\{\frac{1}{2}\int_0^\infty \frac{i\omega\epsilon_2}{\gamma_2}\frac{1+R_{b2}}{1-R_{b2}}\left[\frac{J_0(\lambda\rho_1)-J_0(\lambda\rho_2)}{\lambda}\right]^2 \lambda d\lambda\right.$$
$$\left. - \sum_{m=1}^{\infty}\frac{i\omega\epsilon_i}{\gamma_m}\frac{1+R_{\text{im}}}{1-R_{\text{im}}}\left[\frac{J_0(\lambda_m\rho_1)-J_0(\lambda_m\rho_2)}{\lambda_m b J_1(\lambda_m b)}\right]^2\right. . \tag{8.71}$$

The first term of (8.71) is rewritten as

$$I = \tfrac{1}{2}\int_0^\infty H(\lambda)\lambda\, d\lambda. \tag{8.72}$$

The integral may be approximated by a summation of terms

$$I_n = \tfrac{1}{2}\int_{\lambda_n-\Delta_1}^{\lambda_n+\Delta_2} H(\lambda)\lambda\, d\lambda \approx \tfrac{1}{2}H(\lambda_n)\lambda_n(\Delta_1+\Delta_2), \tag{8.73}$$

where $\Delta_1+\Delta_2 = \frac{1}{2}(\lambda_{n+1}-\lambda_{n-1})$. The zeros of the Bessel function

$$J_0(\lambda_n c) = 0 \tag{8.74}$$

may be chosen for the sequence of λ_ns. In the first approximation $(\lambda_n c) \approx (n-\frac{1}{4})\pi$ and

$$I_n \approx \tfrac{1}{2}H(\lambda_n)\pi^2(n-\tfrac{1}{4})/c^2 \approx H(\lambda_n)/[cJ_1(\lambda_n c)]^2 \tag{8.75}$$

with an error of less than 2 percent for $n \geqslant 1$. Equation (8.75) can be verified using the assymptotic form of $J_1(x)$. This changes the admittance expression (8.71) into

$$Y = \frac{4\pi}{[\log(\rho_2/\rho_1)]^2}\left\{\sum_{n=1}^{\infty}\frac{i\omega\epsilon_2}{\gamma_{n2}}\frac{1+R_{b2}}{1-R_{b2}}\left[\frac{J_0(\lambda_n\rho_1)-J_0(\lambda_n\rho_2)}{\lambda_n c J_1(\lambda_n c)}\right]^2\right.$$
$$\left. - \sum_{m=1}^{\infty}\frac{i\omega\epsilon_i}{\gamma_m}\frac{1+R_{\text{im}}}{1-R_{\text{im}}}\left[\frac{J_0(\lambda_m\rho_1)\, J_0(\lambda_m\rho_2)}{\lambda_m b J_1(\lambda_m b)}\right]^2\right\}, \tag{8.76}$$

where

$$\gamma_{nj} = i(k_j^2-\lambda_n^2)^{1/2}.$$

The admittance, which is given for laterally unbounded plasma layers in (8.71), can be approximated by a sum of waveguide modes for a waveguide of radius c in (8.76) if the approximation (8.73) is valid. Selection of small values of Δ_1 and Δ_2 or of large values of c will improve its accuracy, if singular values of $H(\lambda_n)$ are avoided in the right-hand side of (8.73).

As long as σ_i, σ_1, or $\sigma_2 \neq 0$, the real part of Y will differ from the radiation conductance G_r. However, G_r may be determined after computing the power flow for $z > z_1$. The radiated power P_r and the radiation conductance G_r are given by

$$\begin{aligned} P_r &= G_r V^2 \\ &= \text{Re}\iint E_{\rho 0}\times H^*_{\phi 0}\rho\, d\rho\, d\phi. \end{aligned} \tag{8.77}$$

After $E_{\rho 0}$ and $H_{\phi 0}$ of (8.57) and (8.61) are substituted, the ϕ integration is trivial and the remaining triple integral is reduced to a single integral using (8.64). This gives

$$G_r = \frac{2\pi \mathrm{Re}}{V^2} \int_0 \frac{\gamma_0(\lambda)}{i\omega\epsilon_0} \left| e^{\gamma_0 z} A_0(\lambda) \right|^2 \lambda \, d\lambda,$$

$$\approx 4\pi \mathrm{Re} \sum_n \frac{\gamma_{n0}}{i\omega\epsilon_0} \left| e^{\gamma_{n0} z} \frac{A_{n0}}{V c J_1(\lambda_n c)} \right|^2, \tag{8.78}$$

where (8.73) and (8.75) were utilized in the approximation of the integral by a summation, and where $A_{n0} = A_0(\lambda_n)$. A_{n0} is related to A_{n2} by matching either H_ϕ or E_ρ across the dielectric boundaries at $z = z_1$ and z_2. The amplitude A_{n2} is related to V from (8.65), (8.68), and (8.70). This completes the specification of G_r, and the only values of n that contribute to G_r are those with imaginary γ_{n0}.

8.3.3. *Two-term Approximation of Aperture Fields*

The previous expressions for the slot admittance and radiation conductance were derived for the simple aperture fields (8.70); they can be modified for a two-term trial function

$$E_\rho(\rho) = A f_A(\rho) + B f_B(\rho). \tag{8.79}$$

Substituting this in (8.69), it follows that the slow admittance is given by (7.17), (7.18), and (7.20) where

$$\gamma_{NM} = \gamma^i_{NM} + \gamma^0_{NM}, \tag{8.80}$$

$$\gamma^0_{NM} = 2\pi \sum_{n=1}^{\infty} \frac{i\omega\epsilon_2}{\gamma_{n2}} \frac{1+R_{b2}}{1-R_{b2}} I_{Nn} I_{Mn}, \tag{8.81}$$

$$\gamma^i_{NM} = -2\pi \sum_{m=1}^{\infty} \frac{i\omega\epsilon_i}{\gamma_m} \frac{1+R_{\mathrm{im}}}{1-R_{\mathrm{im}}} I_{Nm} I_{Mm}, \tag{8.82}$$

$$I_{Np} = \int_{\rho 1}^{\rho 2} f_N(\rho) R_{\mathrm{p}}(\rho) \rho \, d\rho, \tag{8.83}$$

$$F_N = \int_{\rho 1}^{\rho 2} f_N(\rho) \, d\rho, \tag{8.84}$$

with $p = n$ or m, and N (or M) $= A$ or B. G_r is given by (8.78), but the ratio A_{n2}/V is computed differently. The trial functions $f_A(\rho)$ and $f_B(\rho)$ of (8.79) will be selected such that the integrals I_{Np} of (8.83) can be readily evaluated.

For electrically small antennas, a highly lossy dielectric layer k_1 will tend to "short circuit" the slot if $z_2 \ll |\rho_2 - \rho_1|$. There will be a strong electric field E between $z = z_2$ and 0 near $\rho = \rho_1$, but **E** will be weak along the z_2 surface at larger radii ρ. A similar variation of $E_\rho(\rho)$ may be expected in the aperture plane $z = 0$: the trial function should be able to account for it. A singularity of the electric field will occur also near ρ_2 in Fig. 8.17. However, the increase of fields near ρ_2 may be negligible for $\rho_2 = b$ and for a finite radius of curvature at ρ_2. A possible form of the trial functions of the antenna aperture fields is

$$f_A(\rho) = 1/\rho \quad \text{for} \quad \rho_1 \leqslant \rho \leqslant \rho_2, \tag{8.85}$$

$$f_B(\rho) = 1/\Delta \quad \text{for} \quad \rho_1 \leqslant \rho \leqslant (\rho_1 + \Delta). \tag{8.86}$$

Considering the limit of $\Delta \to 0, f_B(\rho)$ becomes a delta function

$$f_B(\rho) = \delta(\rho - \rho_1^+), \tag{8.87}$$

where

$$\rho_1^+ = \lim_{\Delta \to 0} (\rho_1 + \Delta).$$

Both forms of $f_B(\rho)$ can be expected to give comparable admittances Y. The form (8.87) obviates a need for experimenting with various Δ values and it will be selected for numerical calculations. I_{Np} and F_N are computed substituting (8.85) and (8.87) in (8.83) and (8.84).

For electrically larger antennas, the admittance Y will be examined using a linear and a quadratic function as $f_A(\rho)$ and $f_B(\rho)$, which are defined by

$$f_A(\rho) = \frac{1}{\rho} - \frac{k}{\rho}(\rho - \rho_1) \tag{8.88}$$

and

$$f_B(\rho) = \frac{k^2}{2\rho}(\rho_2 - \rho)^2 \tag{8.89}$$

for $\rho_1 \leqslant \rho \leqslant \rho_2$.

8.3.4. *Radiation Conductance in Terms of Surface Fields*

The radiation conductance G_r was estimated previously from the power flow at $z = z_1$.

It is also possible to compute the electric field on the surface of the lossy layer, and to relate these surface fields to the power radiated into a laterally unbounded half space.

A spherical coordinate system is centered on the z axis at $z = z_1$ of Fig. 8.17 and the magnetic field component at the circle (r, θ) is related to the ϕ symmetrical surface fields $E_\rho(\rho, \phi, z_1) = E(\rho)$ by (3.28), which is rewritten as

$$H_\phi = \frac{\omega\epsilon_0}{2\pi i} \int_0^{2\pi} \int_0^{\rho_0} E(\rho) \cos\phi \, \frac{e^{ik_0 R}}{R} \rho \, d\rho \, d\phi, \tag{8.90}$$

where $R = (r^2+\rho^2-2r\rho \sin\theta\cos\phi)^{1/2}$ and $k_0 = \omega(\mu_0\epsilon_0)^{1/2}$.

For $r \gg \rho$, the expression for R simplifies and the ϕ integral is evaluated to yield

$$H_\phi = -\omega\epsilon_0 \frac{e^{ik_0 r}}{r} \int_0^{\rho_0} E(\rho) J_1(k_0\rho\sin\theta)\rho\, d\rho. \tag{8.91}$$

After expanding the Bessel function $J_1(x)$ in a power series, H_ϕ becomes

$$H_\phi = -\omega\epsilon_0 \frac{e^{ik_0 r}}{r}\left[\frac{k_0\sin\theta}{2}S_2 - \frac{(k_0\sin\theta)^3}{16}S_4 + \frac{(k_0\sin\theta)^5}{384}S_6 + \ldots\right], \tag{8.92}$$

where

$$S_m = \int_0^{\rho_0} E(\rho)\rho^m\, d\rho. \tag{8.93}$$

Considering H_ϕ to be the RMS value of harmonically varying field component, the power radiated through a large hemisphere is

$$\begin{aligned} P_r &= \lim_{r\to\infty} 2\pi r^2 (\mu_0/\epsilon_0)^{1/2} \int_0^{\pi/2} |H_\phi|^2 \sin\theta\, d\theta, \\ &= \frac{k_0^4}{360}\left\{|S_2|^2 - \frac{k_0^2}{5}\,\mathrm{Re}\,S_2 S_4^* \right. \\ &\quad \left. + \frac{k_0^4}{56}\frac{12}{5}\left[\frac{1}{6}\mathrm{Re}\,S_2 S_6^* + \frac{1}{4}|S_4|^2\right] + \ldots\right\}. \end{aligned} \tag{8.94}$$

The radiation conductance G_r is defined as

$$G_r = P_r/V^2, \tag{8.95}$$

where V is the voltage across the slot. For a narrow slot at $\rho = \rho_0$ and $z_1 = z_2 = 0$, (8.93) simplifies to $S_m \approx -V\rho_0^m$, and G_r is the same as the real part of Y_s in (3.43). In the general case, $E(\rho)$ is determined from (8.57) and (8.61). S_m/V is computed by applying (8.72) and (8.75).

The integrations required for computing the radiated power are only relatively simple for small slots, when they can be confined to the direct vicinity of the antenna ($k_0\rho_0 < 1$).

8.3.5. *Slots with Free Space and Plasma Layers*

The slot admittance is examined first for radiation into a homogeneous large guide of increasing radius c. The plot of the slot conductance is discontinuous at those points where an additional mode starts propagating in the large guide. The slot susceptance exhibits discontinuities at the same points, and large susceptances occur below the cut-off radius of the guide when γ_{n0}

is real and approaches zero. The changes of the waveguide admittance tend to become less for increasing guide radius c, and a randomly selected guide radius will give a fair approximation to the correct admittance if c is of the order of 15–30λ. However, the relatively slow convergence of the sum representing the slot susceptance restricts the choice of c to smaller values.

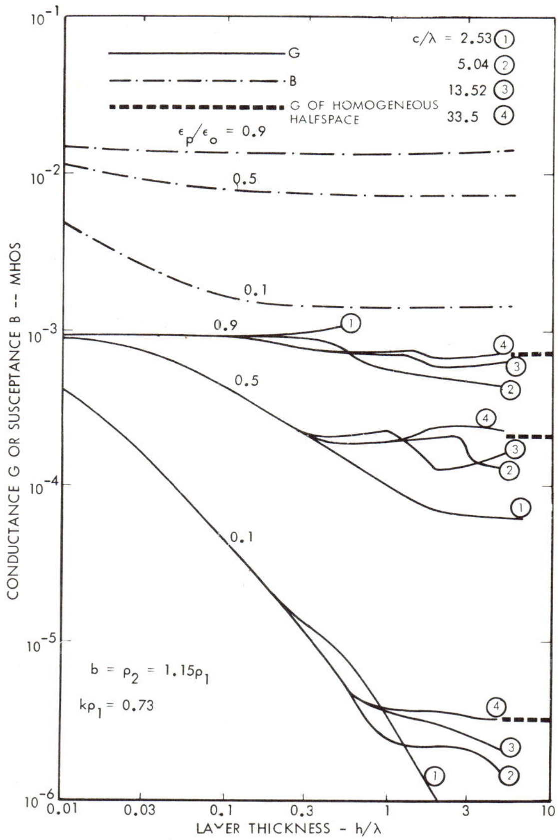

FIG. 8.18. Slot admittance for lossless plasma layers.

For smaller guides, the radius may be always selected such that the waveguide admittance is equal to the slot admittance for a homogeneous halfspace.

Admittance calculations for plasma layers of various thicknesses h are illustrated in Figs. 8.18 and 8.19. The slot conductance G and the radiation conductance G_r decrease with increasing layer thicknesses h, while the slot susceptance is almost constant for $h/\lambda > 0.1$. For plasma layer of $h/\lambda > 0.3$, the conductance depends on the guide radius c. The computed conductance is more accurate for the larger c values when it approaches G of the homogeneous half-space† for $h/\lambda > 1$. A similar effect of increasing waveguide size was also observed for rectangular plasma covered slots [Galejs, 1963c].

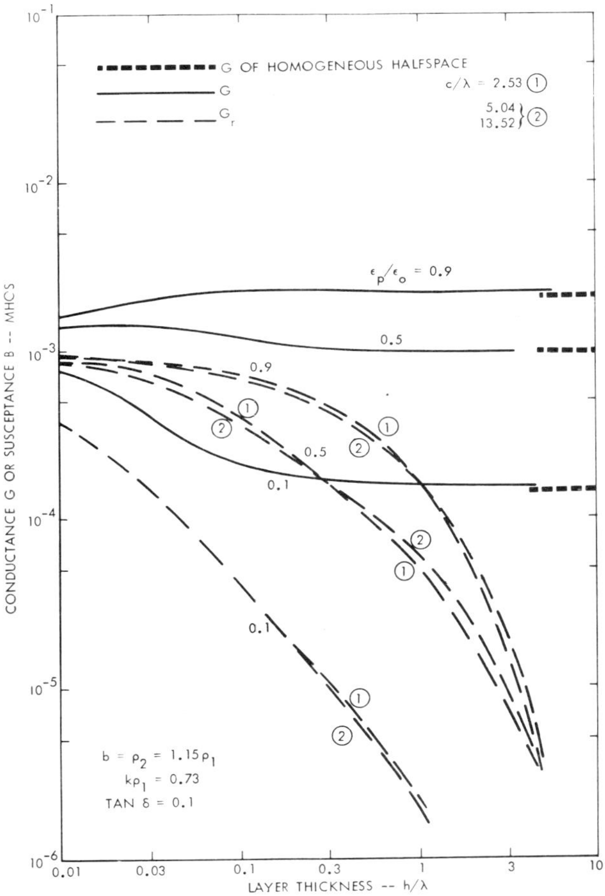

FIG. 8.19. Slot admittance for lossy plasma layers.

For slots covered by thin plasma layers, the admittance is principally determined by the nearby air-to-plasma interface, and is hardly affected by the relatively distant guide walls. The guide walls have an effect for large plasma

†For $k\rho_1 \ll 1$ and $\rho_2/\rho_1 \approx 1$ and for a lossless dielectric half space,

$$G_0 = \left(\frac{\epsilon}{\epsilon_0}\right)^{1/2} \frac{(k\rho)^4}{360}\left[1 - \frac{(k\rho)^2}{5} + \frac{(k\rho)^4}{56} + \ldots\right],$$

where $\rho = (\rho_1 + \rho_2)/2$. For a lossy dielectric half space, $G_1 = G_0 + B \tan \delta$.

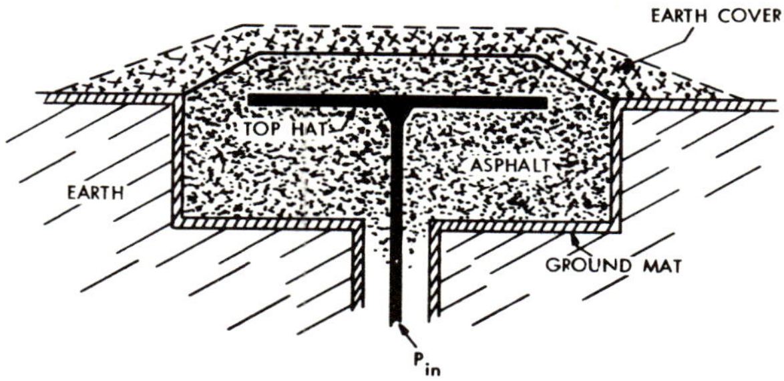

FIG. 8.20. Configuration for a buried electric dipole antenna.

thicknesses, particularly where ϵ_p/ϵ_0 is near unity, or where the electrical discontinuity at the air-to-plasma interface is small.

Only the admittance reflected into the slot plane by the outside space ($z > 0$) has been indicated in Figs. 8.18 and 8.19. The admittance reflected by the cavity or Y_i has been discussed in Section 7.2.2, and is not included in the above admittance figures. All of the above calculations were made for the simple form (8.70) of aperture fields.

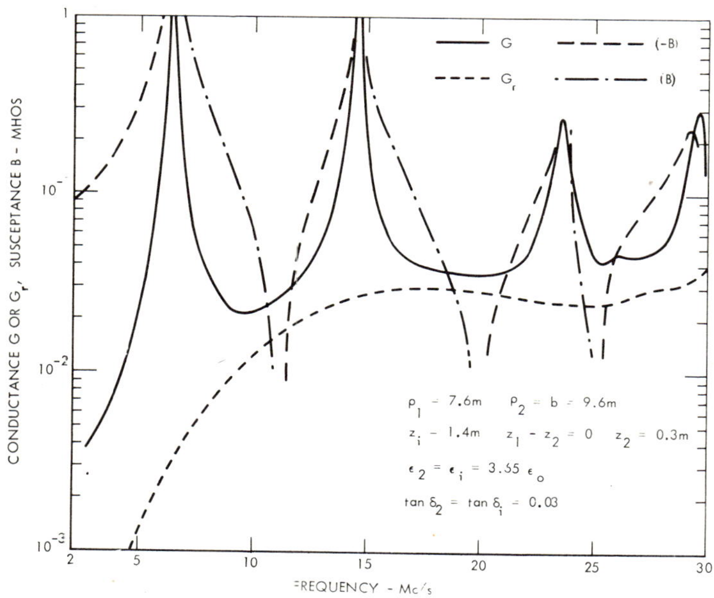

FIG. 8.21. Admittance of the antenna with no ground cover.
$(Z_1 - Z_2 = 0)$.

8.3.6. *Buried Slot Antennas*

A finite electric dipole or a wide top-hat antenna of Fig. 8.20 is embedded in an asphalt pad and is buried below the ground surface. For purposes of the analysis, this antenna geometry is approximated by a cavity-backed annular slot antenna of Fig. 8.17.

The frequency dependence of the admittance is seen in Fig. 8.21 for an antenna with no layer of conducting ground. The conductance G and susceptance B exhibit peaks near frequencies of 6, 14, 23, and 29 Mc/s. A detailed examination of the admittance expression shows that these admittance peaks are caused by resonances of the backup cavity. The conductance G is much lower for intermediate frequencies, where it differs only slightly from the radiation conductance G_r. The antenna efficiency will be relatively high for those frequencies where the difference between G and G_r curves is small. The computed antenna efficiency figures are shown in Fig. 8.22. If there is no ground cover the antenna efficiency exceeds 60 percent for frequencies between 10 and 28 Mc/s, except near the cavity resonances.

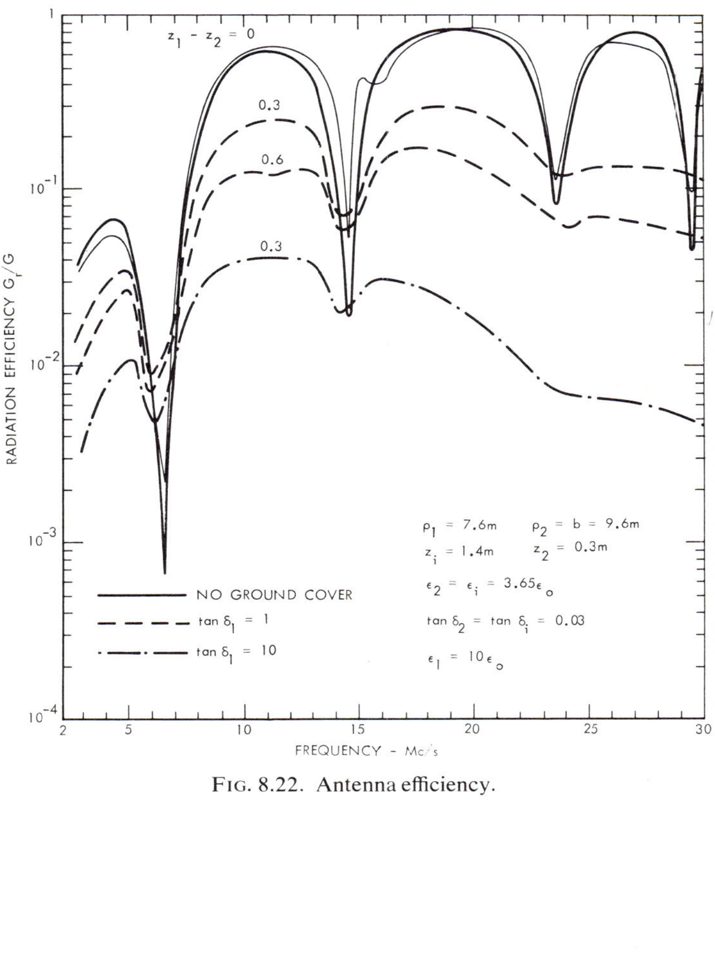

FIG. 8.22. Antenna efficiency.

A conducting ground cover deteriorates the antenna efficiency, but for $\tan \delta_1 = 1$ the efficiency is still above 10 percent for frequencies between 8 and 21 Mc/s, except for the cavity resonance near 14 Mc/s.

In the presence of a ground cover the dips in the efficiency curves near the resonance frequencies are smaller than in the absence of a ground cover. Thus the lossy antenna cover tends to damp out the resonances in the cavity. The thin lined curve is based on the trial functions (8.85) and (8.87), but the other data of Fig. 8.22 have been computed using the trial functions (8.88) and (8.89). With no ground cover ($z_1 - z_2 = 0$), both sets of trial functions give nearly the same efficiency figures. The measured antenna efficiency

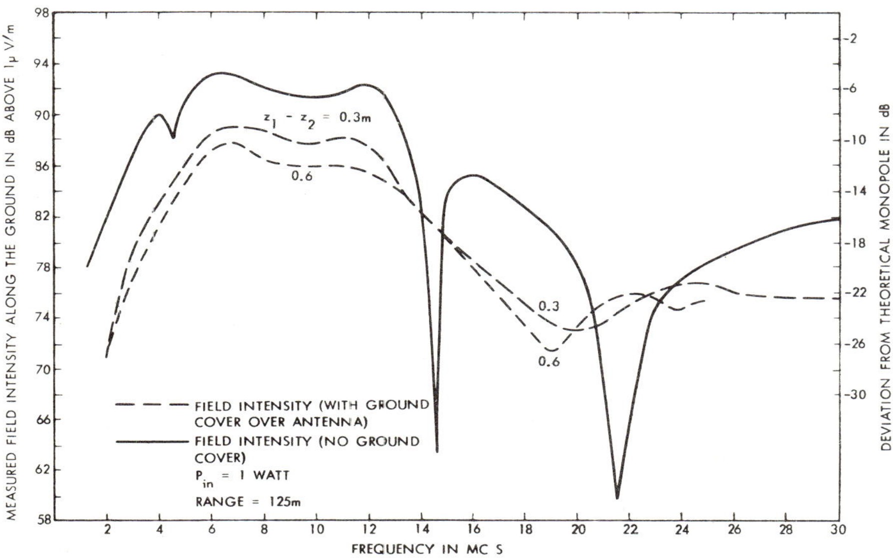

FIG. 8.23. Measured antenna efficiency.

[Wundt, 1964] is shown in Fig. 8.23. The antenna efficiency is estimated by comparing the vertical electrical fields on the ground at a distance of 125 m from the antenna with the fields produced by a short vertical electric dipole above the ground surface for the same input power. With no ground cover, the measured efficiency curve exhibits dips near the frequencies indicated in the theoretical curves of Fig. 8.22. The vertical radiation pattern of the slot antenna exhibits a null in the horizontal direction near 22 Mc/s, which causes this particular dip in the measured efficiency curve to be particularly deep. In the presence of a ground cover, the antenna efficiency is lowered but is also much more slowly varying with frequency and does not show any of the deep dips observed in the absence of the ground cover. A ground cover of 0.3–0.6 m causes a degradation of 4 to 6dB in the antenna efficiency. Still the antenna efficiency is of the order of − 10 dB between 6–12 Mc/s.

8.4. Exercises

8.1. Compute the parameters γ_{MN} for the rectangular slot radiating into a stratified dielectric as indicated in Section 8.1.2.

8.2. Derive the fields $E_y(\xi, \eta)$ of (8.39) for the surface of a slot excited thick plasma layer.

8.3. Compute the ratio (A_{no}/V) in the integral expression (8.78) of the radiation conductance for an annular slot.

8.4. Compute I_{Np} and F_N in the impedance formulation of an annular slot for the trial functions (a) (8.85) and (8.87), (b) (8.88) and (8.89).

8.5. Evaluate the leading terms (S_2/V) and (S_4/V) of the radiation conductance of an annular slot in (8.94) and (8.95).

8.5. References

COHEN, M. H., CROWLEY, T. H. and LEVIS, C. A. (1951) *The Aperture Admittance of a Rectangular Waveguide Radiating into Half Space*, Antenna Laboratory, Ohio State Research Foundation, Columbus, Ohio, Rept. No. 339-22, Contract No. USAF W 33–038 ac 21114.

COMPTON, R. T. (1964) *The Admittance of Aperture Antennas Radiating Into Lossy Media*, Report No. 1691-5, Antenna Laboratory, Ohio State Research Foundation, Columbus, Ohio.

CUTLER, S. (1959) *The Effects of a Plasma Sheath on the Impedance Properties of Some Slot Antennas*. Part I. *Analytical Formulation*, Tech. Research Group, Inc. Syosett, NY., Report No. TRG-127-SR-1, Contract No. AF-19(604)-4574.

ERDELYI, A. (1954) *Tables of Integral Transforms*, McGraw-Hill, New York, NY.

GALEJS, J. (1963a) Admittance of rectangular slot which is backed by a rectangular cavity, *IEEE Trans. on Ant. and Prop.* **AP-11**, 119–26.

GALEJS, J. (1963b) *Slot Antenna Impedance for Plasma Layers*, Scientific Rept. No. 1, Contract No. AF 19(628)-2410, Applied Res. Lab., Sylvania Electronic Systems, Waltham, Mass.

GALEJS, J. (1963c) *Admittance of Waveguide Radiating into Stratified Plasma*, Applied Research Laboratory, Sylvania Electronic Systems Waltham, Mass., Scientific Rept. No. 2, Contract No. AF 19(628)-2410.

GALEJS, J. (1964a), Admittance of annular slot antennas radiating into a plasma layer, *NBS J. Res.* (*Radio Sci.*) **68D** (3), 317–24.

GALEJS, J. (1964b) Slot antenna impedance for plasma layers, *IEEE Trans. on Ant. and Prop.* **AP-12**, (6), 738–45.

GALEJS, J. (1965a) Admittance of waveguide radiating into stratified plasma, *IEEE Trans on Ant. and Prop.* **AP-13**, (1), 64–70.

GALEJS, J. (1965b) Self and mutual admittances of waveguides radiating into plasma layers *NBS J. Res.* (*Radio Sc.*) **69D**, 179–89.

GALEJS, J. (1966) Impedance and radiation efficiency of buried dipole type and loop antennas, Symposium on "Sub-Surface Communications", AGARD NATO, Paris.

GALEJS, J. and MENTZONI, M. (1965) Waveguide admittance for radiation into plasma layers, theory and experiment, Third Symposium on the Plasma Sheath, September 21–23, 1965, Boston.

GALEJS, J. and THOMPSON, T. W. (1962) Admittance of a cavity-backed annular slot antenna, *IRE Trans. on Ant. and Prop.* **AP-10** (6), 671–8.

HARRIS, J. H. (1964) Radiation through cylindrical plasma sheaths, *Electromagnetic Aspects of Hypersonic Flight*, Second Symposium on the Plasma Sheath—Its Effect upon Re-entry Communication and Detection, W. Rotman, H. K. Moore, R. Papa, editors, Spartan Books, Inc., Baltimore, pp. 89–117.

HARRIS, J. H., VILLENEUVE A. T. and BROCK, L. A. (1965) Radiation pattern from plasma enclosed cylindrical hypersonic vehicles, *Radio Science* **69D**, 1335–43.

HASSERJIAN, G. (1965) Fields of a curved plasma layer excited by a slot, *IEEE Trans. on Ant. and Prop.* **AP-13**, 389–96.

KATZIN, M., KOO, B. Y-C. (1964) Effects of re-entry plasma sheath on antenna characteristics, *Electromagnetic Aspects of Hypersonic Flight*, Second Symposium on the Plasma Sheath—Its Effect upon Re-entry Communication and Detection, W. Rotman. H. K. Moore, R. Papa, editors, Spartan Books, Inc., Baltimore, pp. 73–88.

KNOP, C.M. (1961) The radiation fields from a circumferential slot on a metal cylinder coated with a lossy dielectric, *Trans. IRE on Ant. and Prop.* **AP-9** (6), 535–45.

KNOP, C. M. and COHN, G. I. (1964) Radiation from an aperture in a coated plane, *NBS J. Res. (Radio Sci.)* **68D** (4), 363–78.

LEVINE, H. and PAPAS, C. H. (1951) Theory of circular diffraction antenna, *J. Appl. Phys.* **22**, 29–43.

LEVINE, H. and SCHWINGER, J. (1950) On the theory of electromagnetic wave diffraction by an aperture in an infinite plane conducting screen, *Comm. Pure and Appl. Math.* **3**, 355–391.

LEWIN, L. (1951) *Advanced Theory of Waveguides*, Iliffe & Sons, Ltd., London, England.

MARCUVITZ, N. (1951) *Waveguide Handbook*, McGraw-Hill, New York, NY.

MARINI, J. W. (1960) Radiation and admittance of an insulated slotted-sphere antenna surrounded by a strongly ionized plasma sheath, *NBS J. Res. (Radio Prop.)* **64D** (5), 525–32.

OLINER, A. A. and TAMIR, T. (1964) Radiation from semi-infinite slot-excited plasma sheath configurations, *Electromagnetic Aspects of Hypersonic Flight*, The Second Symposium on the Plasma Sheath—Its Effect upon Re-entry Communication and Detection, W. Rotman, H. K. Moore, R. Papa, editors, Spartan Books, Inc., Baltimore, pp. 32–48.

RAEMER, H. R. (1962) Radiation from linear electric or magnetic antennas surrounded by a spherical plasma shell, *IRE Trans. on Ant. and Prop.* **AP-10** (1), 69–78.

ROTMAN, W. and MELTZ, G. (1961) *Experimental Investigation of the Electromagnetic Effects of Re-entry*, US. Air Force Cambridge Research Laboratories, Bedford, Mass., Report No. AFCRL-87.

RUSH, W. V. T. (1963) Radiation from a plasma-clad axially slotted cylinder, *Radio Science*, **67D** (2), 203–14.

STEVENSON, A. F. (1948) Theory of slots in rectangular waveguides, *J. Appl. Phys.* **19**, 24–38.

STRATTON, J. A. (1941) *Electromagnetic Theory*, McGraw-Hill, New York, NY., sections 9.4 and 9.5.

SWIFT, C. T. (1964) Radiation patterns of a slotted-cylinder antenna in the presence of an inhomogeneous lossy plasma, *IEEE Trans. on Ant. and Prop.* **AP-12**, 728–38.

TAI, C. T. (1951) The effect of a grounded slab or the radiation from a line source, *J. Appl. Phys.* **22**, (4), 405–14.

TAMIR, T. and OLINER, A. A. (1962) The influence of complex waves on the radiation field of a slot-excited plasma layer, *IRE Trans. on Ant. and Prop.* **AP-10** (1), 55–65.

VILLENEUVE, A. T. (1965) Admittance of waveguide radiating into plasma environment, *IEEE Trans. on Ant. and Prop.* **AP-13** (1), 115–21.

WAIT, J. R. (1959) *Electromagnetic Radiation from Cylindrical Structures*, pp. 125–41, Pergamon Press, New York, NY.

WAIT, J. R. (1961) The electromagnetic field of a dipole in the presence of a thin plasma sheet, *Appl. Sci. Res.* **8**, sec. B, 397–417.

WATSON, W. H. (1947) *The Physical Principles of Waveguide Transmission and Antenna Systems*, Oxford, Clarendon Press, England, section 10.

WUNDT, R. M. (1964) New results on buried antennas for use at medium and high frequencies, *Northeast Electronics Research and Engineering Meeting (NEREM) Record*, Paper M-5.2, Boston, Mass.

YEE, J. S. (1961) Mutual coupling of two thin infinitely-long slots located on a perfectly conducting plane in the presence of a uniform plasma layer, *Proc. IRE* **49** (12), 1837–45.

CHAPTER 9

IMPEDANCE OF LINEAR ANTENNAS IN THE PRESENCE OF A STRATIFIED DIELECTRIC

THE input impedance of linear antennas in homogeneous lossy media has been discussed by King [1956], King and Harrison [1960], and King and Iizuka [1963]. Treatment of insulated antennas in a highly dissipative medium has been based on transmission line theory [Moore, 1951; Ghose, 1960; Guy and Hasserjian, 1963]. The theory of insulated antennas in an infinite dissipative medium has been further refined by King [1964]. The theory of coupled free space antennas [King and Sandler, 1963] has been adapted to the calculations of the self and mutual impedances of linear antennas immersed in an infinite homogeneous dissipative medium [King *et al.*, 1966]; a good quantitative agreement is indicated between this theory and experiment.

Slot antenna impedance for plasma and dielectric layers was discussed in Chapter 8, and a simular variational impedance formulation can be derived for linear antennas surrounded by a stratified dielectric [Galejs, 1965]. Also, the mutual and self impedance of two linear flat strip antennas has been computed in a similar geometry [Galejs, 1966]. This stratified medium may represent slabs of dielectric or plasma, or lossy dielectric layers. It is also possible to investigate finite insulated antennas in a semi-infinite lossy medium. For antennas in a dielectric layer, the decrease of the radiation efficiency observed with increasing electrical thickness of the layer is explained principally by the increased amounts of surface wave power and also by local dielectric losses. For insulated antennas located in a dissipative medium, this theory yields results that agree better with measurements than the conventional transmission line theory. The transmission line theory is shown to give too large a resistance and reactance at the full-wave resonant peaks of the impedance and too low a resistance for short antennas [Galejs, 1965]. For free space surrounding the antennas, the variational formulation of Galejs [1966] provides a closer agreement with measurements of mutual impedances than past solutions of the iterative type. These calculations tend to agree with transmission line approximations of the antenna impedance for insulated antennas in a highly lossy medium. However, the two sets of calculations differ for small antenna separations, and for lower dielectric losses in particular if the permittivity of the lossy medium is also lower.

9.1. Driving Point Impedance of a Single Antenna

9.1.1. *Impedance Formulation*

The linear antenna is assumed to be in the shape of a rectangular strip of zero thickness. It is excited at its center and sandwiched between parallel dielectric layers, as shown in Fig. 9.1. The dielectric regions are surrounded by free space, but their layers may be unsymmetrical with respect to the

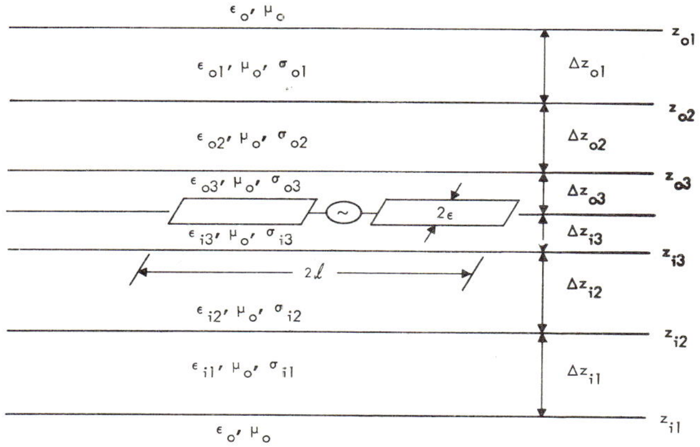

FIG. 9.1. Antenna geometry.

antenna. For semi-infinite media, one of the boundaries between the dielectric and the free space can be made to recede to infinity. For dielectric layers that may be lossy or that can support surface waves, the power radiated into the two lossless half-space regions will generally differ from the power supplied to the antenna. Hence, separate expressions are derived for the input impedance of the antenna and for its radiation resistance.

The fields excited in the geometry of Fig. 9.1. will be expressed as a superposition of TE and TM modes that are derived from scalar functions Ψ_j and Φ_j respectively. The region of $z \geqslant 0$ next to the antenna is designated by a subscript 0, the region of $z \leqslant 0$–by a subscript i. For an exp $(-i\omega t)$ time dependence of the fields, the scalar functions Ψ_0 and Φ_0 are given by (8.3) and (8.4), if $p = 0$. Ψ_i and Φ_i are obtained from (8.3) and (8.4) by letting $i = p$ and $\gamma_i = -\gamma_p$. The field components are related to Ψ_j and Φ_j by (3.10) to (3.15). The current of a linear density J_x is assumed to flow only in the x direction along the antenna, and the field components satisfy the boundary conditions $E_{xi} = E_{x0}$, $E_{yi} = E_{y0}$, $H_{xi} = H_{x0}$, and $H_{yi} = H_{y0} + J_x$, for $z = 0$. Substituting Ψ_j and Φ_j in (3.10) and (3.11), and applying the E_x and E_y boundary conditions gives

$$A_0(1+R_{a0}) = A_i(1+R_{ai}), \tag{9.1}$$

$$\gamma_0 B_0(1-R_{b0}) = \gamma_i B_i(-1+R_{bi}). \tag{9.2}$$

The H_x and H_y boundary conditions lead to

$$\gamma_0 B_0(1-R_{b0})\left[\frac{k_0^2(1+R_{b0})}{\gamma_0(1-R_{b0})}+\frac{k_i^2(1+R_{bi})}{\gamma_i(1-R_{bi})}\right] = \frac{\omega\mu_0 u}{u^2+v^2}\iint J_x \mathrm{e}^{-iux}\,\mathrm{e}^{-ivy}\,\mathrm{d}x\,\mathrm{d}y, \tag{9.3}$$

$$A_0(1+R_{a0})\left[\gamma_0\frac{1-R_{a0}}{1+R_{a0}}+\gamma_i\frac{1-R_{ai}}{1+R_{ai}}\right] = \frac{iv}{u^2+v^2}\iint J_x \mathrm{e}^{-iux}\,\mathrm{e}^{-ivy}\,\mathrm{d}x\,\mathrm{d}y. \tag{9.4}$$

The amplitudes A_j and B_j are related to the linear current density of the antenna J_x by (9.1) to (9.4). The scalar functions Ψ_j and Φ_j and the field components are thus uniquely related to J_x. The driving point impedance of a flat strip antenna may be computed from the expression (1.30). The surface current J_s is assumed to have only an x component J_x, which in addition, is an even function about $x = 0$, $y = 0$; and it follows that

$$Z = \frac{i\omega\mu_0}{[2\pi I(x=0)]^2}\int_{-\infty}^{\infty}\int_{-\infty}^{\infty} F(u,v) \left[\iint J_x(x,y)\cos ux\cos vy\,\mathrm{d}x\,\mathrm{d}y\right]^2 \mathrm{d}u\,\mathrm{d}v, \tag{9.5}$$

where

$$F(u,v) = \frac{v^2}{u^2+v^2}\left[\gamma_{0p}\frac{1-R_{a0p}}{1+R_{a0p}}+\gamma_{iq}\frac{1-R_{aiq}}{1+R_{aiq}}\right]^{-1} - \frac{u^2}{u^2+v^2}\left[\frac{k_{0p}^2}{\gamma_{0p}}\frac{(1+R_{b0p})}{(1-R_{b0p})}+\frac{k_{iq}^2}{\gamma_{iq}}\frac{(1+R_{biq})}{(1-R_{biq})}\right]^{-1}. \tag{9.6}$$

The designation of the dielectric layers by subscripts is seen from Fig. 9.1, and in particular the subscripts $0p$ and iq designate dielectric layers next to $z = 0$ for $z \geqslant 0$ and $z \leqslant 0$ respectively. The other symbols are defined as

$$k_{jl} = (\omega^2\mu_0\epsilon_{jl} + i\omega\mu_0\sigma_{jl})^{1/2},\ \gamma_{jl} = i(k_{jl}^2 - u^2 - v^2)^{1/2},$$

where $j = 0$ or i and $l = 1, 2 \ldots p$ or $1, 2 \ldots q$. The reflection coefficients of the TE modes R_{ajl} and of the TM modes R_{bjl} depend on the dielectric structure for $|z| > 0$. For stratified dielectric layers as shown in Fig. 9.1, the reflection coefficient R_{djl} ($d = a$ or b) in the region of $|z_{l+1}| \leqslant |z| \leqslant |z_l|$ is related to the reflection coefficient $R_{d(l-1)}$ of the region $|z_l| \leqslant |z| \leqslant |z_{l-1}|$ by the expressions (8.18) and (8.19) if the double subscripts jl are substituted for the single subscripts j. Also the absolute value of z_{jl} is considered for $|z_{jl}| < 0$. The antenna current density is assumed to be representable by the same trial function as in Section 2.3, but with an arbitrary value of k. This gives

$$J_x(x, y) = [A \sin [k(l - |x|)] + B\{1 - \cos [k(l - |x|)]\}]f(y) \tag{9.7}$$

with $f(y) = 1/(2\epsilon) = \text{const.}$ Substituting (9.7) in (9.5) and using the stationary character of the impedance expression for determining the complex amplitudes A and B, it follows that (A/B) and Z are given by (2.38) to (2.41), with

$$F_A = \sin kl, \tag{9.8}$$

$$F_B = 1 - \cos kl, \tag{9.9}$$

$$\gamma_{NM} = \frac{4i\omega\mu_0}{\pi^2} \int_0^\infty dv \left(\frac{\sin \epsilon v}{\epsilon v}\right)^2 \int_0^\infty du\, F(u, v) g_N(u) g_M(u), \tag{9.10}$$

$$g_A(u) = \frac{k}{k^2 - u^2} (\cos ul - \cos kl), \tag{9.11}$$

$$g_B(u) = \frac{k}{k^2 - u^2} \left(\frac{k}{u} \sin ul - \sin kl\right). \tag{9.12}$$

The impedance Z depends on the selection of the wave number k of the trial functions (9.7). For a homogeneous dielectric, k should represent the wave number of the dielectric k_d. For antennas in a thick dielectric layer, k should also be equal to the wave number of the dielectric layer k_d, and for very thin dielectric layers, k is approximated by the free space wave number k_0. The antenna impedance computed using these two values of k can be shown to be continuous for dielectric layers of a thickness near $\lambda/100$. The accuracy of the computations may be verified by using

$$k = (k_0^2 + k_d^2)^{1/2},$$

or other intermediate values of k for intermediate thickness of the dielectric layer. For insulated antennas in a dissipative medium, the transmission line theory may provide the first estimates of k. This is usually a satisfactory guess, and small changes of k cause only negligible variations in the computed impedance.

The power radiated into the region outside of the dielectric slabs can be computed as

$$P_r = \text{Re} \int \int \mathbf{E} \times \mathbf{H}^* \cdot d\mathbf{s}, \tag{9.13}$$

where the integration is carried out over the outermost surfaces of the dielectric ($z = z_{0l}$ and z_{il}), and where $d\mathbf{s}$ is in the direction of the outward normal. After the field expressions are substituted, (9.13) becomes a sixfold integral. The x and y integrals are carried out first and result in delta functions. The two subsequent integrations are trivial, and the remaining double integral becomes

$$P_r = \frac{16\pi^2}{\omega\mu_0} \int_0^{k_0} du \int_0^{(k_0^2 - u^2)^{1/2}} dv\,(k_0^2 - u^2 - v^2)^{1/2}(u^2 + v^2) \cdot [|A_{00}|^2 + |A_{i0}|^2 + k_0^2|B_{00}|^2 + k_0^2|B_{i0}|^2], \tag{9.14}$$

where A_{j0} and B_{j0} designate the complex amplitudes of TE and TM modes for $z > z_{01}$ (when $j = 0$), and for $z < z_{i1}$ (when $j = i$). The amplitudes A_{j0} and B_{j0} are related to the corresponding amplitudes in the dielectric region near the antenna A_{js} and B_{js} ($s = p$ for $j = 0$, $s = q$ for $j = i$) by matching the tangential field components across the dielectric interfaces. A_{js} and B_{js} are related to the amplitudes A and B of the trial function for the antenna current density (9.7). The radiation resistance R_r may be defined as

$$R_r = \frac{P_r}{|I(x=0)|^2} = \frac{P_r}{|AF_A + BF_B|^2}, \tag{9.15}$$

where the complex ratio A/B is given by (2.38).

With no losses present in the dielectric layers, the power carried by surface waves guided along the dielectric interfaces can be computed as

$$P_s = \text{Re} \int\int \mathbf{E} \times \mathbf{H}^* \cdot d\mathbf{s}, \tag{9.16}$$

where the integration is carried out over the surfaces $x = \pm\infty$ and $y = \pm\infty$, and where $d\mathbf{s}$ is in the direction of the outward normal. Considering an antenna located at the center of a single lossless dielectric layer ($p = q = 1$) of thickness $2a = z_{01} - z_{i1}$, the integrals representing the fields in (9.16) are singular for those values $\lambda^2 = u^2 + v^2$, where $R_{a01} = R_{ai1} = 1$ or $R_{b01} = R_{bi1} = -1$. This gives

$$\tan\,[ak_0(\epsilon_r - w^2)^{1/2}] = \frac{(w^2 - 1)^{1/2}}{(\epsilon_r - w^2)^{1/2}}, \tag{9.17}$$

and

$$\tan\left[ak_0(\epsilon_r - w^2)^{1/2}\right] = -\frac{(\epsilon_r - w^2)^{1/2}}{\epsilon_r(w^2 - 1)^{1/2}}, \tag{9.18}$$

with $1 < (w = \lambda/k_0) < (\epsilon_r)^{1/2}$ and $\epsilon_r = \epsilon_{01}/\epsilon_0 = \epsilon_{i1}/\epsilon_0$ as the equations for determining the characteristic values λ for TE and TM modes respectively. Equation (9.17) has real solutions for the TE modes for any thickness of the dielectric layer. Equation (9.18) has real solutions for the TM modes only for $ak_0\ (\epsilon_r - 1)^{1/2} > \pi/2$. For lossless dielectric there are no complex solutions that satisfy the radiation condition following Whitmer [1948].

Substituting the field expressions (9.16) becomes a six fold integral. In the integrals over the $x = \pm\infty$ ($y = \pm\infty$) surfaces, the $y(x)$ integrals give delta functions, which makes one of the $v(u)$ integrals trivial. The two $u(v)$ integrals are computed by evaluating the residues at the poles given by (9.17) and (9.18). The z integrations are trivial, and (9.16) is reduced to a sum of four single integrals which are evaluated numerically between the limits of 0 and λ [Galejs, 1965]. The antenna input resistance R_s associated with the surface wave power P_s is computed by analogy with (9.15).

9.1.2. *The Free Space Impedance of a Thin Antenna*

The antenna impedance is expressed in terms of double integrals (9.10), which can be evaluated analytically for the antenna in a homogeneous medium. The free space impedance (2.41) of a thin half-wavelength antenna will be considered as a simple example. For $l = \lambda/4 = \pi/(2k_0)$, k of the trial functions (9.7) is set equal to $k_0 = \omega(\mu_0\epsilon_0)^{1/2}$. With $R_{djl} = 0$, γ_{AA} of (9.10) simplifies to

$$\gamma_{AA} = \frac{2\omega\mu_0}{\pi^2}\int_0^\infty dv\left(\frac{\sin \epsilon v}{\epsilon v}\right)^2 \int_0^\infty \frac{du\, \cos^2[(\pi/2)(u/k)]}{(k^2-u^2-v^2)^{1/2}(k^2-u^2)}. \tag{9.19}$$

With $F_A = 1$, substitution of (9.19) in (2.41) gives the real part of the impedance as

$$\begin{aligned} \mathrm{Re}Z_s &= \frac{2\omega\mu_0}{\pi^2}\int_0^k \frac{du\, \cos^2[(\pi/2)(u/k)]}{k^2-u^2} \int_0^{(k^2-u^2)^{1/2}} \frac{dv}{(k^2-u^2-v^2)^{1/2}} \\ &= \frac{\omega\mu_0}{\pi}\int_0^k \frac{du\, \cos^2[(\pi/2)(u/k)]}{(k-u)(k+u)}. \end{aligned} \tag{9.20}$$

Expanding the denominator in partial fractions and changing the variables of integration to $(k-u)$ and $(k+u)$ gives

$$\begin{aligned} \mathrm{Re}Z_s &= 60\int_0^2 \frac{\sin^2(\pi/2x)}{x}\, dx = 30[\log(2\pi) + 0{\cdot}5772 - Ci(2\pi)] \\ &= 73.1, \end{aligned} \tag{9.21}$$

where $Ci(x)$ is the cosine integral. The imaginary part of the impedance is obtained as

$$\mathrm{Im}Z_s = -\frac{2\omega\mu_0}{\pi^2}\left[\int_0^k \mathrm{d}u \int_{(k^2-u^2)^{1/2}}^{\infty} \mathrm{d}v + \int_k^{\infty} \mathrm{d}u \int_0^{\infty} \mathrm{d}v\right] \cdot \left[\frac{\cos^2[(\pi/2)(u/k)]}{k^2-u^2}\left(\frac{\sin \epsilon v}{\epsilon v}\right)^2 \frac{1}{(v^2+u^2-k^2)^{1/2}}\right]. \tag{9.22}$$

The v integrations become elementary by noting that $(\sin \epsilon v/\epsilon v) \approx 1$ over the range of v, where $(v^2+u^2-k^2)^{1/2} \neq v$. (This applies strictly if $\epsilon \to 0$ and $u \leqslant u_0 =$ finite.) It follows that

$$\mathrm{Im}Z_s = \frac{\omega\mu_0}{\pi^2 k^2}\int_0^{\infty} \frac{\mathrm{d}u \cos^2[(\pi/2)(u/k)]}{1-(u/k)^2}\left[\log\left|1-\frac{u}{k}\right| + \log\left(1+\frac{u}{k}\right)\right]$$

$$= \frac{240}{\pi}\int_0^{\infty} \frac{\mathrm{d}y \log y \sin^2 \pi y}{4-y^2} = -30\, Si(2\pi) = -42.5, \tag{9.23}$$

where $Si(x)$ is the sine integral. The numerical values (9.21) and (9.23) are recognized as results already derived in Section 2.1.

9.1.3. *Antennas in a Homogeneous Dielectric*

Impedance calculations for an antenna of a finite width are first carried out for homogeneous media where other solutions are already available.

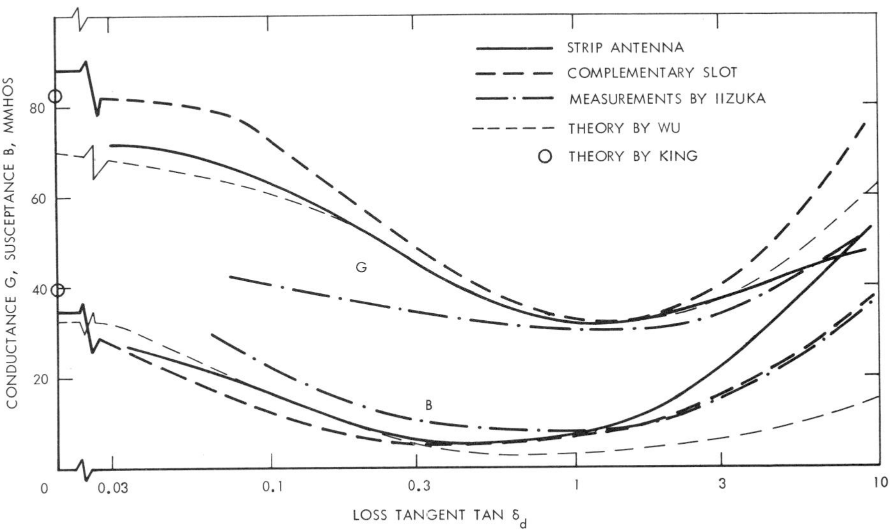

FIG. 9.2. Admittance of a thin antenna in a dissipative medium.

The free space impedance of a $\lambda/2$ antenna was shown in Fig. 2.1 for various values of the width ϵ or of the parameter $\Omega = 2ln(4l/\epsilon)$. The admittance of the strip antenna in a lossy medium is compared in Fig. 9.2 with the admittance of a complementary slot antenna [Galejs, 1964], with measurements of Iizuka and King [fig. 10, 1962], and with the theory of Wu [1961] and Gooch *et al.* [1963]. The antenna admittance calculated from King [1956] for a lossless dielectric is also indicated in Fig. 9.2. The antenna length is equal to one-half wave length for low dielectric losses and $\epsilon_r = \epsilon_d/\epsilon_0$ changes from 78 to 69 with increasing losses as indicated by Iizuka and King [1962]. Good agreement exists between the various theoretical curves for $\tan \delta_d < 2$. The measured conductance appears too low for a low-loss dielectric. The theory of Wu gives lower susceptance figures for high-loss tangents. The admittance of the complementary slot has been computed using a complex value of $k = k_0(\epsilon_d/\epsilon_0)^{1/2}(1+i \tan \delta_d)^{1/2}$ in the trial functions (9.7), while the admittance of the strip antenna was computed using real values of $k = k_0(\epsilon_d/\epsilon_0)^{1/2}$. This may account for some of the differences between the two sets of calculations for larger values of $\tan \delta_d$.

9.1.4. *Antenna in Dielectric Layers*

The impedance of a linear antenna located either in the center or on the surface of dielectric layers of various thicknesses $2d$ is shown with the corresponding efficiency figures in Figs. 9.3–9.8. All the calculations of these figures are made for antennas of $\Omega = 8$ (the heavy horizontal lines indicated on the sides of the figures denote the impedance for $d = 0$ or ∞ computed from a complementary slot antenna). The radiation efficiency is computed as a ratio between the power radiated normally through the dielectric surface as indicated in (9.13) and the power supplied to the antenna. The surface wave efficiency denotes the ratio between the total power carried by the surface wave for a lossless dielectric in the direction parallel to the dielectric surface and the power supplied to the antenna. For a lossy dielectric, that part of surface wave power that flows outside of the dielectric layer will eventually be dissipated inside the dielectric, and will not appear as part of the power radiated outside the lossy dielectric in (9.13). The sum of the surface wave efficiency and the radiating efficiency are a little less than unity for a low-loss dielectric ($\tan \delta_d = 0.03$) in Figs. 9.3 and 9.4, and appreciably less than unity for higher dielectric losses ($\tan \delta_d = 0.1$ and 1) in Fig. 9.5, when significant dissipation occurs in the immediate vicinity of the antenna. For antennas located in the center of the dielectric layer, the radiation efficiency is decreased by increasing the dielectric constant of the layer (Fig. 9.4). It is unchanged, however, for the several antenna lengths shown in Fig. 9.3.

It was indicated following equation (9.12) that the impedances are

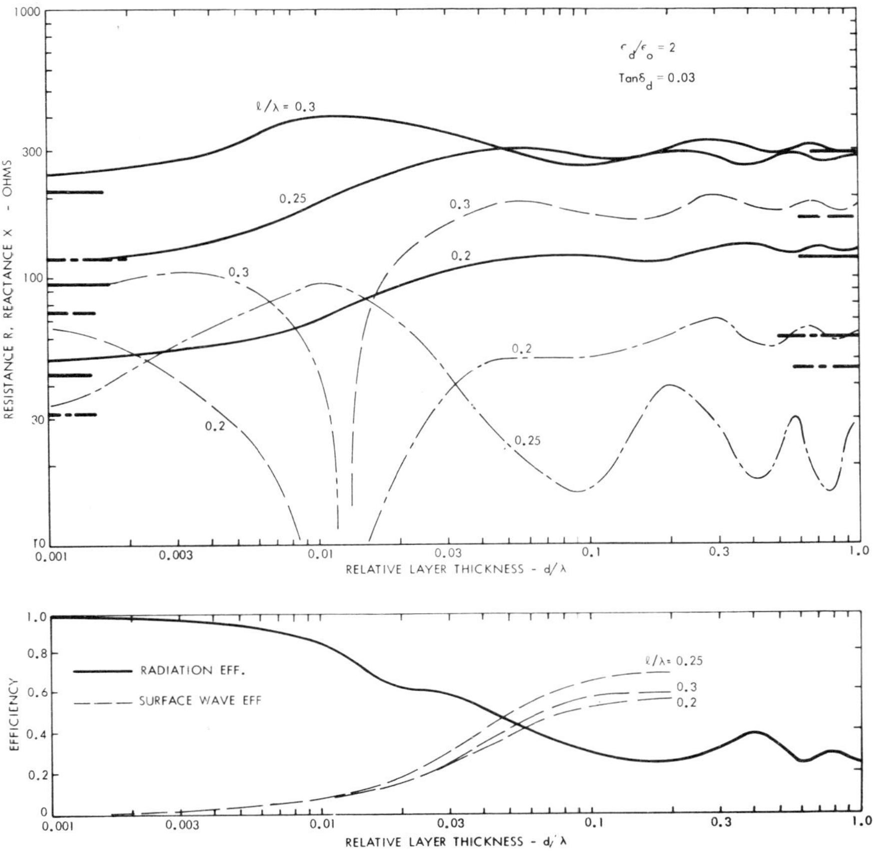

FIG. 9.3. Impedance and efficiency of antennas in the center of a dielectric layer. Effects of antenna length.

relatively insensitive to changes in the wave number k for intermediate thickness of the dielectric layer. Subsequently the radiated power P_r and the surface wave power P_s are computed using a current distribution (9.7) with A/B and k determined from the impedance calculation. The expressions (9.13) or (9.16) used for calculating P_r and P_s are not stationary. (This follows after substituting (1.10) in (1.39) and interchanging the dot and cross product signs.) Therefore P_r and P_s should be more sensitive to changes in the wave number k than the impedance Z of (9.5). It is possible to have inaccurate values of k where (R_r+R_s) is larger than $\text{Re}\,Z$, particularly for the larger values of ϵ_d/ϵ_0. More accurate values of k make $(R_r+R_s) < \text{Re}\,Z$, and only such calculations have been shown in the numerical results.

For antennas located on the surface of the dielectric layer, the radiation efficiency is practically unaffected by changes of the antenna length, and it is also decreased for higher values of ϵ_d as seen from Fig. 9.6. The calculations of surface wave efficiency have not been carried out in the case of antennas located on the surface of the dielectric layer.

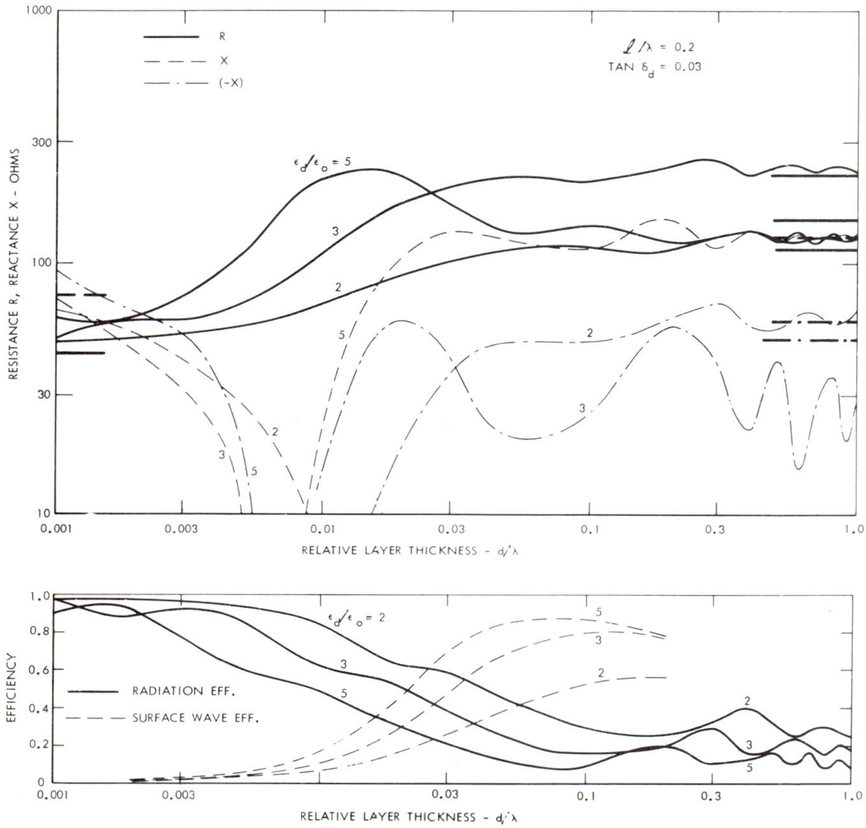

FIG. 9.4. Impedance and efficiency of antennas in the center of a dielectric layer. Effects of the dielectric constant.

For dielectric layers of $\epsilon_d/\epsilon_0 < 1$ (approximation of a plasma layer by an isotropic dielectric), the impedance curves of Fig. 9.7 are much smoother than for $\epsilon_d/\epsilon_0 > 1$, when the layers could support surface waves. Also the radiation efficiency is decreased more gradually with increasing layer thickness.

Variational approximations to the current distributions of the antenna configurations shown in Figs. 9.3–9.6 are computed from equation (9.7) and (2.38), and are shown in Fig. 9.8. There are significant deviations from the usually assumed half-wave sinusoidal current distribution, these deviations becoming more significant with increasing thickness of the dielectric layer d, the antenna length l, or the dielectric constant ϵ_d. Thick layers of a lossy dielectric make the current distribution approach a triangular one. A

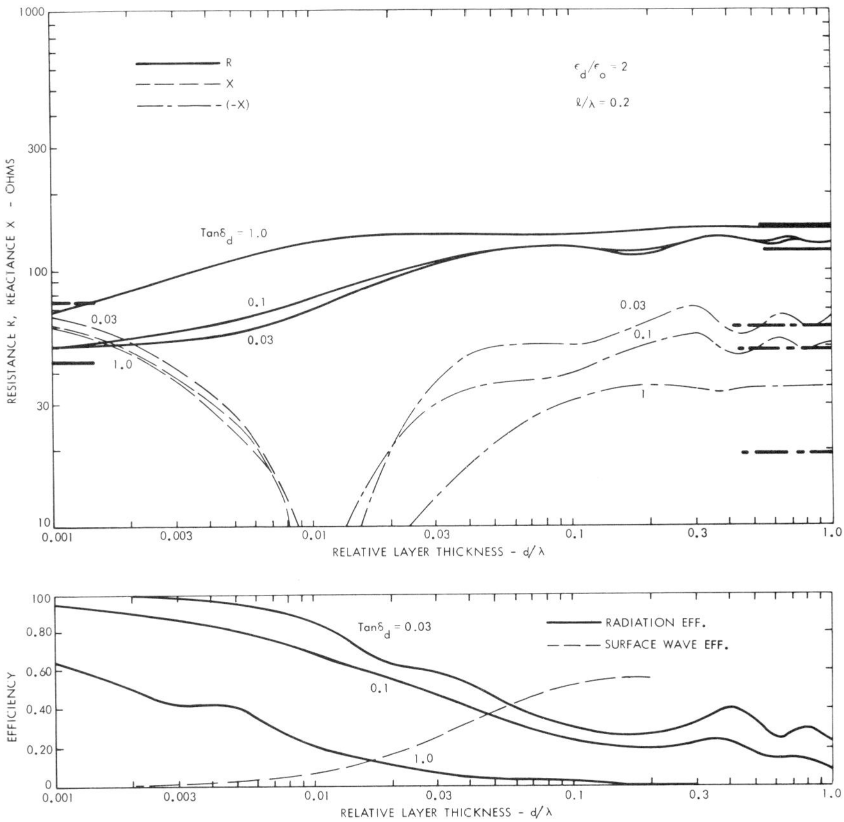

FIG. 9.5. Impedance and efficiency of antennas in the center of a dielectric layer. Effects of dielectric losses.

similar change is observed with increasing layer thickness d if $\epsilon_d < \epsilon_0$, and with the decreasing ϵ_d when the antenna tends to become electrically shorter.

The above approximate current distributions are the best possible fits to the actual current distribution using the trial function (9.7) and should be indicative of the trends observed in the actual current distributions with changing antenna parameters.

9.1.5. *Antennas in Dielectric or Plasma Cylinders*

The impedance curves of Fig. 9.3–9.7 have been computed for flat strip antennas in or on planar dielectric slabs, and it may be of interest to compare these results with computations for dielectric cylinders. The formulation

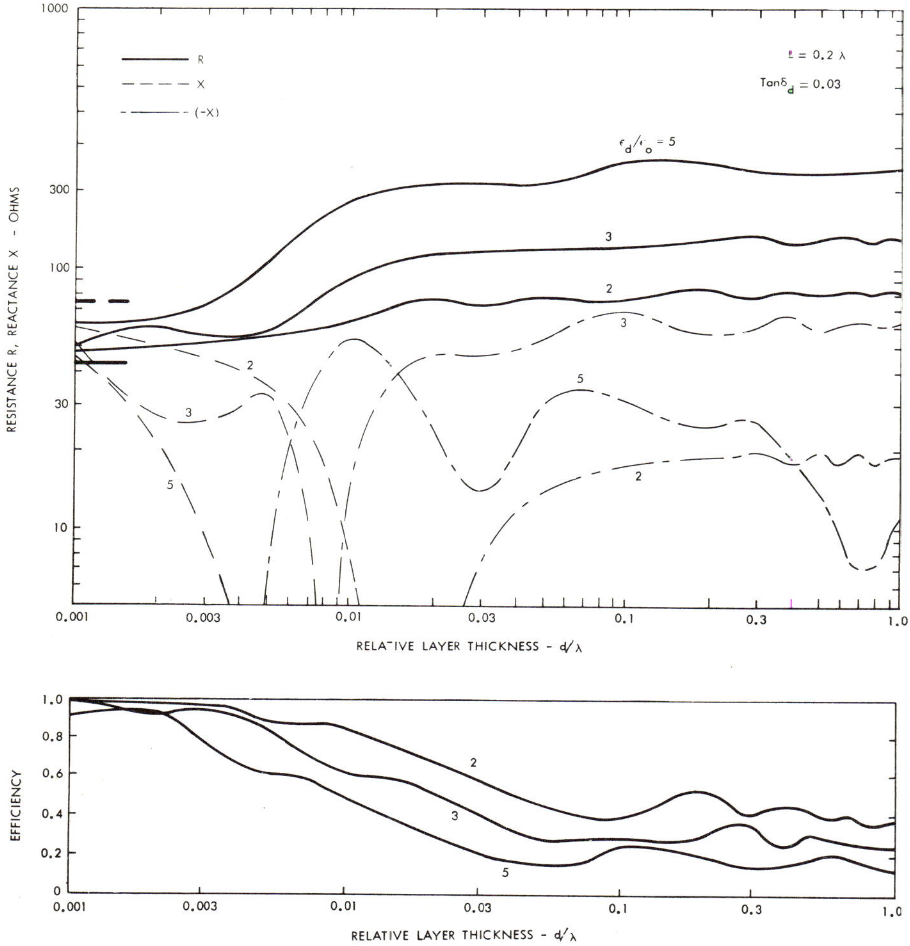

FIG. 9.6. Impedance and efficiency of antennas on the surface of a dielectric layer. Effects of the dielectric constant.

of Section 12.1 can be specialized to cylindrical antennas of radius a embedded in a dielectric cylinder of radius b, which is surrounded by free space.

The antenna impedance for a dielectric cylinder is shown in Fig. 9.9; it can be compared with the data shown in Fig. 9.4 for a dielectric slab. The impedances exhibit about the same variation with increasing thickness of the dielectric layer $(b-a)$ or d, except that the impedance curves are more oscillatory in the cylindrical geometry of Fig. 9.9 for larger thicknesses of the

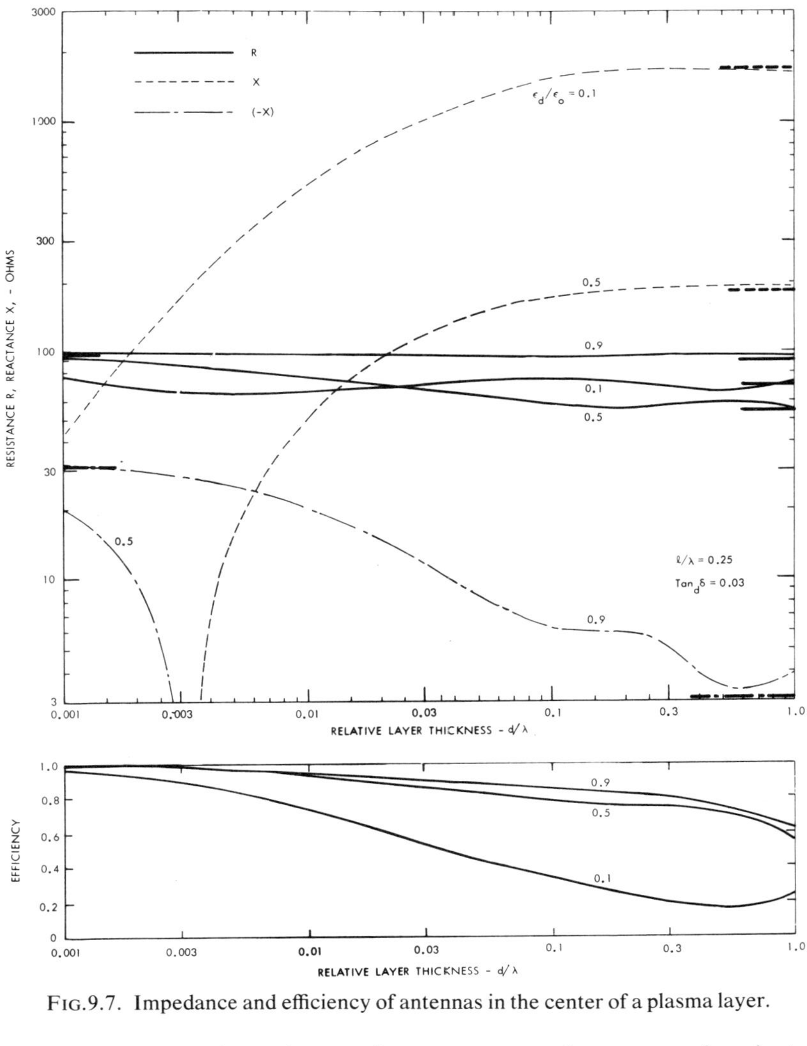

FIG. 9.7. Impedance and efficiency of antennas in the center of a plasma layer.

dielectric layer. The impedance of an antenna at the center of an isotropic cylindrical plasma column in Fig. 9.10 is nearly the same as the antenna impedance for a plasma slab in Fig. 9.7.

In these two examples the antenna impedance in a cylindrical geometry of

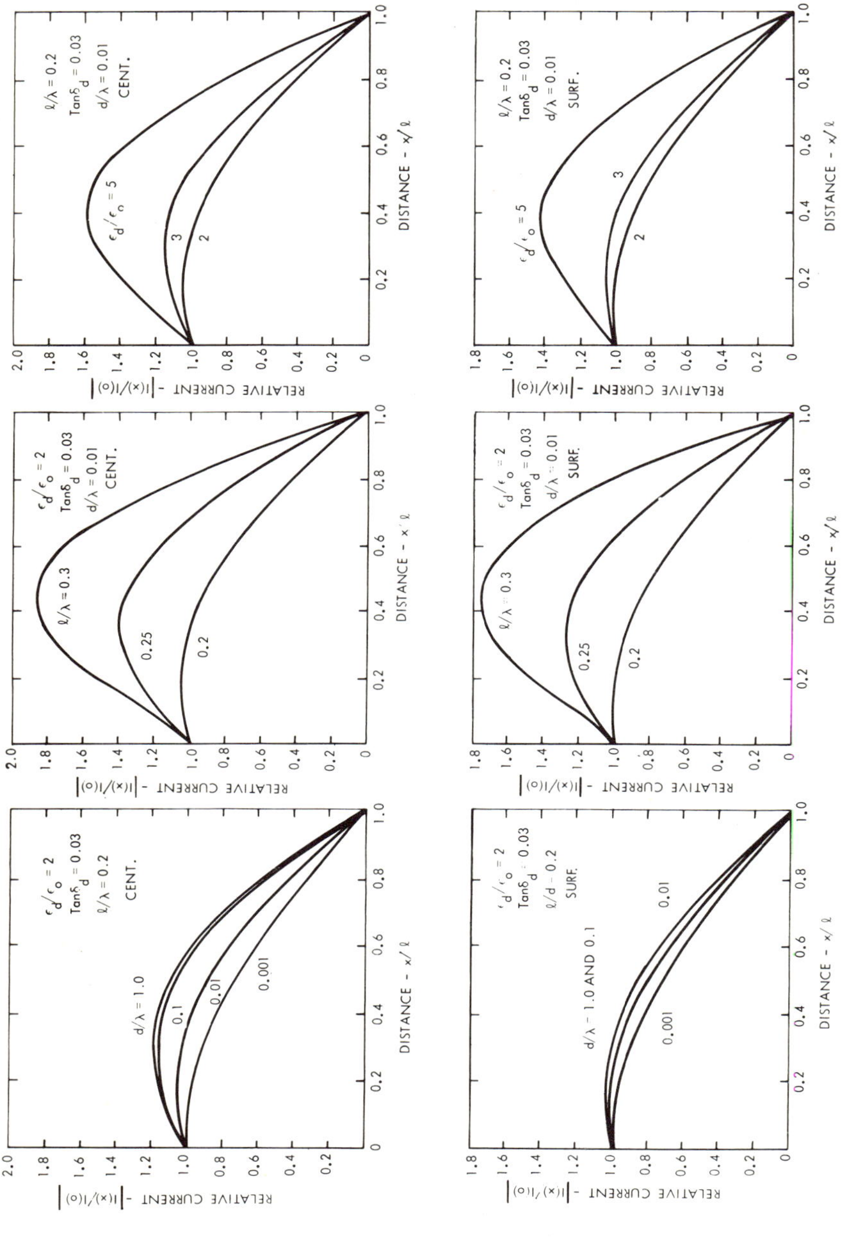

FIG. 9.8. Variational approximation to the current distribution of antenna in the center (CENT.) or on the surface (SURF.) of dielectric layers of thickness 2*d*.

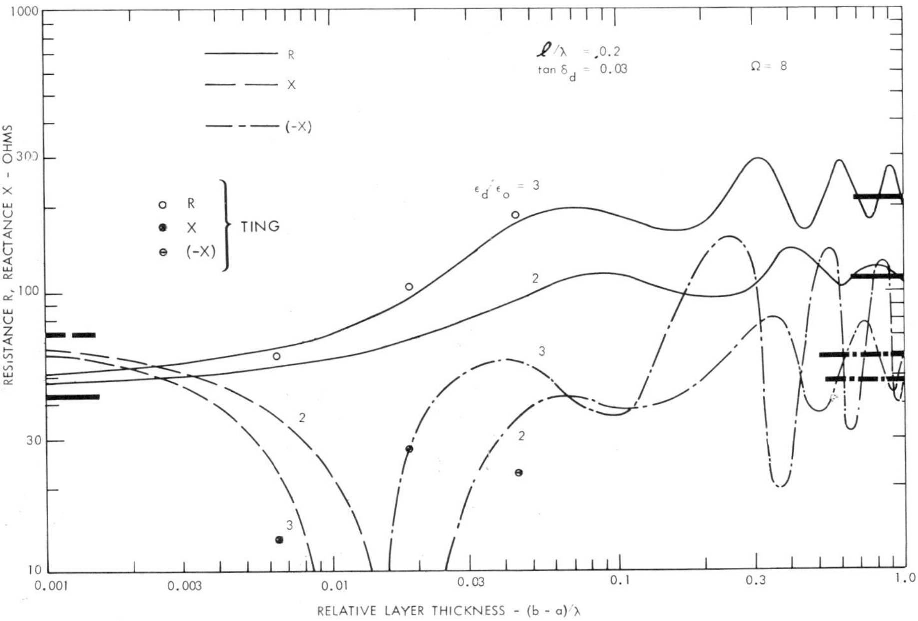

FIG. 9.9. Impedance of an antenna in the center of a dielectric cylinder.

the antenna and of the surrounding dielectric was shown to correspond closely to the antenna impedance in a planar geometry for comparable thicknesses of the dielectric layers.

Ting [1966] has investigated the impedance of a dielectric-coated cylindrical antenna using a numerical method. He indicates that the imaginary part of the current is well represented by a sine term, the real part by a shifted cosine term, both with Goubau surface-wave numbers [Goubau, 1950]. The radius of this antenna is 15 percent larger than in our calculations, but the impedance data of the two calculations are comparable as can be seen from his data points indicated in Fig. 9.9 for $\epsilon_d/\epsilon_0 = 3$.

9.1.6. *Insulated Antenna in Dissipative Medium*

In Fig. 9.11 the impedance of an insulated cylindrical wire measured by Guy [Figs. 3.24 and 3.25, 1962] is compared with transmission line calculations following Section 5.1, with calculations for a cylindrical antenna following Section 12.1, and with an approximately equivalent model of the strip antenna. The calculated resistance data agree very well with measurements, but there is only fair agreement with the measured reactance. The transmission line theory gives too low resistance figures for small antenna lengths and it also produces too large resonance peaks in resistance and

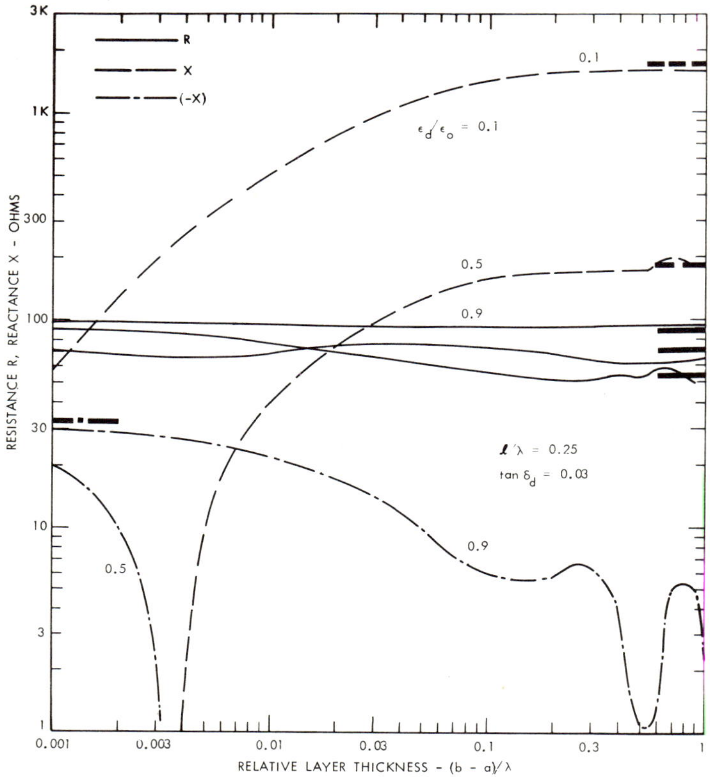

FIG. 9.10. Impedance of an antenna in the center of a plasma cylinder.

reactance. These observations are particularly pronounced in the present example where tan δ_g varies from 2.4 to 0.3 in the frequency range indicated in Fig. 9.11. The same basic differences between transmission line theory and the measured data can be seen from other measurements with higher loss tangents (figs. 3.12–3.29 of Guy [1962], or figs. 3 and 5 of Fenwick and Weeks [1963]).

The impedance of an insulated strip antenna located near the interface between free space and a lossy dielectric half space (ground) is shown in Fig. 9.12 for several values of the loss tan δ_g. The transmission line theory is applied to an approximately equal geometry with a cylindrical conductor in Fig. 5.1. For tan $\delta_g = 100$, there is fair agreement between the two sets of calculations, but for tan $\delta_g = 10$, the transmission line theory gives an increase in the size of the resonant peaks, while they are definitely lower for the calculations shown in Fig. 9.12. Also there is a significant increase of the resistance for short antenna lengths and there is a relatively small variation

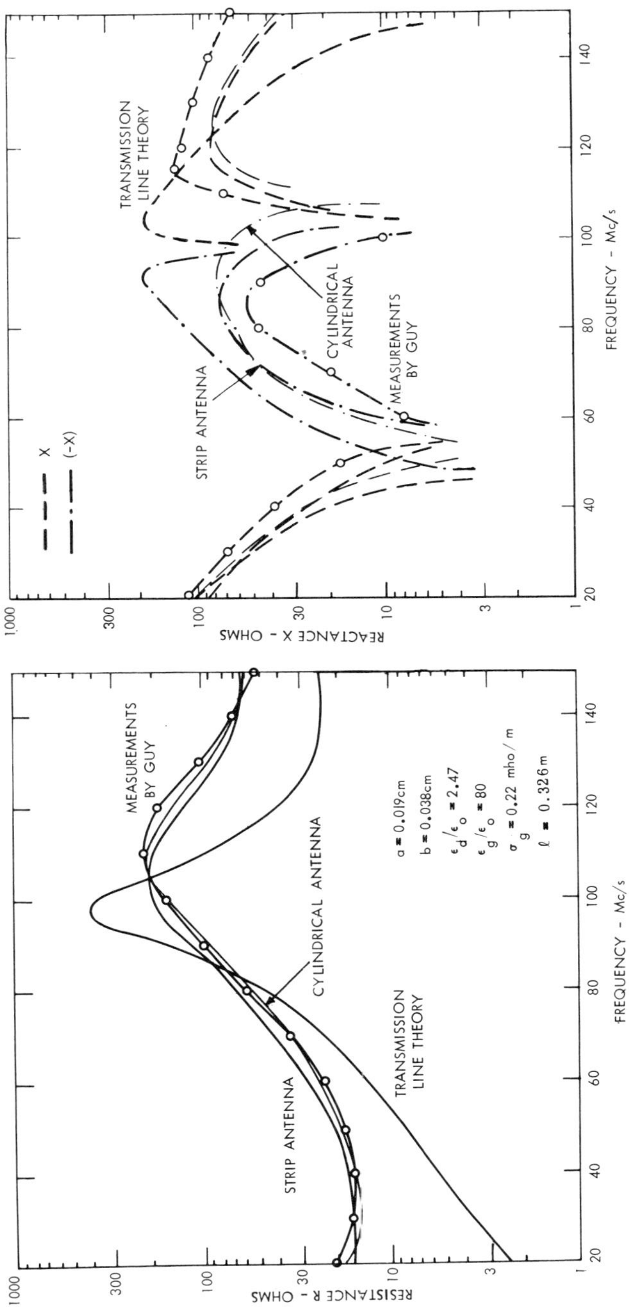

FIG. 9.11. Impedance of an insulated antenna in a semi-infinite dissipative medium. Antenna of length l above a conducting ground plane.

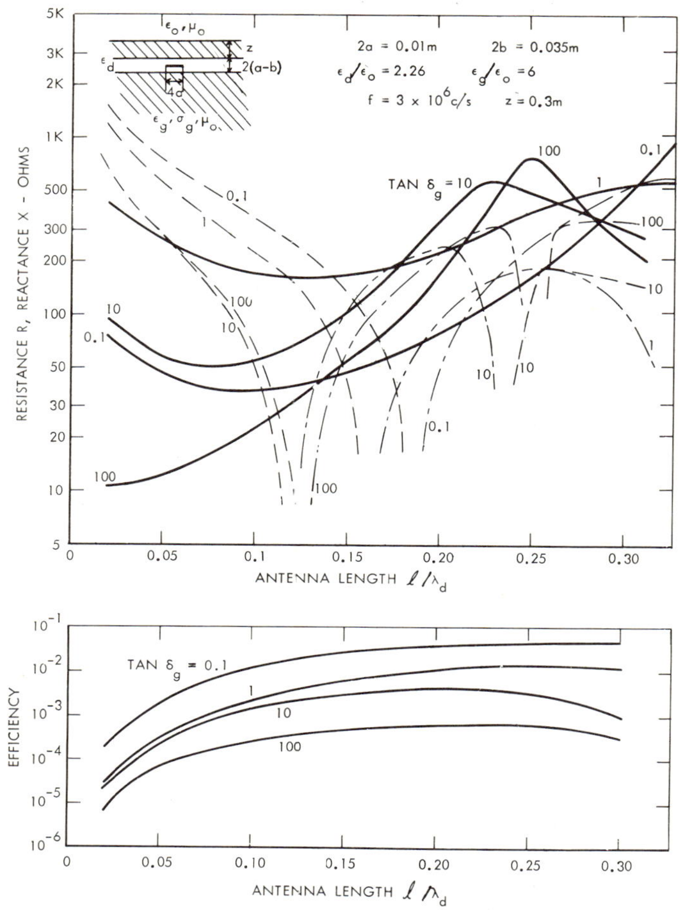

FIG. 9.12. Impedance $Z = R + iX$ of a buried strip antenna and its radiation efficiency.

of the resistance component with the changing antenna length for tan $\delta_g = 1$. The high values of resistance for a small antenna length indicate significant power dissipation in the vicinity of the antenna. For a dipole in direct contact with the ground, the resistance should approach infinity, but insulation and the finite antenna length have the effect of keeping this resistance finite. The radiation efficiency should be improved by decreasing the values of tan δ_g, but for short antenna lengths and for tan $\delta_g = 10$ the efficiency is nearly the same as for tan $\delta_g = 1$, because of the very high resistance components shown in Fig. 9.12 for tan $\delta_g = 1$. The resistance components shown for a short antenna in Fig. 9.12 with $\epsilon_g/\epsilon_0 = 6$ are higher than those of Fig. 9.11 with $\epsilon_g/\epsilon_0 = 80$ for the lower frequencies, because in the latter case there is a large difference in dielectric constant between the dielectric and ground. As a consequence, less of the near fields is stored in the ground with smaller attendant ohmic losses.

9.2. Mutual Impedance

An approximate mutual impedance figure is simplest to compute for an assumed current density along two antennas [Carter, 1932]. A more detailed analysis shows that the mutual and self impedance can be determined from a solution of two coupled integral equations. For two identical antennas the driving point impedances can be computed for symmetrical (Z_s) and anti-symmetrical excitation (Z_a) following Section 5.2. These two driving point impedances are uniquely related to the self-impedance Z_{11} and the mutual impedance Z_{12} of the antennas. For two antennas of different size, a variational formulation has been provided by Levis and Tai [1956], and a unique solution of the self impedances Z_{11}, Z_{22} and of the mutual impedance Z_{12} requires the examination of the two coupled antennas under three different excitation conditions.

The considerations of this Section are restricted to two antennas of equal size. With $Z_{11} = Z_{22}$, two excitation conditions are sufficient and the symmetrical and antisymmetrical antenna excitations are selected for convenience of the latter calculations. This formulation of the antenna impedance is reduced for assumed antenna current distributions to the standard expressions of mutual impedance [Carter, 1932; Harrington, 1961]. Numerical examples are discussed for antennas in free space and in dielectric slabs, and for insulated antennas in lossy dielectric media. For antennas in lossy media, the present formulation is also compared with the more conventional transmission line calculations.

9.2.1. *Impedance Formulation*

The equations used for calculating the self-impedance of a single antenna can be adapted for computing the driving point impedance of one antenna in the presence of a symmetrically or antisymmetrically excited second antenna. In the impedance expression

$$Z = -\frac{\int\int E_x J_{x1}\,dx\,dy}{[I_1(0)]^2} = -\frac{\int\int (E_{x11}+E_{x12})\,J_{x1}\,dx\,dy}{[I_1(0)]^2}, \tag{9.24}$$

E_x represents now the fields E_{x11} induced at (x, y) on antenna 1 by the surface current density $J_{x1}(x', y')$ of antenna 1, plus the fields E_{x12} induced on antenna 1 by the surface current density $J_{x2}(x', y')$ of antenna 2. The center of the second antenna is located at $x = 0$ and $y = h$, and the surface current densities J_{x1} and J_{x2} are related as

$$J_{x1}(x', y') = J_{x2}(x', y' + h) \tag{9.25}$$

for symmetrical excitation, and as

$$J_{x1}(x', y') = -J_{x2}(x', y' + h) \tag{9.26}$$

for antisymmetrical excitation. Substituting (9.25) and (9.26) into the field expressions, it follows that

$$Z_{s,a} = \frac{i\omega\mu_0}{[2\pi I_1(0)]^2} \int_{-\infty}^{\infty}\int_{-\infty}^{\infty} F(u, v)(1 \pm \cos vh) \cdot \left[\iint J_x(x, y) \cos ux \cos vy \, dx \, dy\right] du \, dv, \tag{9.27}$$

where $F(u, v)$ is defined by (9.6). This is a stationary expression that can be used to determine Z_s and Z_a, and also the approximations to the current distributions appropriate to the two excitations. The antenna current density is assumed to be representable by the trial function (9.7). Substituting (9.7) in (9.27) and using the stationary character of the impedance expression for determining the complex amplitudes A and B, it follows that $A/B|_{s,a}$ and $Z_{s,a}$ are given by (2.28) to (2.39) with

$$\gamma_{NM} = \frac{4i\omega\mu_0}{\pi^2} \int_0^{\infty} dv \left(\frac{\sin \epsilon v}{\epsilon v}\right)^2 \int_0^{\infty} du F(u, v) g_N(u) g_M(u) [1 \pm \cos vh] \tag{9.28}$$

and with F_N and $g_N(u)$ defined in (9.8), (9.9), (9.11), and (9.12). The parameter γ_{NM} can be divided corresponding to its two terms in (9.28) as

$$\gamma_{NM} = \gamma_{NM}^0 \pm \gamma_{NM}^h. \tag{9.29}$$

Δ of (2.40) is written similarly as

$$\Delta = \Delta^0 \pm \Delta^h, \tag{9.30}$$

where Δ^0 and Δ^h contain terms proportional to γ_{NM}^0 and γ_{NM}^h, respectively. In (9.28), (9.29), and (9.30), the plus sign refers to symmetrical excitation (subscript s) and the minus sign refers to antisymmetrical excitation (subscript a).

9.2.2. *Limiting Impedance Values*

In the limiting case of $h \to 0$, $\gamma_{NM}^h = \gamma_{NM}^0$. It follows that $Z_a = 0$ and $Z_s = 2Z_{11} = 2Z_{12}$. With $h \to \infty$, the integrand of γ_{NM}^h is highly oscillatory and $\gamma_{NM}^h \to 0$. It follows that $Z_s \approx Z_a \approx Z_{11}$ and $Z_{12} \to 0$. (A/B) is the same for $h \to 0$ and ∞. The current distributions on two coincident symmetrically excited antennas $(h = 0)$ are the same as on two widely separated antennas $(h \to \infty)$.

Substituting (2.41) in (5.21) and (5.22), it follows that for a single trial function

$$Z_{11} = \frac{\gamma^0_{AA}}{F^2_A}, \tag{9.31}$$

$$Z_{12} = \frac{\gamma^h_{AA}}{F^2_A}. \tag{9.32}$$

Z_{11} is recognized as the self-impedance of a single antenna, and Z_{12} of (9.45) can be obtained from a standard expression for mutual impedances [Harrington, 1961]:

$$Z_{12} = -\frac{\int E_{x12} J_{x1} \, dx \, dy}{I_1(0) \, I_2(0)} \tag{9.33}$$

by requiring that J_{x1} and J_{x2} exhibit the same x and y variations. A single term trial function was not sufficiently accurate for self-impedance calculations of a single antenna, and the impedance expressions (9.31) and (9.32) can be expected to be similarly inaccurate. A different approximation can be obtained by assuming the current distribution required in (9.33) to be the same as for a single antenna. Such an approximation is obviously correct for two closely spaced antennas and also for two widely separated antennas. Starting out with the field expressions listed earlier, this gives

$$Z_{12} \approx \frac{(A/B)^2 \gamma^h_{AA} + 2(A/B)\gamma^h_{AB} + \gamma^h_{BB}}{[(A/B)F_A + F_B]^2}, \tag{9.34}$$

where (A/B) is obtained from (2.38) by letting

$$\gamma_{NM} = \gamma^0_{NM}. \tag{9.35}$$

Z_{12} is now a function of both γ^0_{NM} and γ^h_{NM}. In the calculations using Z_s and Z_a, Z_{11} and Z_{12} depend also on γ^0_{NM} and γ^h_{NM} as can be verified by substituting (9.29) and (9.30) in (2.39) and (2.39) in (5.21) and (5.22).

9.2.3. *Free Space Impedance*

Numerical calculations made for a half-wave antenna of several widths 2ϵ or several values of the parameter $\Omega = 2 ln(4l/\epsilon)$ are shown in Fig. 9.13. The impedance curves follow the general behavior reported in past calculations, but the resistance components R_{11} and R_{12} are increased for small antenna separations. Two closely spaced and strongly coupled antennas should be equivalent to a single antenna having a larger width and exhibiting a higher resistance (Fig. 2.1). However, this increase of R_{12} and of R_{11} has not been observed in past calculations by King [1956], Harrington [1961], or Kraus [1950].

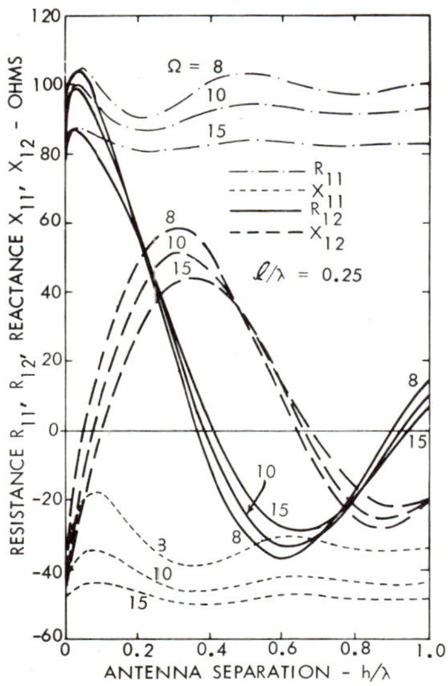

FIG. 9.13. Free space impedance of two antennas.

The free space mutual impedance has also been computed using the approximation (9.34). These results are not distinguishable from the

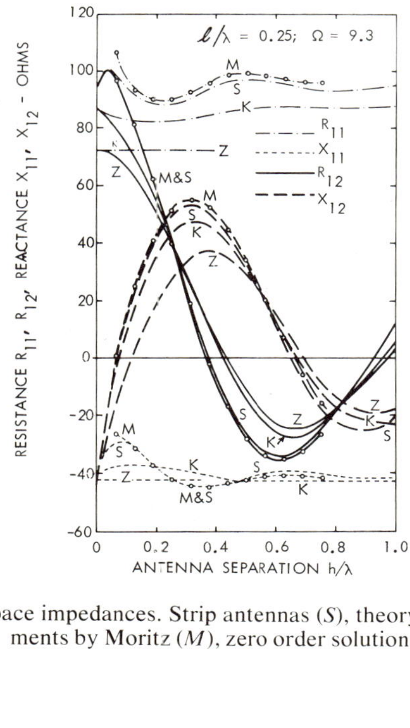

FIG. 9.14. Free space impedances. Strip antennas (S), theory by King (K), measurements by Moritz (M), zero order solution (Z).

calculations based on the more accurate impedance expression within the accuracy of the plots shown in Fig. 9.13.

The present calculations are shown in Fig. 9.14 to agree very closely with measurements of [Moritz, 1952; King, 1956, fig. III.13.1] made for two cylindrical antennas of comparable dimensions. The solution obtained by using the sinusoidal term of the current distribution, (9.31) and (9.32), are numerically the same as the zero order solution reported by King [1956]. It differs considerably from the more accurate calculations using the two-term trial functions. The second order calculations of King [1956] are also indicated in Fig. 9.14 for purposes of the comparison. The resistance components R_{11} of third order computations by King [1956] are higher by approximately 3 ohms, but they follow the same general shape of the second order curve.

9.2.4. *Antenna in Dielectric Layers*

The self and mutual impedance of antennas embedded in a center of a dielectric slab of thickness $2d$ are shown in Figs. 9.15 and 9.16. The mutual impedance is increased by increasing the thickness of a dielectric slab in Fig. 9.15 and by increasing the dielectric constant in Fig. 9.19. Increasing

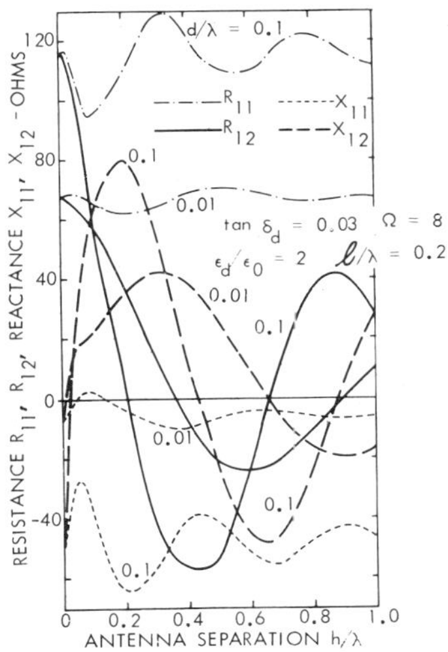

FIG. 9.15. Impedances for antenna's in center of a dielectric layer. Effects of layer thickness.

dielectric losses, however, significantly decrease the mutual impedance, especially for larger antenna separations shown in Fig. 9.16.

For a homogeneous isotropic plasma slab of a relative dielectric constant within the range $1 > \epsilon_d/\epsilon_0 > 0$, the mutual impedance is decreased with

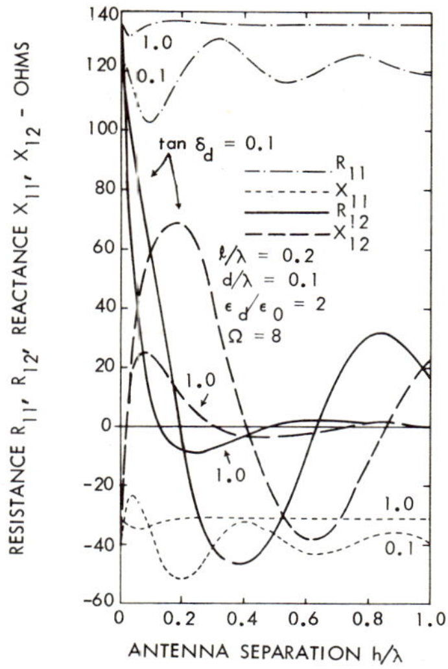

FIG. 9.16. Impedance for antennas in center of a dielectric layer. Effects of dielectric losses.

decreasing values of ϵ_d in Fig. 9.17. This is similar to the behavior of the mutual admittance observed for waveguides radiating into a homogeneous plasma layer [Figs. 8.12–8.14].

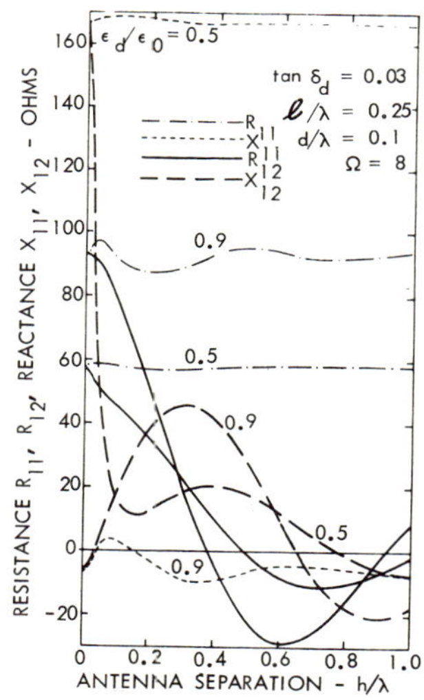

FIG. 9.17. Impedances of antennas in the center of a plasma layer.

9.2.5. *Insulated Antennas in Dissipative Medium*

Two linear strip antennas of width $4a$ and of length $2l$ are separated from the surrounding lossy ground of permittivity ϵ_g and a loss tangent

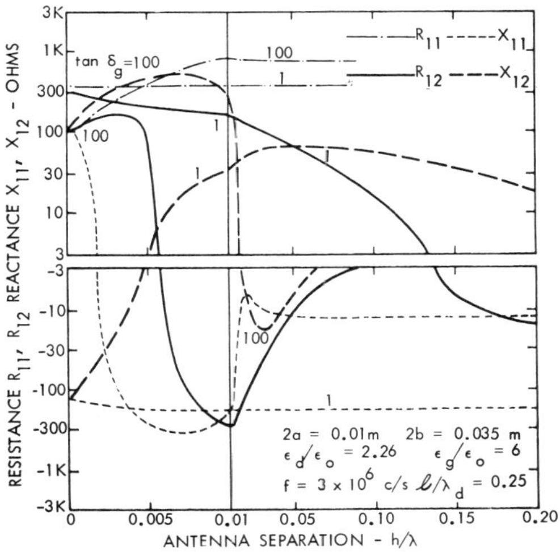

FIG. 9.18. Impedances of insulated strip antennas buried at a depth $z = 0.3$ m.

$\tan \delta_g = \sigma_g/(\omega\epsilon_g)$ by an insulating layer of permittivity ϵ_d and of total thickness $2(b-a)$. The self-impedances and mutual impedances are computed

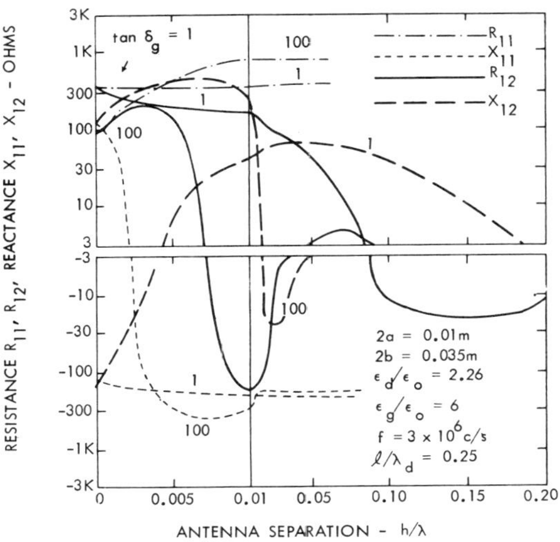

FIG. 9.19. Impedances of insulated strip antennas for infinite burial depth.

for shallow ($z = 0{\cdot}3$ m) and large burial depth ($z = \infty$) of antennas with length $l/\lambda_d = 0{\cdot}25$ (Figs. 9.18 and 9.19), where λ_d is the wavelength in the dielectric. For small antenna separations, the mutual impedances are about the same in Figs. 9.18 and 9.19 but tend to decay more rapidly for the large burial depth as shown in Fig. 9.19. The mutual impedances of antennas with length $l/\lambda_d = 0{\cdot}1$ are shown in Fig. 9.20 for infinite burial depth. The resistance

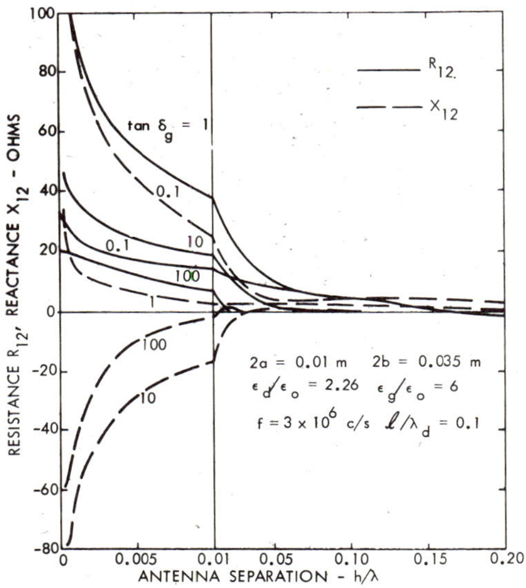

FIG. 9.20. Impedance of insulated strip antennas for infinite burial depth.

component R_{12} is smaller and also the reactance component X_{12} is significantly decreased for the smaller values of ground losses in Fig. 9.20. The corresponding values of the self-impedance Z_{11} are shown in Table 9.1

TABLE 9.1 SELF-IMPEDANCE OF INSULATED BURIED ANTENNAS

z	h/λ	Tan δ_g	100	10	1	0.1
0.3 m	10^{-3}	$Z_{11} =$	$24.5 + i68$	$58 + i80$	$168 + i218$	$36.6 + i347$
	∞ or 0		$22.4 + i70$	$54 + i82$	$168 + i216$	$36.6 + i347$
∞	10^{-3}		$20.2 + i71.8$	$48 + i52$	$141 + i181$	$33.6 + i295$
	∞ or 0		$18.6 + i72.5$	$44 + i54$	$141 + i180$	$33.5 + i295$

$l/\lambda_d = 0.1$; $2a = 0.01$ m; $2b = 0.035$ m; $\epsilon_d/\epsilon_0 = 2.26$;
$f = 3 \times 10^6$ c/s.

for two antenna separations h. Z_{11} is nearly independent of h for the smaller values tan δ_g. The self-impedance changes by less than 20 percent between shallow and large burial depths of the antennas.

The mutual and self impedance of two cylindrical antennas of radii a and separated from the surrounding homogeneous ground by a dielectric cylinder of radius b can be also formulated in a transmission line approximation. The resulting impedance curves are shown in Fig. 5.2. For a highly lossy ground with tan $\delta_g = 100$, the transmission line approximation agrees closely with the variationally computed impedance values of Fig. 9.19. For tan $\delta_g = 10$, the agreement is fair, but the transmission line approximation becomes inaccurate for tan $\delta_g = 1$. The resonances shown in Fig. 5.2 for tan $\delta_g = 1$ and $h/\lambda_d \approx 0{\cdot}0075$ are not present in Fig. 9.19.

A larger value of relative dielectric constant of the surrounding medium improves the accuracy of the transmission line calculations as can be seen from the impedance data shown in Fig. 9.21. Antennas of such para-

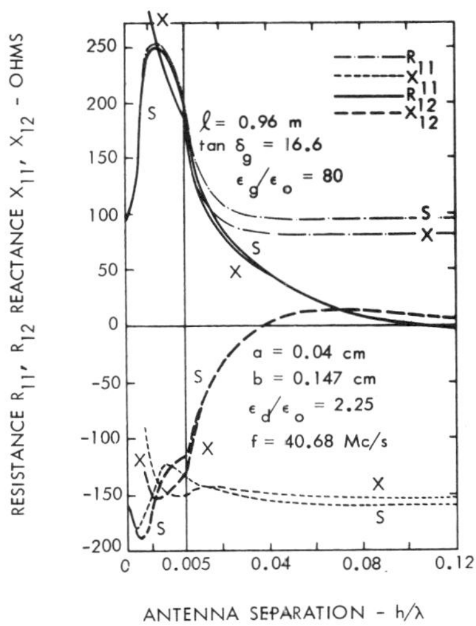

FIG. 9.21. Impedances of insulated buried antennas for infinite burial depth. Strip antennas in a variational formulation (S) and cylindrical wires in a transmission line approximation (X).

meters have been investigated by Guy [1962]. The principal discrepancy between the two sets of computations occurs for very small antenna separations where the assumption of an h-independent y_{11} of (5.9) may be violated in the transmission line theory.

The above limitations of the transmission line approximations are similar to the ones observed when calculating the impedance of a single antenna.

9.3. Exercises

9.1 Compute the power radiated outside the dielectric slab by relating the amplitudes A_{j0} and B_{j0} of the TE and TM modes in (9.14) to the amplitudes of the trial functions A and B defined by (9.7).

9.2 Compute the power carried by surface waves along a dielectric slab; substitute the field expressions in (9.16) and reduce the sixfold integral to a sum of four single integrals.

9.3 Examine the amplitude ratios $(A/B)_{s,a}$ of mutual impedance calculations:

(a) Substitute (9.7) and (9.27) in (5.21) and (5.22) and determine the amplitude ratios $(A/B)_{s,a}$ for symmetrical and antisymmetrical excitations by postulating

$$\partial Z_{11}/\partial I_{1s} = 0 \quad \text{and} \quad \partial Z_{11}/\partial I_{1a} = 0$$

or

$$\partial Z_{12}/\partial I_{1s} = 0 \quad \text{and} \quad \partial Z_{12}/\partial I_{1a} = 0.$$

(b) Derive the ratios $(A/B)_{s,a}$ by substituting (9.7) into (9.27) and postulating

$$\partial Z_{s,a}/\partial (A/B)_{s,a} = 0.$$

(c) Compare the results of the two derivations.

9.4. References

CARTER, P. S. (1932) Circuit relations in radiating systems and applications to antenna problems, *Proc. IRE* **20**, (6), 1004–41.

FENWICK, R. C. and WEEKS, W. L. (1963) Submerged antenna characteristics, *IEEE Trans. on Ant. and Prop.* **AP-11**, 296–305.

GALEJS, J. (1964) Slot antenna impedance for plasma layers, *IEEE Trans. on Ant. and Prop.* **AP-12**(6), 738–45.

GALEJS, J. (1965) Driving point impedance of linear antennas in the presence of stratified dielectric, *IEEE Trans. on Ant. and Prop.* **AP-13**, 725–37.

GALEJS, J. (1966) Mutual impedance of linear antennas in the presence of the stratified dielectric, *IEEE Trans. on Ant. and Prop.* **AP-14**.

GHOSE, R. N. (1960) The radiator-to-medium coupling in an underground communication system, *Proc. Nat. Elec. Conf.* **16**, 279–89.

GOOCH, D. W., HARRISON Jr., C. W., KING, R. W. P. and WU, T. T. (1963) Impedances of long antennas in air and dissipative media, *NBS J. Res (Radio Science)* **67D**, 355–60; and Cruft Lab., Harvard University, Cambridge, Mass., Tech. Rept. 353.

GOUBAU, G. (1950) Surface waves and their application to transmission lines, *J. Appl. Phys.* **21** (9), 1119–28.

GUY, A. W. (1962) *Experimental Data on Buried Antennas*, The Boeing Company, Seattle, Wash., Rept. D2-90051.

GUY, A. W. and HASSERJIAN, G. (1963) Impedance properties of large sub-surface antenna arrays, *IEEE Trans. on Ant. and Prop.* **AP-11** (3), 232–40.

HARRINGTON, R. F. (1961) *Time-harmonic Electromagnetic Fields*, McGraw-Hill, New York, NY, Section 7.9.

IIZUKA, K. (1963) An experimental study of the insulated dipole antenna immersed in a conducting medium, *IEEE Trans. on Ant. and Prop.* **AP-11**, 518–32.

IIZUKA, K. and KING, R. W. P. (1962) The dipole antenna immersed in a homogeneous conducting medium, *IEEE Trans. on Ant. and Prop.* **AP-10**, 384–92.

KING, R. W. P. (1956) *Theory of Linear Antennas*, Harvard University Press, Cambridge, Mass., Sections III. 1 to 8.

KING, R. W. P. (1964) Theory of the terminated insulated antenna in a conducting medium, *IEEE Trans. on Ant. and Prop.* **AP-12**, 305–18.

KING, R. W. P. and HARRISON JR., C. W. (1960) Half-wave cylindrical antenna in a dissipative medium: current and impedance, *NBS J. Res.* **64D**, 365–80.

KING, R. W. P. and IIZUKA, K. (1963) The complete electromagnetic field of half-wave dipole in dissipative media, *IRE Trans. on Ant. and Prop.* **AP-11**, 275–85.

KING, R. W. P. and SANDLER, S. S. (1963) Linear arrays: current impedances and fields, *Electromagnetic Waves*, ed. E. C. Jordan, Pergamon Press, New York, NY, Part II, 1307–23.

KING, R. W. P., SANDLER, S. S. and IIZUKA, K. (1966) Coupled dipoles in a dissipative medium, *IEEE Trans. on Ant. and Prop.* **AP-14** (3), 275–82.

KRAUS, J. D. (1950) *Antennas*, McGraw-Hill, New York, NY, Section 10. 5.

LEVIS, C. A. and TAI, C. T. (1956) A method of analyzing coupled antennas of unequal sizes, *IRE Trans. on Ant. and Prop.* **AP-4** (2), 128–32.

MOORE, R. K. (1951) Theory of radio communication between submerged submarines, Ph.D. thesis, Cornell University, Ithaca, NY.

MORITZ, C. (1952) *The Coupled Receiving Antenna*, Tech. Reports Nos. 146 and 147, Cruft Laboratory, Harvard University Cambridge, Mass.

TING, C. Y. (1966) *A Theoretical Study of Dielectric-Coated Cylindrical Antennas*, Tech. Rept. 506, Cruft Laboratory, Harvard University, Cambridge, Mass.

WHITMER, R. M. (1948) Fields in nonmetallic waveguides, *Proc. IRE* **36**, 1105–9.

WU, T. T. (1961) Theory of the dipole antenna and the two-wire transmission line, *J. Math. Phys.* **2** (4), 550–74.

CHAPTER 10

INSULATED LOOPS IN LOSSY MEDIA

Loop antennas are simplest to analyze with an assumed uniform current flow; such impedance calculations for insulated loops surrounded by a lossy medium have been carried out by Wait [1957], Kraichman [1962], Galejs [1963], and Row [1964]. The current distributions on loop antennas surrounded by a homogeneous medium have been computed by solving integral equations by Storer [1956], Wu [1962], Chen and King [1963], and King *et al.* [1964]. This work shows that, for loops surrounded by a lossless dielectric, the uniform current distribution is justified for loops of sizes $\beta b < 0.1$, where b is a radius of the loop, and β is the real part of the complex phase constant $k = \beta + i\alpha$ (see Section 2.4). For loops surrounded by a lossy medium, the assumption of a constant current may give a nearly correct susceptance for $\beta b < 0.3$, but the conductance is of an incorrect order of magnitude for βb as small as 0.05. However, following Chapter 5 the propagation along a straight wire depends critically on the thickness of the insulation. A similar dependence may be expected also for insulated loops surrounded by a lossy medium. It has been shown that loop antenna analyses based on uniform current distributions are valid over a wider range of loop diameters than is predicted for insulating layers of negligible thickness [Galejs, 1965].

A variational expression of the antenna impedance will be derived for a loop in the shape of a flat ring that is at a finite distance from two semi-infinite layers of a lossy dielectric as is shown in Fig. 10.1. The validity of the usually assumed uniform current distribution is determined using a simple two-term trial function for the loop current. A function of the form $I = I_0 + I_1 \cos \phi$, which was also used by Chen and King [1963], will indicate the range of loop parameters where I_0 is much larger than I_1. However, this trial function will not give accurate current distributions if this inequality is not strictly satisfied. It is shown that loop impedances computed for dielectrics of negligible thickness with an assumed uniform current flow are nearly the same as a loop impedance computed for dielectric of finite thickness, where a uniform current flow is justified based on the two-term trial function. This procedure will establish the range of validity of analyses, similar to that of Kraichman [1962] or Galejs [1963], more accurately than is possible based on more exact results for loops in direct contact with the

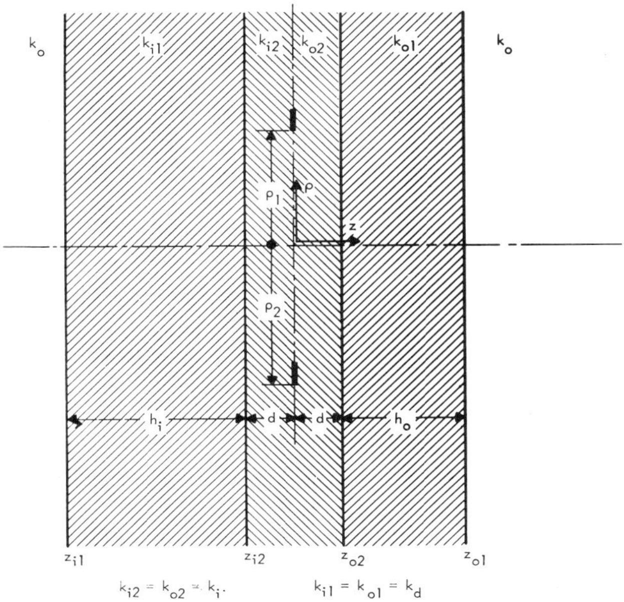

FIG. 10.1. Geometry of the loop antenna.

lossy medium [King *et al.*, 1964]. The impedance and radiation efficiency of loops surrounded by thin layers of lossy dielectric and of buried insulated horizontal loops are alsc investigated in the geometry of Fig. 10.1. For thin lossy layers, the radiation efficiency of the loop is compared with the closed forms expressions of Section 4.3 for loops in spherical shells. A buried horizontal loop exhibits about the same efficiency as the buried horizontal dipole of Section 9.1.6 if its length $2l$ is the same as the mean circumference of the loop.

10.1. Impedance Formulation

The loop impedance is computed in the geometry of Fig. 10.1, where the individual layers are denoted similarly as in Fig. 9.1 for a flat strip antenna. A concentric perfectly conducting infinitely long cylinder of diameter a that is large relative to the loop is assumed to surround the antenna and the dielectric layers. The cylinder is introduced to change the continuous spectrum of eigenvalues of the unbounded problem to a set of discrete eigenvalues that characterize the permissible waveguide modes. Such an artificial boundary has been used in admittance calculations of rectangular slots in Section 8.1 and of annular slot antennas in Section 8.3. The antenna current is represented by a ϕ-directed surface current density $J_{s\phi}(\rho, \phi)$. The fields of a small loop of uniform current are characterized by $E_z = 0$ and can be expressed in TE modes. For non-uniform current along the loop, the loop will exhibit also an electric dipole moment and $E_z \neq 0$ outside the $z = 0$ plane. The surface current $J_{s\phi}(\rho, \phi)$ will excite field components in a superposition of TE and TM modes that can be derived for region m from two scalar functions Ψ_m and Φ_m, defined as

$$\Psi_m = \sum_{n,q} A_{nq}^m \cos n\phi J_n(\lambda_{qa}\rho)\,(\mathrm{e}^{\gamma_{am}z} + R_{am}\mathrm{e}^{-\gamma_{am}z}), \tag{10.1}$$

$$\Phi_m = \sum_{n,q} B_{nq}^m \sin n\phi J_n(\lambda_{qb}\rho)\,(\mathrm{e}^{\gamma_{bm}z} + R_{bm}\,\mathrm{e}^{-\gamma_{bm}z}), \tag{10.2}$$

where m is used for the double subscripts shown in Fig. 10.1. The function Φ_m of (10.2) can be obtained from (8.60) by considering discrete values of λ, and Ψ_m can be derived in a similar way starting out with (8.3). The neglect of terms proportional to sin $n\phi$ in Ψ_m and proportional to sin $n\phi$ in Φ_m will be justified in the discussion following. The transverse electric and magnetic field components are computed from (1.21), (1.22), (1.24), and (1.25) as

$$E_\rho = \frac{i\omega\mu_0}{\rho}\frac{\partial}{\partial\phi}\Psi_m + \frac{\partial^2}{\partial\rho\partial z}\Phi_m, \tag{10.3}$$

$$E_\phi = -i\omega\mu_0\frac{\partial}{\partial\rho}\Psi_m + \frac{1}{\rho}\frac{\partial^2}{\partial\phi\partial z}\Phi_m, \tag{10.4}$$

$$H_\rho = \frac{\partial^2}{\partial\rho\partial z}\Psi_m - \frac{i\omega\epsilon_m}{\rho}\frac{\partial}{\partial\phi}\Phi_m, \tag{10.5}$$

$$H_\phi = \frac{1}{\rho}\frac{\partial^2}{\partial\phi\partial z}\Phi_m + i\omega\epsilon_m\frac{\partial}{\partial\rho}\Phi_m, \tag{10.6}$$

The reflection coefficients R_{am} and Γ_{bm} are determined by requiring the tangential field components (10.3) to (10.6) to be continuous across the dielectric interfaces. In the outermost regions ($m = 00$ and $i0$) there are no reflected waves and $R_{am} = R_{bm} = 0$.

In the other regions the reflection coefficients R_{am} and R_{bm} are determined similarly as in Section 9.1.1. The eigenvalues λ_{qa} and λ_{qb} are determined from the conditions

$$\frac{\mathrm{d}}{\mathrm{d}\rho}J_n(\lambda_{qa}\rho)\bigg|_{\rho=a} = 0, \tag{10.7}$$

$$J_n(\lambda_{qb}a) = 0, \tag{10.8}$$

where $q = 1, 2, 3 \ldots$ designates the consecutive roots of the equations (10.7) or (10.8) for a given value of n. The propagation constants γ_j are defined as

$$\gamma_{jm} = i(k_m^2 - \lambda_{qj}^2)^{1/2},$$

where

$$j = a \quad \text{or} \quad b; \qquad k_m = (\omega^2 \mu_0 \epsilon_m + i\omega\mu_0\sigma_m)^{1/2},$$

or

$$k_m = k_0(\epsilon_m/\epsilon_0)^{1/2}\,(1+i\tan\delta_m)^{1/2},$$

with

$$k_0 = \omega(\mu_0\epsilon_0)^{1/2} = 2\pi/\lambda.$$

The continuity of the tangential electric fields at $z=0$ leads to

$$A_{nq}^{02}(1+R_{a02}) = A_{nq}^{i2}(1+R_{ai2}) \tag{10.9}$$

$$B_{nq}^{02}\gamma_{b02}(1-R_{b02}) = -B_{bq}^{i2}(1-R_{bi2}) \tag{10.10}$$

which are similar to (9.1) and (9.2). After substituting (10.1) and (10.2) in (10.5) and (10.6), the orthogonality of the transverse magnetic field components of different modes can be used to compute the mode amplitudes A_{nq}^3 and B_{nq}^3. It follows that

$$A_{nq}^m = \frac{\iint (i_\rho H_\rho + i_\phi H_\phi)\cdot\left[i_\rho \cos n\phi \frac{\mathrm{d}}{\mathrm{d}\rho}J_n(\lambda_{qa}\rho) - i_\phi \frac{n\sin n\phi}{\rho}J_n(\lambda_{qa}\rho)\right]\rho\,\mathrm{d}\rho\,\mathrm{d}\phi}{\gamma_a(1-R_{am})\frac{2\pi}{\epsilon_n}I_n(\lambda_{qa})} \tag{10.11}$$

$$B_{nq}^m = \frac{\iint (i_\rho H_\rho + i_\phi H_\phi)\cdot\left[i_\rho \frac{n\cos n\phi}{\rho}J_n(\lambda_{qb}\rho) - i_\phi \sin n\phi \frac{\mathrm{d}}{\mathrm{d}\rho}J_n(\lambda_{qb}\rho)\right]\rho\,\mathrm{d}\rho\,\mathrm{d}\phi}{\left(-\frac{ik^2}{\omega\mu_0}\right)(1+R_{bm})\pi I_n(\lambda_{qb})}, \tag{10.12}$$

with

$$I_n(\lambda_{qj}) = \int_0^a \left[\frac{\mathrm{d}}{\mathrm{d}\rho}J_n(\lambda_{qj}\rho)\right]^2 \rho\,\mathrm{d}\rho + n^2\int_0^a \left[J_n(\lambda_{qj}\rho)\right]^2\frac{\mathrm{d}\rho}{\rho}, \tag{10.13}$$

$$\epsilon_n = \begin{cases} 1 \text{ for } n=0 \\ 2 \text{ for } n\neq 0 \end{cases}. \tag{10.14}$$

Noting that $H_{\rho 02} - H_{\rho i2} = J_{s\phi}$ and $H_{\phi 02} = H_{\phi i2}$ at $z=0$, it follows from (10.11) that

$$[A_{nq}^{0q}\gamma_{a02}(1-R_{a02}) + A_{nq}^{i2}\gamma_{ai2}(1-R_{ai2})]\,\frac{2\pi}{\epsilon_n}I_n(\lambda_{ga}) = -\iint J_{s\phi}(\rho,\phi)\,\frac{\partial}{\partial\rho}J_n(\lambda_{qa}\rho)\cos n\phi\,\rho\,\mathrm{d}\rho\,\mathrm{d}\phi, \tag{10.15}$$

$$[B_{nq}^{02}\epsilon_{02}(1+R_{b02}) - B_{nq}^{i2}\epsilon_{i2}(1+R_{bi2})]\,i\omega\pi I_n(\lambda_{qb}) = -\iint J_{s\phi}(\rho,\phi)\,J_n(\lambda_{qb}\rho)\,n\cos n\phi\,\mathrm{d}\rho\,\mathrm{d}\phi. \tag{10.16}$$

The surface current density of the loop is approximated by the trial function

$$J_{s\phi}(\rho,\phi)=\frac{I_0+I_1\cos\phi}{\rho\log(\rho_2/\rho_1)}, \tag{10.17}$$

for $\rho_1 \leqslant \rho \leqslant \rho_2$ and with $J_{s\rho}(\rho,\phi)$ assumed to be negligible relative to $J_{s\phi}(\rho,\phi)$. With $J_{s\phi}(\rho,\phi)$ of the form (10.17), the only nonzero amplitude coefficients are A_{0q}, A_{1q}, and B_{1q}. Also terms of Ψ_m proportional to $\sin n\phi$ and terms of Φ_m proportional to $\cos n\phi$, which have not been shown in (10.1) and (10.2), will not be excited for this form of $\mathbf{J}_s$. The Bessel function integrals of (10.13), (10.15), and (10.16) can be evaluated using relations listed by Luke [1962]. Application of (10.9) and (10.10) shows that

$$A_{0q}^{02}=\frac{I_0}{\log(\rho_2/\rho_1)}\frac{J_0(v)\Big|_{v_1}^{v_2}}{\gamma_a F(R_a)(1+R_{a02})[v_0J_0(v_0)]^2}, \tag{10.18}$$

where

$$v_m=\lambda_{qa}\rho_m \quad\text{and}\quad J_1(\lambda_{qa}a)=J_1(v_0)=0,$$

$$A_{1q}^{02}=\frac{I_1}{\log(\rho_1/\rho_1)}\frac{J_1(w)\Big|_{w_1}^{w_2}}{\gamma_a F(R_a)(1+R_{a02})[-1+w_0^2]J_1^2(w_0)}, \tag{10.19}$$

where

$$w_m=\lambda_{qa}\rho_m \quad\text{and}\quad J_1'(\lambda_{qa}a)=J_1'(w_0)=J_0(w_0)-\frac{1}{w_0}J_1(w_0)=0,$$

$$B_{1q}^{02}=\frac{I_1}{\log(\rho_2/\rho_1)}\frac{i\omega\mu_0\left[\int_{v_1}^{v_2}J_0(v)\,\mathrm{d}v-J_1(v)\Big|_{v_1}^{v_2}\right]}{k^2F(R_b)(1-R_{b02})[v_0J_0(v_0)]^2}, \tag{10.20}$$

where

$$v_m=\lambda_{qb}\rho_m \quad\text{and}\quad J_1(\lambda_{qb}a)=J_1(v_0)=0.$$

$F(R_a)$ and $F(R_b)$ of (10.18) to (10.20) are given by

$$F(R_a)=\frac{1}{2}\left[\frac{1-R_{a02}}{1+R_{a02}}+\frac{1-R_{ai2}}{1+R_{ai2}}\right], \tag{10.21}$$

$$F(R_b)=\frac{1}{2}\left[\frac{1+R_{b02}}{1-R_{b02}}+\frac{1+R_{bi2}}{1-R_{bi2}}\right]. \tag{10.22}$$

The ϕ component of the electric field is computed from (10.4) after substituting Ψ_m and Φ_m of (10.1) and (10.2) with A_{nq} and B_{nq} of (10.18) to (10.22). With E_ϕ determined and $J_{s\phi}$ given by (10.17), the loop impedance may be computed from (1.30) as

$$Z=-\frac{\int\int\mathbf{E}\cdot\mathbf{J}_s\rho\,\mathrm{d}\rho\,\mathrm{d}\phi}{[I(\phi=0)]^2}, \tag{10.23}$$

which is stationary with respect to small changes of the surface current density $\mathbf{J}_s$ or of the current $I(\phi)$ about its correct value. After evaluating the integral, (10.23) is expressed as

$$Z = K\frac{I_0^2 S_0 + I_1^2 S_1}{(I_0 + I_1)^2} \tag{10.24}$$

where

$$K = i\omega\mu_0\pi/[\log(\rho_2/\rho_1)]^2, \tag{10.25}$$

$$S_0 = \sum_q \left\{\frac{1}{F(R_a)}\frac{2}{\gamma_a}\frac{\left[J_0(v)\Big|_{v_1}^{v_2}\right]^2}{[v_0 J_0(v_0)]^2}\right\}, \tag{10.26}$$

$$S_1 = \sum_q \left\{\frac{1}{F(R_a)}\frac{\left[J_1(w)\Big|_{w_1}^{w_2}\right]^2}{\gamma_a(-1+w_0^2)\,J_1^2(w_0)} - \frac{1}{F(R_b)}\frac{\gamma_b}{k^2}\frac{\left[\int_{v_1}^{v_2} J_0(v)\,\mathrm{d}v - J_1(v)\Big|_{v_1}^{v_2}\right]^2}{[v_0 J_0(v_0)]^2}\right\}, \tag{10.27}$$

and where v_0, v_m, w_0, and w_m are defined as in (10.18) to (10.20). Considering the ratio $u = I_1/I_0$ as a variable in (10.24), the stationarity of Z or $\mathrm{d}Z/\mathrm{d}u = 0$ leads to

$$u = \frac{I_1}{I_0} = \frac{S_0}{S_1}, \tag{10.28}$$

or

$$\frac{1}{Z} = \frac{1}{K}\left[\frac{1}{S_0} + \frac{1}{S_1}\right]. \tag{10.29}$$

For

$$|I_1| \ll |I_0|,\ |S_1| \gg |S_0|$$

and

$$Z_0 = KS_0, \tag{10.30}$$

which can be also obtained directly from (10.24). The variation of the complex current along the loop may be computed as

$$I(\phi) = I_0\left(1 + \frac{I_1}{I_0}\cos\phi\right) = \frac{V}{Z}\frac{1 + (I_1/I_0)\cos\phi}{1 + (I_1/I_0)}, \tag{10.31}$$

where V is assumed to be real, but Z and I_1/I_0 are generally complex.

The computation of Z in (10.29) requires the evaluation of two infinite summations S_0 and S_1. The numerical calculations are facilitated by approximating the summations by integrals in the range of $Q < q \leqslant \infty$, where all the Bessel functions can be represented by their asymptotic expansions, and

$$v_m = \lambda_{qa}\rho_m \approx (q + \tfrac{1}{4})\pi\rho_m/a$$

and

$$w_m = \lambda_{qb}\rho_m \approx (q - \tfrac{1}{4})\pi\rho_m/a$$

exceed 20. Q is then computed as $Q \approx 20a/(\pi\rho_m)$. R_a and R_b will be nearly zero in this range of q.

Although this development has been restricted to a simple two-term approximation (10.17) of the current distribution, it can be readily generalized to consider trial functions

$$J_{s\phi}(\rho, \phi) = \frac{\Sigma I_n \cos n\phi}{\rho \log(\rho_2/\rho_1)}. \tag{10.32}$$

The integrals required for evaluating A_{nq}, B_{nq}, and S_n are similar. After considering the ratios I_n/I_0 as variables in a generalization of (10.24), the stationary character of this expression is used to generalize (10.29). As a result, Z is computed by adding further terms $(1/S_n)$ to (10.29).

For finite values of h_0 and h_i, the radiation resistance of the loop R_r is computed by integrating the Poynting's vector over the outer dielectric surfaces at $z_{01} = d + h_0$ and $z_{il} = -d - h_i$. The radiation resistance is defined by

$$R_r = \mathrm{Re}\frac{1}{[I(0)]^2}\iint \mathbf{E} \times \mathbf{H}^* \cdot da \tag{10.33}$$

After computing the field expressions from (10.1) to (10.6), it follows that

$$R_r = \mathrm{Re}\frac{i\omega\mu_0\pi}{|I(0)|^2}\sum_q \Big\{[|A_{0q}^{00}|^2 + |A_{0q}^{i0}|^2]\gamma_{a00}^* I_0(\lambda_{qa}) + [|A_{1q}^{00}|^2 + |A_{1q}^{i0}|^2]\gamma_{a00}^* I_1(\lambda_{qa}) - \frac{\epsilon_0}{\mu_0}\gamma_{b00}[|B_{1q}^{00}|^2 + |B_{1q}^{i0}|^2] I_1(\lambda_{qb})\Big\}, \tag{10.34}$$

where $I_n(\lambda_{gj})$ is defined by (10.13). A_{nq}^{j2} and B_{nq}^{j2} are related to the current amplitudes I_0 or I_1 by (10.18) to (10.20), which can be expressed in terms of the total current $I(0) = I_0 + I_1$ using (10.28). A_{nq}^{j0} and B_{nq}^{j0} are related to A_{nq}^{j2} and B_{nq}^{j2} by matching the tangential field components across the dielectric interfaces. The summation is restricted to those values of q that represent propagating modes in the free space region with imaginary values of γ_{a00} and γ_{b00}.

10.2. Loops in a Homogeneous Lossy Dielectric or in Free Space

The dependence of the loop impedance on changes of the waveguide diameter has been examined and was found to be similar to that of the annular slot admittance considered in Section 8.3.5. The loop resistance is a discontinuous function of the guide diameter, and the discontinuities occur at those guide diameters, where an additional mode starts propagating in the waveguide. However, the guide diameter can be selected such that the loop resistance for an empty guide is the same as the free space loop resistance with a uniform current flow computed from the admittance of a complementary slot (3.43) using (3.6).

The reactance of the loop depends less on the guide diameter, and the selection of the guide diameter for a correct resistance is sufficient for obtaining also a valid reactance. Guide radii $a/\lambda = 6.28$ and 2.27 have been used in the numerical computations. A smaller guide radius is preferable because the number of terms $Q \sim a$ and less terms are required in the summations for the smaller values of a. (This is further discussed in Section 8.3.5 in connection with radiation from annular slots.)

The guide diameter a is noncritical in presence of lossy dielectric ($\tan \delta_d \neq 0$) even for loop insulation of finite thickness ($d \neq 0$, $\tan \delta_i = 0$). The admittance computed for $\tan \delta_d = 0.06$ and 0.2 changes by less than 7 and 1 percent, respectively, when the guide diameter is gradually varied from $a/\lambda = 1.5$ to 3. Even smaller variations may be expected for larger values of $\tan \delta_d$, and the guide radius a, which is selected for an empty guide ($\epsilon_i = \epsilon_d = \epsilon_0$, $\tan \delta_i = \tan \delta_d = 0$), can be legitimately used for a dissipative surrounding medium ($\epsilon_i \neq \epsilon_d \tan \delta_d > 0.06$).

The loop admittances ($Y = 1/Z$) and current distributions computed with no dielectric layers ($\epsilon_i = \epsilon_d = \epsilon_0$) are shown in Figs. 2.4 and 2.5, and are discussed in Section 2.4.

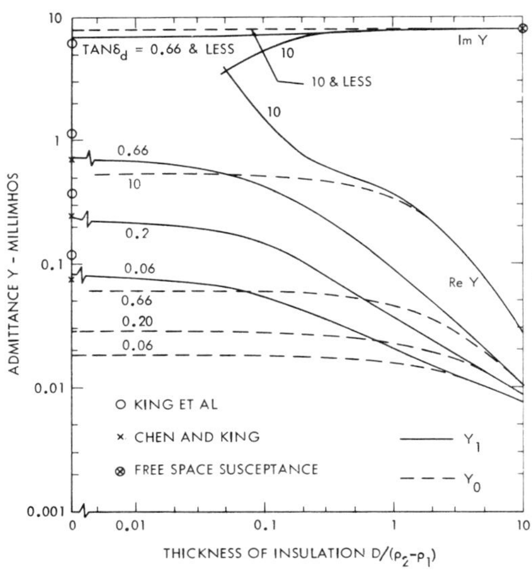

FIG. 10.2. Loop admittance $k_i\rho_0 = 0.1$, $\epsilon_d/\epsilon_i = 5$, $\Omega = 2 \log [8\pi\rho_0/(\rho_2 - \rho_1)] = 10$, $a = 2.27\lambda$.

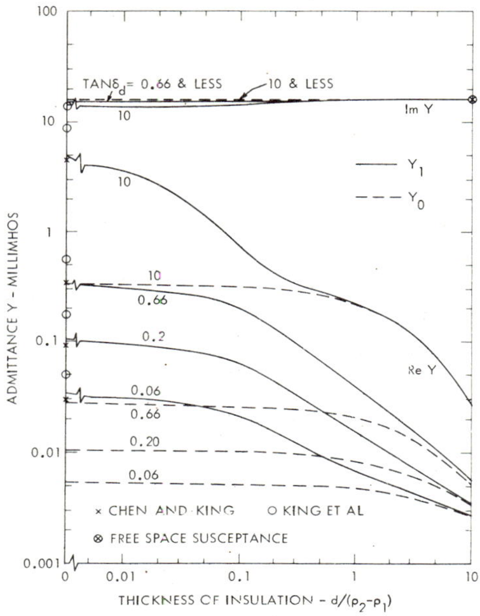

FIG. 10.3. Loop admittance $k_i\rho_0 = 0.05$, $\epsilon_d/\epsilon_i = 5$, $\Omega = 2 \log [8\pi\rho_0/(\rho_2 - \rho_1)] = 10$, $a = 2.27\,\lambda$.

The effects of a lossy dielectric are shown in Figs. 10.2 and 10.3 for loop radii $k_i\rho_0 = 0.1$, and 0.05. The loss tangent of the dielectric tan δ_d is selected in such a way as to afford a direct comparison with the results of King *et al.* [1964] or Chen and King [1963] for a homogeneous lossy dielectric. The loss tangents tan $\delta_d = 0.06$, 0.2, 0.66, and 10 correspond to the parameters $\alpha/\beta = \mathrm{Im}k_d/\mathrm{Re}k_d = 0.03$, 0.1, 0.3, and 0.9 respectively. For $d = 0$ the computed admittance Y_1 is nearly the same as that in the computations of Chen and King [1963]. It differs more from the more accurate results of King *et al.* [1964], which are indicated in Figs. 10.2 and 10.3, only for the range of ρ_0 given by

$$\beta\rho_0 < 0.3,$$

where

$$\beta = \mathrm{Re}k_d = k_0[(0.5\epsilon_d/\epsilon_0)([1 + \tan^2\delta_d]^{1/2} + 1)]^{1/2}.$$

This should indicate the accuracy of the present method in the presence of a lossy dielectric, when Y_1 differs significantly from Y_0.

Similar admittance data are shown in Figs. 10.4 and 10.5 for $k_i\rho_0 = 0.1$ and 0.05 and for the loss tangents tan $\delta_d = 10$, 100, and 1000, with parameters that approximate an insulated wire ($\epsilon_i/\epsilon_0 \approx 2.5$) in sea water ($\epsilon_d/\epsilon_0 \approx 80$, tan $\delta_d \gg 1$).

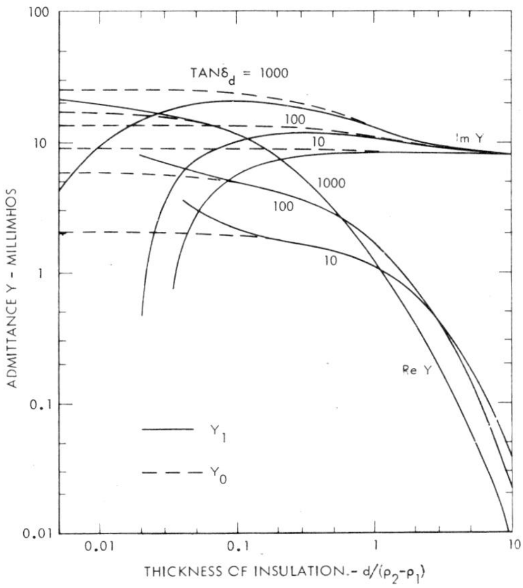

FIG. 10.4. Loop admittance $k_i\rho_0 = 0.1$, $\epsilon_d/\epsilon_i = 32$, $\Omega = 2\log[8\pi\rho_0/(\rho_2-\rho_1)] = 10$, $a = 2.27\ \lambda$.

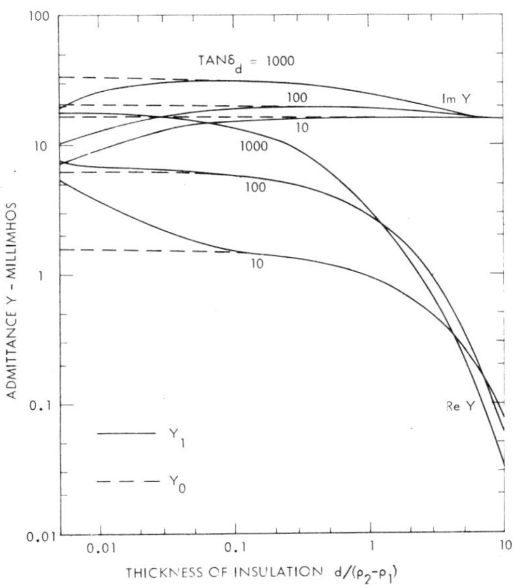

FIG. 10.5. Loop admittance $k_i\rho_0 = 0.05$, $\epsilon_d/\epsilon_i = 32$, $\Omega = 2\log[8\pi\rho_0/(\rho_2-\rho_1)] = 10$, $a = 2.27\ \lambda$.

The calculations show that Y_0 differs only slightly from Y_1 for thick layers of insulation. The assumption of a uniform current flow is therefore justified for larger thicknesses d of the insulating layer. Very thin layers of insulation ($d \approx 0.2(\rho_2 - \rho_1)$) are adequate for maintaining a uniform current flow around the loop with smaller loop radii ρ_0 and larger losses $\tan \delta_d$. The loop becomes electrically large with further decreasing values of d for the larger values of $\tan \delta_d$, and the admittance can be expected to become inaccurate as Im Y_1 is decreased and the first resonance of the loop is approached. The assumption of uniform current flow (Y_0) gives incorrect conductance estimates in the absence of insulation ($d = 0$), as has been pointed out by Chen and King [1963] and King *et al.* [1964]. The admittance Y_0 computed with $d = 0$, however, is nearly the same as Y_0 and Y_1 computed with $d \neq 0$ at the lower limit of d, where $Y_0 \approx Y_1$, or where the assumption of a uniform current is still justified. Insulating layers of a larger thickness will affect the admittance as can be seen from Figs. 10.2 to 10.5 or from analyses that

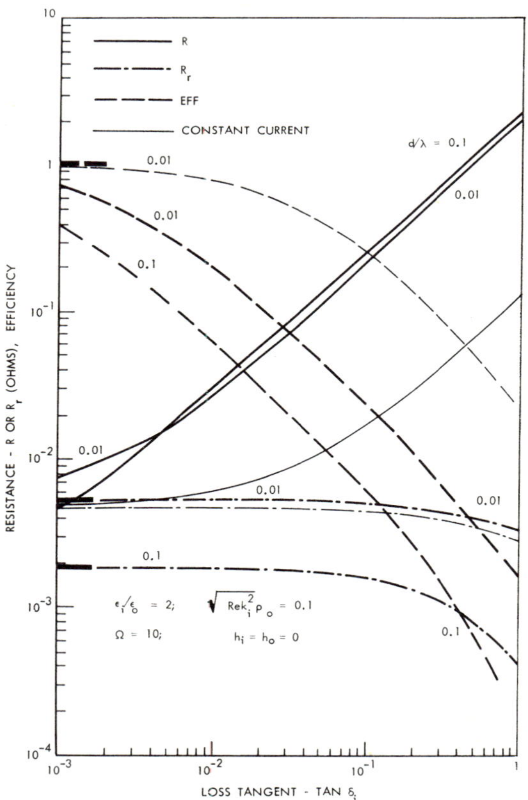

FIG. 10.6. Impedance and efficiency of a loop in a lossy dielectric slab.

consider the finite distance between the loop and the lossy dielectric for an assumed uniform loop current [Wait, 1957; Row, 1964].

10.3. Loops in Finite Layers of Lossy Dielectric

The calculations of Fig. 10.6 are made for a dielectric slab of $d/\lambda = 0.01$ and 0.1, and for various values of the loss tangent $\tan \delta_i$. The width of the loop $(\rho_2 - \rho_1)$ is such that $\Omega = 2 \log [8\pi\rho_0/(\rho_2 - \rho_1)] = 10$. The radiation resistance and the efficiency for $\tan \delta_i = 0$ are indicated by heavy horizontal lines on the left-hand side of the figure. The radiation resistance is nearly constant for $\tan \delta_i < 0.1$, the resistance R increases, and the radiation efficiency that is defined as the ratio between the radiation resistance R_r and the input resistance R decreases correspondingly. For larger losses, there is a decrease in the radiation resistance, particularly for larger thicknesses of the dielectric slab. The zero order $(I = I_0)$ resistance and efficiency are indicated by thin lines. The zero order radiation resistance is nearly the same

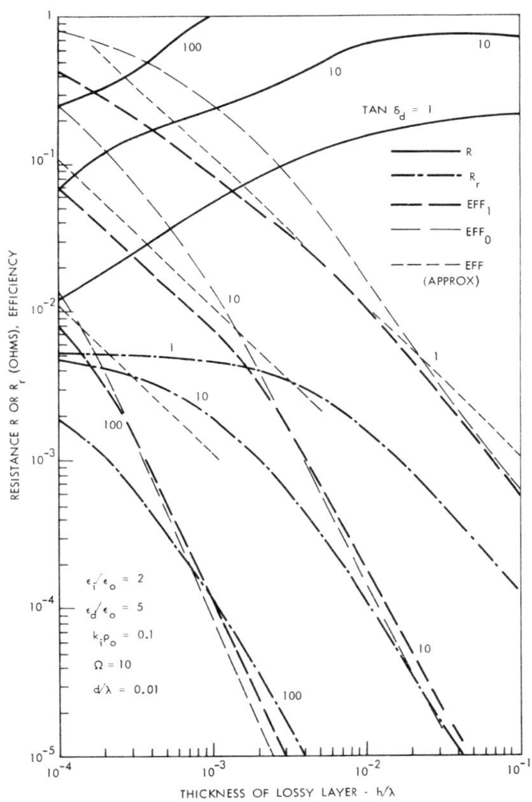

FIG. 10.7. Impedance and efficiency of a loop between layers of lossy dielectric.

as the first order radiation resistance, but the zero order input resistance is significantly lower. Hence the zero order efficiency exceeds the more accurate first order efficiency of the antenna. The loop reactance remains constant for the parameters shown in Fig. 10.6, the first order reactance is −90.6 ohms, and the zero order reactance is smaller by 2 ohms in magnitude.

Figure 10.7 shows the resistance and efficiency data for a loop that is inside a lossless dielectric slab of similar dimensions, and is surrounded by lossy layers. The resistance is increased, the radiation resistance is decreased, and the efficiency decreases correspondingly with increasing thickness of the lossy layer. The antenna reactance is nearly constant for the parameters indicated in Fig. 10.7. The first order reactance changes from −90.8 to −90.6 ohms, and the zero order reactance from −88.6 to −88.4 ohms. In the present example of thickness of the insulating layer $d = 0.01\lambda$ exceeds the loop width $(\rho_2 - \rho_1)$ by more than five times. For thick layers of lossy dielectric, the zero order impedance of the loop should be nearly the same as the first order impedance following Figs. 10.2 or 10.3. The zero order efficiency indicated by thin dashed lines can be seen to differ little from the first order efficiency figures for larger values of h. The geometry of the loop surrounded by layers of lossy dielectric may approximate a loop in a lossy dielectric shell. The radiation efficiency of the loop in the latter geometry was given by (4.67) as

$$EFF = \frac{\omega\epsilon_0}{\sigma_d} k_0 b \frac{b}{h} \frac{1 + (k_0\rho_0)^4/1400 + \ldots}{1 + (9/56)(\rho_0/b)^4 + \ldots}, \tag{10.35}$$

where σ_d is the conductivity of the lossy shell, h is the thickness of the shell, and ρ_0 and b are the radii of loop and shell, respectively. The derivation is strictly valid for $\sigma_d \gg \omega\epsilon_0$, $|k_d h| \ll 1$, and $|k_d b| \ll 1$. For the planar geometry of Fig. 10.7, an equivalent spherical shell is introduced by defining $b = d$ and letting $\sigma_d/\omega\epsilon_0 = \epsilon_d \tan \delta_d$. The resulting efficiency figures are indicated by thin dotted lines in Fig. 10.7. For small values of h, the approximate efficiency data lie between the first and zero order efficiency figures, but the efficiency approximation becomes inaccurate for larger values of h and larger values of $\tan \delta_d$. There are obvious differences between spherical and planar loop geometries, but the previous comparisons indicate that the present numerical calculations give efficiency figures of the same magnitude as simple closed form expressions for antenna parameters, where the latter derivations are valid.

The final example considers a buried insulated horizontal loop. Loops of $k_i\rho_0 = 0.05$, 0.1, and 0.15 are investigated for various amounts of ground losses. As shown in Fig. 10.8, the antenna resistance reaches maximum values in the range of $0.5 < \tan \delta_d < 3$. For low losses, the absorption of energy in the ground is low; for high losses, the ground acts as a reflector and the antenna resistance is low in both extremes. Small loops are inductive ($X < 0$), while larger loops are capacitive ($X > 0$) for low dielectric losses, but the reactance changes sign with increasing losses for $k_i\rho_0 > 0.1$. Indicated by thin lines is the impedance of the loop in a lossy dielectric

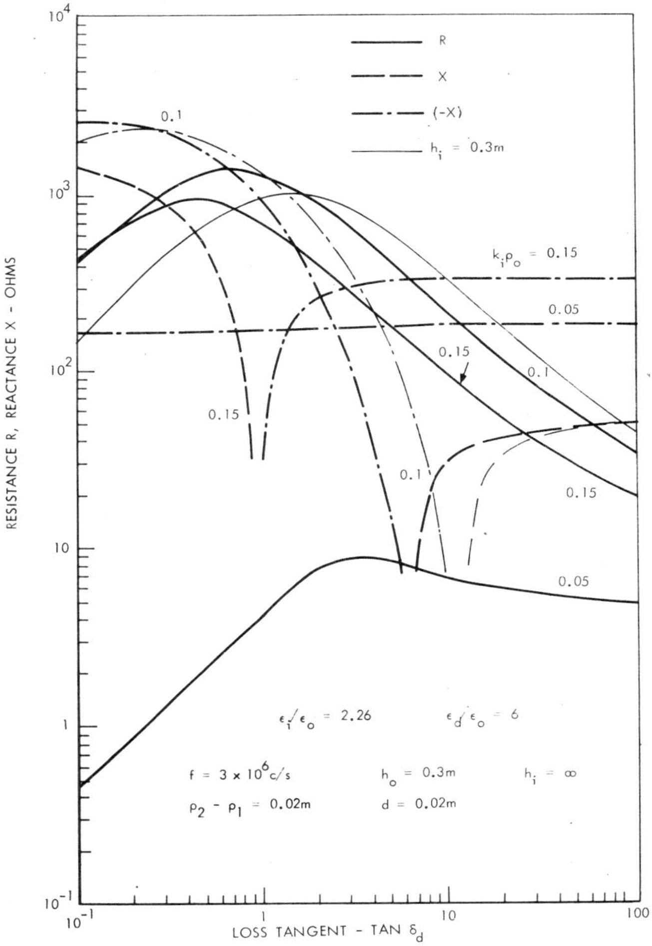

FIG. 10.8. Impedance of a buried insulated loop.

slab of $h_i = h_0 = 0.3$ m, and it shows the same impedance behavior as buried loops with $h_i = \infty$.

The radiation efficiency of the buried loops are shown in Fig. 10.9. A decreasing radius of the loop decreases the efficiency. Also indicated by dashed lines is the radiation efficiency of a horizontal dipole antenna, which has a total length of $2l = 2\pi\rho_0$ and is buried in a similar geometry (Fig. 9.12). The efficiency data of loops and horizontal dipoles of equal length differ by a factor of less than 2, with the two sets of efficiencies being nearly equal for the smaller loss tangents with $k_i\rho_0 = 0.1$ or 0.15. The antenna efficiencies are increased by increasing the thickness d of the insulating layer, except for larger loops of $k_i\rho_0 = 0.15$, where the present two-term approximation of the loop current will be less accurate. The efficiency of the loop in a dielectric slab ($h_i = h_0 = 0.3$ m) is almost 2 times larger than the efficiency of the buried loop ($h_i = \infty$).

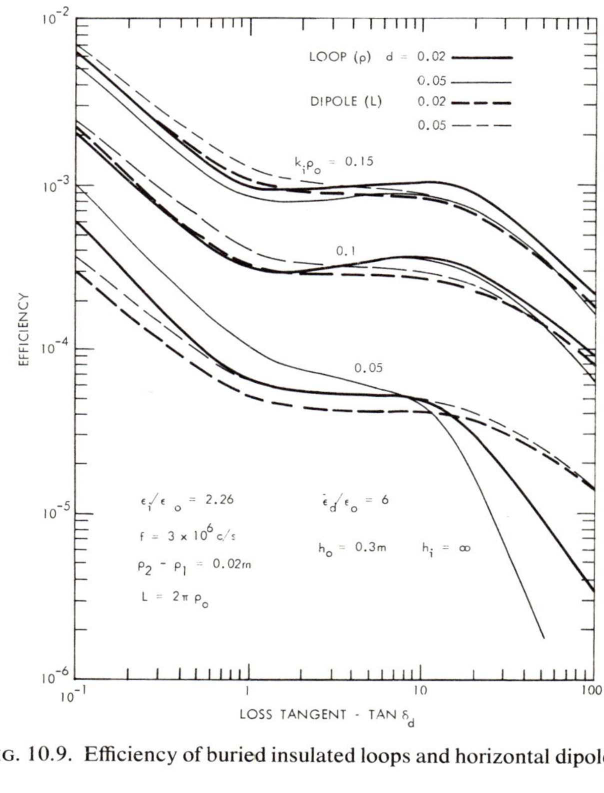

FIG. 10.9. Efficiency of buried insulated loops and horizontal dipoles.

The previous comparison indicates that the radiation efficiency, which is defined here as the ratio of energy radiated outside the ground to the energy dissipated inside the ground, is nearly the same for horizontal loops and dipoles. There are of course obvious differences in the antenna impedances, radiation patterns, and polarization of the transmitted signals. The numerical examples of Fig. 10.9 were confined to antennas of a small size. A large loop can be considered as a superposition of horizontal dipole antennas, and can be anticipated to have a comparable efficiency.

10.4. Exercises

10.1. In the calculations of the loop impedance substitute asymptotic expansions of the Bessel functions in the summations S_0 and S_1 and approximate the summations by an integral in the range $Q < q \leqslant \infty$. Evaluate the integrals for $R_a = R_b = 0$.

10.2. Assume a further term $I_2 \cos 2\phi$ for the loop current (10.32) and compute the function S_2 of the loop impedance.

10.3. Compute the radiation resistance of the loop (10.34) by relating the coefficients A_{nq}^{j0} and B_{nq}^{j0} to current amplitudes I_0 and I_1.

10.5. References

CHEN, C. L. and KING, R. W. P. (1963) The small bare loop antenna immersed in a dissipating medium, *IEEE Trans. on Ant. and Prop.* **AP-11** (3), 266--9.

GALEJS, J. (1963) Small electric and magnetic antennas with cores of a dielectric, *NBSJ. Res. (Radio Prop.)* **67D** (4), 445–51.

GALEJS, J. (1964) Admittance of annular slot antennas radiating into a plasma layer, *NBS J. Res. (Radio Science)* **68D** (3).

GALEJS, J. (1965) Admittance of insulated loop antennas in a dissipative medium, *IEEE Trans. on Ant. and Prop.* **AP-13** (2), 229–35.

KING, R. W. P., HARRISON, W. and TINGLEY, D. G. (1964) The admittance of bare circular loop antennas in a dissipative medium, *IEEE Trans. on Ant. and Prop.* **AP-12**, 434–8.

KRAICHMAN, M. B. (1962) Impedance of a circular loop in an infinite conducting medium, *NBS J. Res. (Radio Prop.)* **66D** (4), 499–503.

LUKE, Y. L. (1962) *Integrals of Bessel Functions*, McGraw-Hill, New York, NY.

ROW, R. V. (1964) "Radiation Efficiency of Electric and Magnetic Dipole Antennas Surrounded by a Small Spherical Shell of Lossy Dielectric," *IEEE Trans. on Ant. and Prop.*, **AP-12** (5), 646–647.

STORER, J. E. (1956) Impedance of thin-wire loop antenna, *Commun. and Electronics, Trans. AIEE* **75** (27), 606–19.

WAIT, J. R. (1957) Insulated loop antenna immersed in a conducting medium, *NBS J. Res.* **59**, 133–7.

WU, T. T. (1962) Theory of thin circular loop antenna, *J. Math. Phys.* **3** (6), 1301–4.

CHAPTER 11

ANTENNAS IN COMPRESSIBLE ISOTROPIC PLASMA

EXPERIMENTAL observations by Whale [1962, 1963, 1964] suggest that antennas excite both electromagnetic (EM) and plasma (P) waves in the ionosphere. Majumdar [1960], Cohen [1961, 1962a, 1962b], Hessel and Shmoys [1962], and Seshadri [1965a] have considered radiation due to electric current sources in a homogeneous plasma. For an infinitesimal dipole the power radiated in the P wave is proportional to $(c/u)^3$ where c is the free space velocity of the EM waves and u is the acoustic velocity [Hessel and Shmoys, 1962; Seshadri, 1965a]. Chen [1964], in examining the dipole antenna and the finite antenna with an assumed current distribution of the EM wave, indicates that the power radiating in the acoustic wave is independent of the u/c ratio. Wait [1964a] and Wait and Spies [1966] examine a spherical antenna that is excited by a circumferential slot of an infinitesimal width. For an antenna radius a that is large compared with the plasma wave length ($k_p a \gg 1$), the power radiated by the P wave is shown to be proportional to the acoustic velocity u, and the presence of an insulating layer or of an ion sheath tends to decrease the power radiated by the acoustic waves [Wait and Spies, 1966; Seshadri, 1965b]. Fejer [1964] treats a charged oscillating sphere: he approximates a dipole antenna by an anti-phase excitation of two such spheres. His results are similar to those of Wait [1964a] for $k_p a \gg 1$. A dipole antenna of a finite radius is considered by Balmain [1965], and the electro-acoustic radiation resistance approaches zero with decreasing acoustic velocity u. For dipole antennas separated from compressible plasma by an insulating cylinder or by an ion sheath, the power transferred to the P mode is significantly decreased with increasing frequency and thickness of the insulating layers [Seshadri, 1965d].

Surface waves supported by a halfspace of compressible plasma have been investigated by Hessel *et al.* [1962] and Seshadri [1964] in cases where the plasma is bounded by either free space or a conducting plane. For a lossy plasma the phase constant acquires a small imaginary part [Hessel *et al.*, 1962]. A lossless plasma slab is capable of supporting a large number of guided waves [Stewart and Caron, 1965; Caron 1965; Wenger, 1966]. For a long cylinder surrounded by a compressible plasma, the propagation constant of the current distribution contains terms that may be identified as EM and P waves [Seshadri, 1965c; Galejs 1965c]. Wait [1965a] indicates that the radially evanescent electroacoustic surface waves are not significant for computing the antenna far fields, which are nearly the same as in a cold plasma for antennas of a finite diameter. More recently, Cook and Edgar [1966] show that considerable amounts of power are radiated by the antenna if its current distribution has the wave number of the electro-acoustical surface waves.

In boundary value problems involving compressible plasmas it is necessary to apply acoustic boundary conditions in addition to the electromagnetic boundary conditions discussed in Section 1.2. Different boundary conditions apply to the interface between plasma and the metallic antenna surface or between the various plasma layers and the ion sheath that is surrounding the antenna. The various boundary conditions have been discussed by Bachynski [1966] in a comprehensive review paper dealing with sources in unbounded plasmas. In the simplest approach the normal component of the electron velocity is assumed to be equal to zero at the rigid plasma boundary. This rigidity condition is used in most of the above papers. This can be improved by requiring the normal velocity and the scalar pressure to be continuous at the interface between the various plasma regions [Wait, 1965b]. The ion sheath near a metallic boundary can be collapsed with a positive d–c bias; the electron density is uniform up to the boundary, but electrons are continuously absorbed by the metallic surface, which is referred to as an absorptive boundary condition [Balmain, 1966; Wait, 1966]. A hybrid admittance boundary condition has been developed by Cohen [1961, 1962b], in which the normal component of the electron velocity depends on the normal electric field and the perturbation of the electron density. Tidman and Boyd [1962] have developed a boundary condition which considers the presence of an electrostatic field which is generated by the discontinuity of plasma density in a plasma of constant temperature. A number of other boundary effects have been also discussed [Bachynski, 1966]. The near fields of the radiating source have an influence on the resulting sheath. Moving metallic bodies will collect ion currents by collecting ions in their path. Long antennas on satellites cut through the earth's magnetic field, which develops a potential gradient along the system and which causes a redistribution of the particles and a nonsymmetrical sheath. There is also an interaction effect between a moving plasma and the magnetic field. However, no consideration has been paid to the effects of such sheaths on the radiation properties of sources.

The preceding summary listed a number of pertinent radiation problems which have been treated in the literature. It also outlines some of the processes to be considered in establishing the boundary conditions. Waves guided along the antenna structures have been examined, but source current distributions have not been established in detail. In the case of the linear antenna, the radiation characteristics may be quite different for current distributions that consider the presence of electromagnetic waves [Chen, 1964] or the plasma waves [Cook and Edgar, 1966]. The principal problem investigated in this chapter is the antenna impedance, when the current (or field) distributions along it consist of parts that can be identified as electromagnetic and plasma waves. The relative amplitudes of the two wave components are estimated from variational formulations [Galejs, 1966a, 1966b]. The characteristics of guided waves supported by representative antenna structures [Galejs, 1966c] will be discussed prior to the actual impedance computations. Only the conditions of the rigid plasma boundary or of continuous normal velocity and scalar pressure will be considered in this discussion. All of the antenna geometries discussed in this chapter are in rectangular coordinates. The reader is referred to Wait [1967] for a detailed treatment of cylindrical and spherical antennas.† The perturbations of antenna current by plasma waves are noted also in the recent study of Wunsch [1967].

11.1. Guided Waves Supported by Layers of a Lossy Compressible Plasma

11.1.1. *Field Expressions*

A lossy plasma slab of thickness h is bounded by free space at $z = 0$. The slab is excited by sources located at $z = 0$, as is shown in Fig. 11.1.

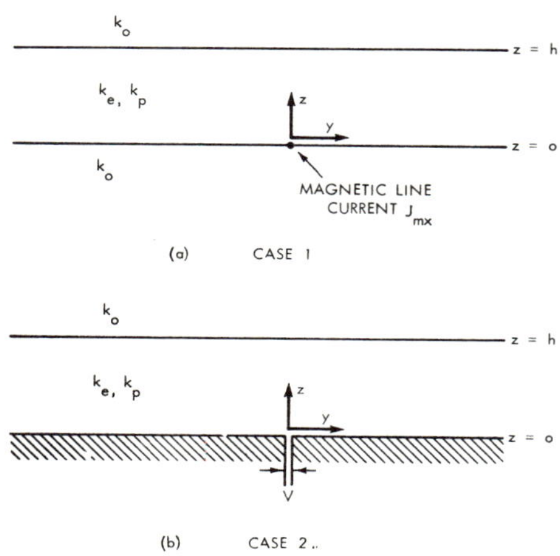

FIG. 11.1. Plasma geometry.

†On spherical antennas [Wait, 1964; 1967] it is possible to compute the current distribution including the electromagnetic and plasma wave components over the surface of the antenna for an assumed field distribution across the exciting slot. However the relative amplitudes of the plasma waves in the slot of a finite width are not determined in these formulations.

The field equations can be written for a single fluid electron gas following Oster [1960] as

$$\times \mathbf{E} = i\omega\mu_0 \mathbf{H}, \tag{11.1}$$

$$\nabla \times \mathbf{H} = -i\omega\epsilon_0 \mathbf{E} + Ne\mathbf{V}, \tag{11.2}$$

$$(-i\omega + \nu)\, mN\mathbf{V} = +Ne\mathbf{E} - \nabla p, \tag{11.3}$$

$$u^2 mN\ \nabla \cdot \mathbf{V} = i\omega p, \tag{11.4}$$

where **E** and **H** are the electric and magnetic field vectors; where N, **V**, and ν are the average values of electron density, velocity, and collision frequency respectively; where p is the time varying component of the scalar pressure; and where u, e, and m are the acoustic velocity in the electron gas, the particle charge, and the mass respectively. The time dependence $\exp(-i\omega t)$ is suppressed throughout.

Eliminating **V** from (11.2) and (11.3) gives

$$\mathbf{E} = -\frac{1}{i\omega\epsilon_0\epsilon} \nabla \times \mathbf{H} - \frac{1-\epsilon}{\epsilon Ne} \nabla p, \tag{11.5}$$

with

$$\epsilon = 1 - \frac{\omega_p^2}{\omega(\omega + i\nu)} = \epsilon_r + \frac{i\nu}{\omega}(1 - \epsilon_r), \tag{11.6}$$

$$\omega_p^2 = Ne^2/(\epsilon_0 m), \tag{11.7}$$

where $\epsilon_r = \operatorname{Re} \epsilon$, and where N denotes the electron density. The divergence of the left-hand-side of (11.1) is equal to zero by definition; in a medium of constant permeability μ_0 the divergence of (11.1) is computed as $\nabla \cdot \mathbf{H} = 0$. Substituting (11.5) in (11.1) and noting that $\nabla \times \nabla \times \mathbf{H} = \nabla(\nabla \cdot \mathbf{H}) - \nabla^2 \mathbf{H}$ results in

$$\nabla^2 \mathbf{H} + k_e^2 \mathbf{H} = 0 \tag{11.8}$$

with $k_e^2 = \omega^2 \mu_0 \epsilon_0 \epsilon$. Only the transverse magnetic (TM) fields that have a non-zero electric field component in the direction of propagation are coupled to the acoustic waves and are affected by the finite temperature of the plasma [Galejs, 1966a]. There is no loss in generality by restricting the considerations to TM fields (an x-directed magnetic line source, or an infinitesimal slot, can be assumed for the sake of simplicity). The **H** vector has only a single component H_x, and consequently (11.8) is reduced to a scalar equation. Eliminating **E** from (11.2) and (11.3) gives

$$\mathbf{V} = -\frac{1-\epsilon}{Ne\epsilon} \left[\nabla \times \mathbf{H} + \frac{i\omega\epsilon_0}{Ne} \nabla p \right]. \tag{11.9}$$

Substituting (11.9) in (11.4) gives

$$\nabla^2 p + k_p^2 p = 0, \tag{11.10}$$

with $k_p^2 = [\omega(\omega+i\nu) - \omega_p^2]/u^2 \approx \omega^2(\epsilon_r + i\nu/\omega)/u^2$. H_x and p can be expressed in terms of the integrals

$$H_x = \frac{1}{2\pi}\int_{-\infty}^{\infty} B(\lambda)\, e^{i\lambda y}(e^{\gamma_e z} + R_b e^{-\gamma_e z})\, d\lambda, \tag{11.11}$$

$$p = \frac{1}{2\pi}\int_{-\infty}^{\infty} C(\lambda)\, e^{i\lambda y}(e^{\gamma_p z} + R_c\, e^{-\gamma_p z})\, d\lambda, \tag{11.12}$$

where $\gamma_e = i(k_e^2 - \lambda^2)^{1/2}$ and $\gamma_p = i(k_p^2 - \lambda^2)^{1/2}$ are defined for real values of λ and $z \geqslant 0$ in the second quadrant of the complex plane, including the positive imaginary and the negative real axes. The reflection coefficients R_b and R_c are equal to zero in semi-infinite regions, where the above definition of γ_e and γ_p assures outgoing phase fronts and possibly decaying amplitudes for $z \to \infty$. For $z < 0$, γ_e and γ_p are defined in the fourth quadrant. **E** and **V** are obtained from (11.5) and (11.9) by making use of (11.12).

A uniform plasma slab is assumed to have rigid boundaries at $z = 0$ and $z = h$, and to be bounded by free space for $z > h$. The continuity of tangential electric and magnetic fields at $z = h$ and the vanishing normal particle velocity at the plasma boundaries $z = 0$ and $z = h$ are satisfied by the following expression for the reflection coefficients introduced in (11.11) and (11.12)

$$R_b = e^{2\gamma_e h}\frac{(\gamma_e/\gamma_0 - \epsilon)\sinh\gamma_p h - \dfrac{(1-\epsilon)\lambda^2}{\gamma_e\gamma_p}(\cosh\gamma_p h - e^{-\gamma_e h})}{(\gamma_e/\gamma_0 + \epsilon)\sinh\gamma_p h + \dfrac{(1-\epsilon)\lambda^2}{\gamma_\epsilon\gamma_p}(\cosh\gamma_p h - e^{-\gamma_e h})}, \tag{11.13}$$

and

$$R_c = \frac{(e^{\gamma_e h} + R_b\, e^{-\gamma_e h}) - e^{\gamma_p h}(1+R_b)}{(e^{\gamma_e h} + R_b\, e^{-\gamma_e h}) - e^{-\gamma_p h}(1+R_b)}, \tag{11.14}$$

where $\gamma_0 = (k_0^2 - \lambda^2)^{1/2}$ and $k_0 = \omega(\mu_0\epsilon_0)^{1/2}$. The boundary condition of zero normal particle velocity can be replaced by the more elaborate boundary conditions of continuous scalar pressure and of continuous normal particle velocity at an interface between plasma and ion sheath or non-ionized gas. Although this leads to increased algebraic complexity of the derivations, essentially the same results are obtained using the simpler and less accurate boundary condition of zero normal particle velocity [Galejs, 1966a].

The ratio of the amplitudes $B(\lambda)$ and $C(\lambda)$ introduced in (11.11) and (11.12) is determined through the $V_z = 0$ boundary condition at $z = 0$ to be

$$\frac{C(\lambda)}{B(\lambda)} = \frac{Ne\lambda}{\gamma_p\omega\epsilon_0}\frac{1+R_b}{1-R_c}. \tag{11.15}$$

The amplitude $B(\lambda)$ depends further on the excitation geometry and on the electro-magnetic boundary conditions at $z=0$.

For the plasma slab bounded by free space at $z=0$ and excited by a magnetic line current source J_{mx} at $z=0$ and $y=0$, the tangential magnetic field component H_x is continuous at $z=0$, while the tangential electric field component E_y is discontinuous by the amount $J_{mx}\delta(y)$. After applying these boundary conditions, the amplitude $B(\lambda)$ is related to the magnetic line current J_{mx} by

$$\frac{B(\lambda)}{J_{mx}}=-i\omega\epsilon_0\epsilon\left[\gamma_e\frac{1-R_b}{1+R_b}+\epsilon\gamma_0-\lambda^2\frac{1-\epsilon}{\gamma_p}\frac{1+R_c}{1-R_c}\right]^{-1}. \tag{11.16}$$

For the plasma slab bounded by a conducting plane at $z=0$ and excited by a voltage V_s across a slot of infinitesimal width at $y=0$, the electric field in the $z=0$ plane is simply $E_y=V_s\delta(y)$, and the amplitude $\beta(\lambda)$ is related to V_s by

$$\frac{B(\lambda)}{V_s}=-i\omega\epsilon_0\epsilon\left[\gamma_e\frac{1-R_b}{1+R_b}-\lambda^2\frac{1-\epsilon}{\gamma_p}\frac{1+R_c}{1-R_c}\right]^{-1}. \tag{11.17}$$

This completes the formal determination of the field components. **E** and **V** are related by (11.5) and (11.9) to H_x and p of (11.11) and (11.12), which again are uniquely specified in terms of R_b and R_c, $C(\lambda)$ and $B(\lambda)$ of (11.13) to (11.17).

11.1.2. *Dispersion Relations*

11.1.2.1. *Plasma slab in free space.* For a plasma slab surrounded on both sides by free space, the phase constant k of the guided waves is computed from the zeros λ of the denominator of $B(\lambda)$ in (11.16). Applying (11.13) and (11.14) gives k as a solution of the equation

$$\begin{aligned}&-2k^2\gamma_e\gamma_p(1-\epsilon)(1-\cosh\gamma_e h\cosh\gamma_p h)\\&-[(1-\epsilon)^2k^4+\gamma_p^2(\gamma_e^2+\epsilon^2\gamma_0^2)]\sinh\gamma_e h\sinh\gamma_p h\\&-2\epsilon\gamma_0\gamma_p[(1-\epsilon)k^2\sinh\gamma_e h\cosh\gamma_p h-\gamma_p\gamma_e\cosh\gamma_e h\sinh\gamma_p h]=0,\end{aligned} \tag{11.18}$$

where

$$\gamma_p=i(k_p^2-k^2)^{1/2},\quad \gamma_e=i(k_e^2-k^2)^{1/2},\quad \gamma_0=i(k_0^2-k^2)^{1/2},\quad k_0=\omega(\mu_0\epsilon_0)^{1/2},$$

and h is the thickness of the plasma slab. When allowing wave number k to be complex, the propagation constants γ_0, γ_e, and γ_p are defined in the second and third quadrants of the complex plane including the positive imaginary axis. Hence $(k_p^2-k^2)^{1/2}$, $(k_e^2-k^2)^{1/2}$, and $(k_0^2-k^2)^{1/2}$ are defined in the upper half of the complex plane.

Assuming that $k \gg k_0$ or k_e, that $c = 1/(\mu_0\epsilon_0)^{1/2} \gg u$, and that h is of the order of a free space wavelength $2\pi/k_0$, it follows that γ_e, $\gamma_0 = -k$, $\cosh \gamma_e h = -\sinh \gamma_e h \gg 1$, and (11.18) simplifies to

$$\frac{1+\epsilon}{1-\epsilon}\frac{\gamma_p}{k}\sinh \gamma_p h - \cosh \gamma_p h = \pm 1, \tag{11.19}$$

which can be expressed as

$$\frac{1-\epsilon}{1+\epsilon}\frac{k}{(k_p^2-k^2)^{1/2}} = A(k) = \begin{cases} -\tan\left[(k_p^2-k^2)^{1/2}h/2\right] \\ \text{or} \\ \cot\left[(k_p^2-k^2)^{1/2}h/2\right] \end{cases}. \tag{11.20}$$

Equation (11.20) can be rewritten alternately as

$$e^{i(k_p^2-k^2)^{1/2}h} = \pm\frac{1-iA(k)}{1+iA(k)}. \tag{11.21}$$

For a lossless plasma ($\nu = 0$) and $k < k_p$, $A(k)$ is real, and both sides of (11.21) have a magnitude of unity. These equations will have a number of real solutions. For $k > k_p$, $A(k)$ is imaginary, both sides of (11.21) have a magnitude of less than unity, and there will be real solutions k.

For a lossy plasma ($\nu \neq 0$), $A(k)$ becomes complex and the complex solution k can be considered as a perturbation of the former real solution k_r. Considering (11.21) as a function of the collision frequency ν and of the complex phase constant $k = k_r + \Delta$, (11.21) is expanded in a Taylor series as

$$f(0+\nu, k_r+\Delta) = f(0, k_r) + \nu\frac{\partial}{\partial \nu}f(0, k_r) + \Delta\frac{\partial}{\partial k}f(0, k_r), \tag{11.22}$$

where $f(0,k_r)$ represents the difference between the two sides of (11.21) for real parameters. The term $f(0,k_r)$ is zero by the definition of the solution k_r. After carrying out the differentiations indicated in (11.22), Δ is related to ν as

$$\Delta = \frac{i\nu}{2\omega\epsilon_r}\frac{k_{pr}hF \pm \left[-8k_r\epsilon_r\dfrac{1-\epsilon_r}{(1+\epsilon_r)^2} + \dfrac{2k_{pr}^2k_r(1-\epsilon_r)}{(1+\epsilon_r)(k_r^2-k_{pr}^2)}\right]}{k_rhF \pm 2\dfrac{1-\epsilon_r}{1+\epsilon_r}\dfrac{k_{pr}^2}{k_r^2-k_p^2}}, \tag{11.23}$$

where $F = (1+iA)^2 \exp\left[i(k_{pr}^2-k_r^2)^{1/2}h\right]$, and where the real parts of ϵ, k_p, and k denoted as ϵ_r, k_{pr}, and k_r are computed with $\nu = 0$. For finite values of F (11.23) can be simplified by substituting (11.21). This gives

$$\Delta = \frac{i\nu}{2\omega}\frac{k_{pr}^2h[(1+\epsilon_r)^2k_{pr}^2 - 4\epsilon_rk_r^2] - 8k_r\epsilon_r(1-\epsilon_r)(k_{pr}^2-k_r^2) - 2(1-\epsilon_r^2)k_rk_{pr}^2}{\epsilon_rk_rh[(1+\epsilon_r)^2k_{pr}^2 - 4\epsilon_rk_{pr}^2] - 2\epsilon_r(1-\epsilon_r^2)k_{pr}^2}. \tag{11.24}$$

Δ of (11.24) is imaginary and the complex solution k is obtained by adding a small imaginary part $\Delta = ik_i$ to the former real solution k_r. For $k_p h \gg 1$ and $\epsilon_r \ll 1$, (11.24) simplifies to

$$\Delta = \frac{i\nu}{2\omega}\frac{1}{\epsilon_r k_r}\left\{k_{pr}^2 + \frac{2}{k_r h}[k_{pr}^2 - k_r^2(1+4\epsilon_r - 4\epsilon_r k_r^2/k_{pr}^2)]\right\}. \qquad (11.25)$$

In the limit of large values of h, the second term of (11.25) approaches zero, resulting in

$$(k_p^2 - k^2)^{1/2} \approx \left[\left(\frac{\omega}{u}\right)^2 \epsilon_r - k_r^2\right]^{1/2} = (k_{pr}^2 - k_r^2)^{1/2}, \qquad (11.26)$$

which is real for $k_r < k_{pr}$. The plasma losses make both k_p and the wave number k of surface waves complex. In the first approximation, however, $\text{Im}(k) = k_i$ is such that $(k_p^2 - k^2)^{1/2}$ and $\gamma_p = i(k_p^2 - k^2)^{1/2}$ remain the same as for a lossless plasma $[\text{Im}(k^2) \approx \text{Im}(k_p^2)]$. For smaller values of h and $k_r < k_{pr}$, Δ of (11.25) is increased in magnitude and $\text{Im}(k^2) > \text{Im}(k_p^2)$. For $k_r < k_{pr}$, the square root of (11.26) is defined in the second quadrant of the complex plane and the propagation constant γ_p is in the third quadrant near the negative imaginary axis. For $k_r < k_{pr}$, there are standing waves of p in the transverse dimension of the slab, and p can be shown to have the same magnitude near the boundaries $z = 0$ and $z = h$ even for a lossy plasma [Galejs, 1966c].

For $k_r > k_{pr}$, (11.26) is imaginary; for smaller values of h, Δ of (11.25) is decreased in magnitude. For $\text{Im}(k^2) < \text{Im}(k_p^2)$, $(k_p^2 - k^2)^{1/2}$ is in the first quadrant, and $\gamma_p = i(k_p^2 - k^2)^{1/2}$ is in the second quadrant near the negative real axis. For $k_r > k_{pr}$, the pressure $|p|$ has the same maximum values at the boundaries of $z = 0$ and $z = h$, but decays as z approaches the center of the plasma slab [Galejs, 1966c].

Under conditions where $(k_p^2 - k^2)^{1/2}$ has a small imaginary part, the left-hand side of (11.21) approaches zero for sufficiently large values of h, and $A(k) = -i$. Substituting this in (11.20) gives

$$k^2 = k_p^2 \frac{(1+\epsilon)^2}{4\epsilon}, \qquad (11.27)$$

with $|k| > |k_p|$. This may be recognized as the phase constant of a surface wave guided along the air-to-plasma interface of a semi-infinite plasma [equation (54) of Hessel *et al.*, 1962; equation (55) Galejs, 1966a]. For a lossy plasma, the imaginary part of k computed from (11.27) is the same as k_i computed from (11.23) in the limit of $F \to 0$.

The factor ϵ^2 of (11.18) corresponds to a factor ϵ in the dispersion relations listed by Caron [1965]. Equations (9) and (10) of Caron [1965] can not be reduced to (11.27) in the limit of $h \to \infty$, and are believed to be misprinted.

11.1.2.2. *Plasma slab on a conducting plane.* For a conducting plane covered by a plasma slab, the phase constant k of the guided waves is computed from the zeros λ of the denominator of $B(\lambda)$ in (11.17). Applying (11.13) and (11.14) gives k as a solution of the equation

$$\begin{aligned}-2k^2\gamma_e\gamma_p(1-\epsilon)(1-\cosh\gamma_e h\cosh\gamma_p h)-[(1-\epsilon)^2k^4+\gamma_p^2\gamma_e^2]\\ \sinh\gamma_e h\sinh\gamma_p h-\epsilon\gamma_0\gamma_p[(1-\epsilon)k^2\sinh\gamma_e h\cosh\gamma_p h\\ -\gamma_p\gamma_e\cosh\gamma_e h\sinh\gamma_p h]=0.\end{aligned} \tag{11.28}$$

Equation (11.28) is in agreement with (25) of Stewart and Caron [1965], or with (18), (23) to (25), and (52) of Galejs [1966a].

Assuming that $k \gg k_0$ or k_e, $c \gg u$, and h is of the order of the free-space wavelength, it follows that

$$\tan[(k_p^2-k^2)^{1/2}h]=\frac{(1-\epsilon)(2+\epsilon)k(k_p^2-k^2)^{1/2}}{(1-\epsilon)^2k^2+(1+\epsilon)(k^2-k_p^2)}=B(k), \tag{11.29}$$

which can be written alternately as

$$e^{2ih(k_p^2-k^2)^{1/2}}=\frac{1+iB(k)}{1-iB(k)}. \tag{11.30}$$

For a lossless plasma and $k < k_p$, (11.29) or (11.30) has a number of real solutions k, and (11.29) or (11.30) have also real solutions for $k > k_p$. For a lossy plasma, the complex solutions k will be examined numerically.

Under conditions where $(k_p^2-k^2)^{1/2}$ has a positive imaginary part, the left-hand side of (11.30) approaches zero for sufficiently large values of h. Hence $B(k)=i$. This gives

$$4(2-\epsilon)k^4-(6\epsilon+3\epsilon^2-\epsilon^4)k_p^2k^2+(1+\epsilon)^2k_p^4=0. \tag{11.31}$$

The two solutions

$$k^2=\frac{k_p^2}{\epsilon(2-\epsilon)} \tag{11.32}$$

and

$$k^2=k_p^2\frac{(1+\epsilon)^2}{4\epsilon} \tag{11.33}$$

exhibit $|k| > |k_p|$. Equation (11.32) represents the phase constant of a surface wave guided along the conducting plane [equation (44), Seshadri 1964; equation (54), Galejs 1966a]. Equation (11.33) is identical to (11.27) and represents the phase constant of a surface wave guided along the air-to-plasma interface. For $k_r < k_{pr}$, there are standing waves of p within the plasma layers. However, the fields near the air-to-plasma boundary of $z = h$ may become very weak for a lossy plasma, and even for the lossless plasma if the wave number of the surface waves $k_r > k_{pr}$.

11.1.3. *Numerical Results*

The numerical calculation of the complex wave numbers k for the guided waves proceeds in two steps. First, a real root k is determined from the solution of (11.20) or (11.29) for a lossless plasma. Secondly, for a lossy plasma, the complex root $k = k_r + ik_i$ is found by applying Newton's iteration method and by using the real root determined in the first step for the initial estimate of k_r. The imaginary part k_i is set equal to $\mathrm{Im}(k_p^2)/(2k_r)$ initially, making the initial value of

$$(k_p^2 - k_2^2)^{1/2}$$

real for complex values of k_p also. For a plasma slab in free space, the perturbation solution (11.24) is most accurate for k near k_p, where it is found to agree within 4 digits, with the result of Newton's iteration method. However, (11.24) gives an accuracy of a few percent only for $k \ll k_p$.

The calculations were made first for a plasma layer in free space with $h/\lambda = 0{\cdot}1$ and $c/u = 100$. The computed values of the ratio between the free space velocity of the electromagnetic waves c and the phase velocity of the guided waves $c/v_{ph} = k_r/k_0$ are shown in the top part of Fig. 11.2. The wavenumbers k_r increase with increasing values of ϵ_r. It follows from (11.6) that ω is increased monotonically with increasing k_r. Consequently the phase velocity $v_{ph} = \omega/k_r$ is of the same sign as the group velocity $v_g = \partial\omega/\partial k_r$, and there are no backward waves. The dotted line represents the ratio k_{pr}/k_0. It separates the solution for guided waves with an oscillatory p variation in the plasma slab $(k_r < k_{pr})$ from the solution for surface waves guided along the plasma boundaries $(k_r > k_{pr})$ exhibiting maximum p values at the boundaries. The number of waves supported by the plasma slab is increased with increasing values of ϵ_r. A given mode of $k_r < k_{pr}$ exhibits large values of v_{ph} or small values of (c/v_{ph}) near its cut-off point for smaller values of ϵ_r. The phase velocity is decreased with increasing values of ϵ_r when k_r tends towards k_{pr} or when (c/v_{ph}) tends towards

$$k_{pr}/k_0 \approx c\epsilon_r^{1/2}/u.$$

It was indicated earlier that $k_i = \mathrm{Im}k$ is in the first approximation chosen such as to make

$$(k_p^2 - k^2)^{1/2}$$

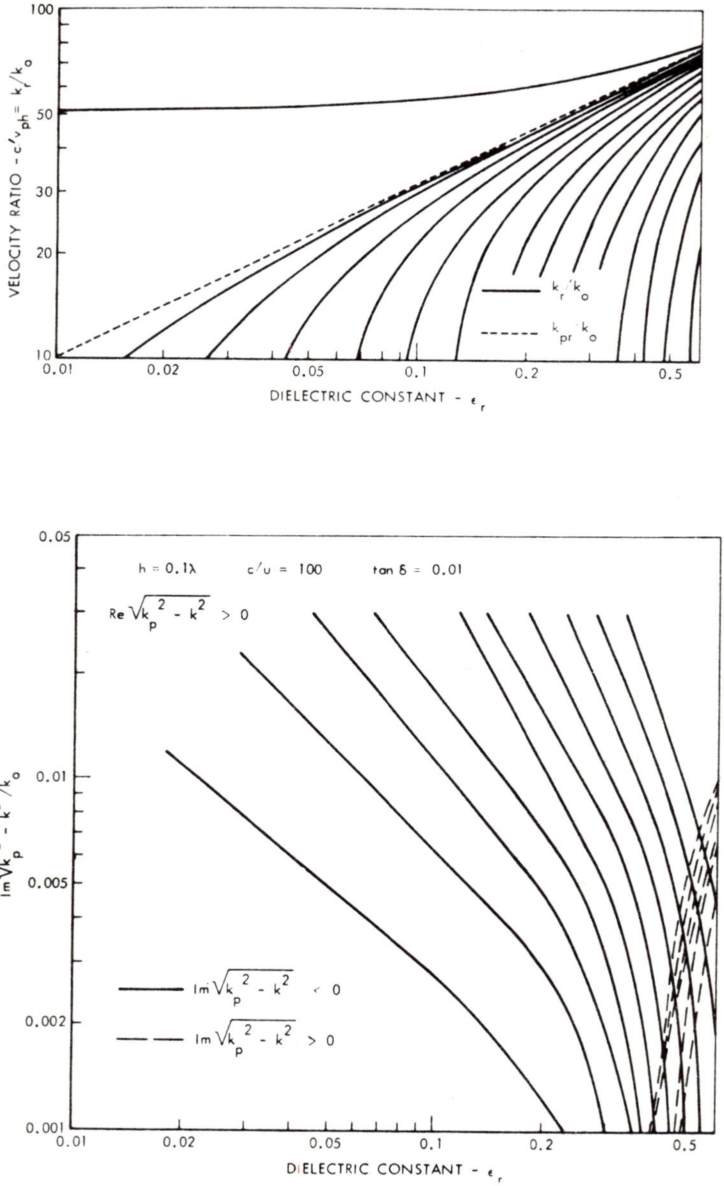

FIG. 11.2. Dispersion relations for a plasma slab in free space.

nearly real also for a lossy plasma. The difference between $\mathrm{Im}(k_p^2)$ and $\mathrm{Im}(k^2)$ becomes more apparent by plotting

$$\mathrm{Im}\,(k_p^2 - k^2)^{1/2},$$

which can be seen to be less than $k_r \tan\delta$ in magnitude.

Plasma layers of larger thickness support an increased number of waves of $k_r < k_{pr}$, and the number of guided waves is increased from 13 to 140 by changing h/λ from 0.1 to 1 for $\epsilon_r = 0.5$. The dispersion curves have the same shape as indicated in Fig. 11.2, are spaced more closely, and are bounded by the same dotted line that represents $k_r = k_{pr}$. For larger c/u ratios, the number of solutions of $k_r < k_{pr}$ is also increased, but c/v_{ph} is increased proportionately for all guided waves. The dispersion curves for guided waves supported by a plasma slab bounded by a conducting plane are similar for $k_r < k_{pr}$ to the curves indicated in Fig. 11.2 for a plasma slab bounded on both sides by free space. Increasing thickness h of the plasma layer makes the dispersion curves more dense, and an increased c/u ratio increases c/v_{ph} proportionately for all the dispersion curves.

All the dispersion curves may be normalized by multiplying the ordinate of Fig. 11.2 with (u/c). The resulting curves are indicated in Fig. 11.3. The solutions of $k_r < k_{pr}$ occupy the space below the dotted line and follow the shape of the curves indicated in Fig. 11.2. In the shaded area the solutions exhibit $\mathrm{Im}(k^2) > \mathrm{Im}(k_p^2)$, and on the dashed thin lines exhibit $\mathrm{Im}(k^2) = \mathrm{Im}(k_p^2)$. For plasma layers having a thickness larger than the plasma wave length

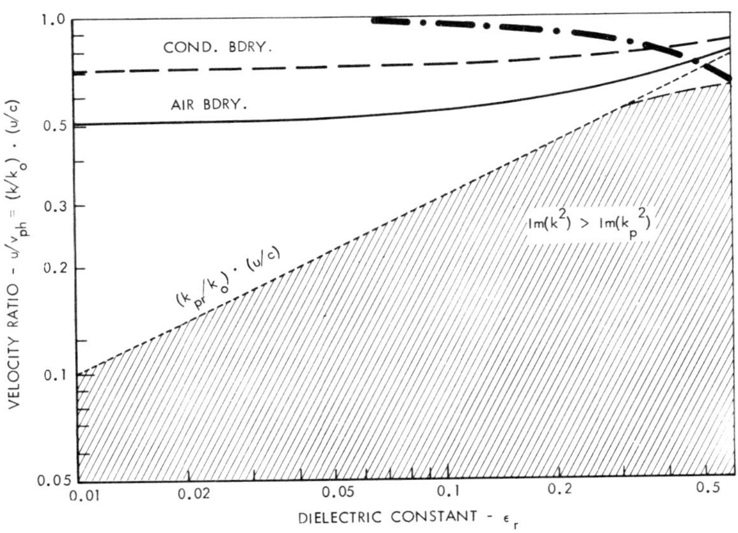

FIG. 11.3. Normalized dispersion curves.

$\lambda_p = 2\pi/k_{pr}$, the solutions for waves guided along the plasma boundaries ($k_r > k_{pr}$) are independent of the slab thickness. The solid curve of Fig. 11.3 characterizes waves guided along the air-to-plasma boundary, and the dashed curve characterizes the waves guided along the plasma-conductor boundary.

Electrons with a thermal speed close to the phase velocity of the longitudinal plasma wave tend to absorb energy, causing the so-called Landau damping. The waves are heavily damped if their wavelength becomes comparable to the Debye length λ_D, or $2\pi/k > 2\pi(\gamma)^{1/2}\lambda_D$, where γ is the ratio of specific heats, and where $\lambda_D = u/(\omega_p\gamma^{1/2})$ [Holt and Haskel, 1965]. This condition for Landau damping is rearranged into $k > \omega_p/u$ or $(k/k_0)\cdot(u/c) > (\omega_p/\omega) = (1-\epsilon)^{1/2}$. This boundary is indicated as the dotted and dashed line in Fig. 11.3. There will be severe damping above this line, but the damping will be negligible for small values of ϵ.

11.1.4. *Complex Power of Guided Waves*

The complex power supplied per unit length of the source is computed by integrating the z component of the complex power flow $\mathbf{E}\times\mathbf{H}^* + p\mathbf{V}^*$ over the $z = 0^+$ plane, and integrating the $(-z)$ component of the complex power flow over the $z = 0^-$ plane for plasma slabs excited by a line source and bounded on both sides by free space (geometry 1). For a slot excited plasma slab bounded on one side by a conducting plane, the integral is carried out only over the $z = 0^+$ plane (geometry 2). For the assumed rigid boundary, $V_z = 0$ and the total power supplied by the source is

$$P_j = \int_{-\infty}^{\infty} \mathrm{d}y\left[E_y H_x^*\big|_{z=0^+} - \delta_{1j} E_y H_x^*\big|_{z=0^-}\right], \tag{11.34}$$

where the two geometries are denoted by subscripts $j = 1$ or 2, and the Kronecker delta δ_{1j} is defined as 1 for $j = 1$ and 0 for $j \neq 1$. H_x of (11.11) is substituted in (11.34) (it is to be noted that H_x is continuous across the $z = 0$ boundary.) The boundary condition

$$E_y(0^+) - E_y(0^-) = J_{mx}\delta(y)$$

is applied in geometry 1 and the boundary condition $E_y(0^+) = V_s\,\delta(y)$ is applied in geometry 2. Applying the equations (11.16) and (11.17), evaluating the y integrals first, and assuming that $J_{mx} = 1$ and $V_s = 1$ gives

$$P_j = \frac{i\omega\epsilon_0\epsilon^*}{2\pi}\int_{-\infty}^{\infty} \frac{\mathrm{d}\lambda}{\left[\gamma_e\dfrac{1-R_b}{1+R_b} - \lambda^2\dfrac{1-\epsilon}{\gamma_p}\dfrac{1+R_c}{1-R_c} + \delta_{1j}\epsilon\gamma_0\right]^*}. \tag{11.35}$$

For arbitrary values of J_{mx} and V, the right-hand side of (11.35) should be multiplied with $|J_{mx}|^2$ or $|V_s|^2$ for $j = 1$ or 2 respectively. This integral exhibits singularities for the same values of the integration variable λ as the field components derived from (11.11), (11.12), (11.16), and (11.17). Hence, it is appropriate to talk about the contributions of surface wave poles to the complex power flow.

The integrand is an even function of λ, and a surface wave pole at $\lambda = k$ makes the same contribution to the integral as the pole at $\lambda = -k$. For a lossless plasma, the surface wave poles are on the real axis and the integration contour is indented above the poles for $\lambda > 0$ when evaluating the field components (11.11) or (11.12). The poles of the power integral are complex conjugates of the poles of the field components in (11.11) or (11.12). For a slightly lossy plasma, the poles occur above the real axis for $\lambda < 0$ and below the real axis for $\lambda > 0$. In the power computation the integration path

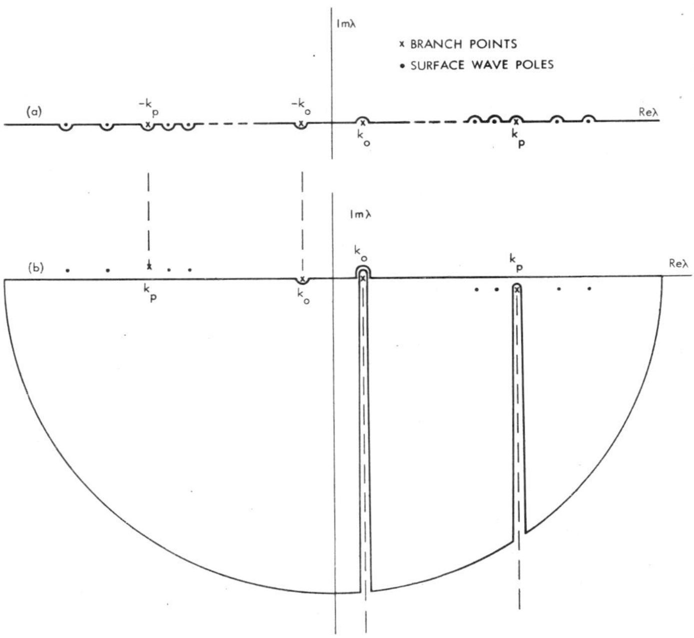

FIG. 11.4. Contour of integration in the λ plane.

is therefore indented below the real axis for $\lambda < 0$ and above the real axis for $\lambda > 0$, as is shown in Fig. 11.4a. After evaluating the integrals along a pair of indentations, the power contributed by surface wave poles at $\lambda = \pm k$ is given by

$$P_{jk} = -\pi i[R(-k) + R(k)], \tag{11.36}$$

where $R(x)$ designates the residue of the integrand (11.35) at the pole $\lambda = x$. For a lossy plasma, the surface wave poles occur for complex values of h. Their contributions to the integral (11.36) can be computed by evaluating the residues of the integrand after closing the contour of integration in the lower half of the complex plane and deforming the contour of integration along the branch cuts as shown in Fig. 11.4b. The integral along the half circle can be shown to give 0, and the pole contribution can be evaluated by computing the residue of the pole included in the integration contour. For poles $\lambda = \pm k$,

$$P_{jk} = -2\pi i R(k), \tag{11.37}$$

which is valid for real as well as complex poles. The branch cut contributions will not be considered in the present computations, and the calculations are further restricted to poles characterized by $k \gg k_0, k_e$. Under this restriction, the expression P_j of (11.35) simplifies to

$$P_j = \frac{i\omega\epsilon_0\epsilon^*}{2\pi} \int_{-\infty}^{\infty} \frac{N^*(\lambda)\, d\lambda}{D_j^*(\lambda)}, \tag{11.38}$$

where

$$N(\lambda) = \frac{\gamma_p}{\lambda}[(1+\epsilon)\gamma_p \tanh(\gamma_p h) - (1-\epsilon)\lambda], \tag{11.39}$$

$$D_1(\lambda) = [(1+\epsilon)^2\gamma_p^2 + (1-\epsilon)^2\lambda^2] \tanh(\gamma_p h) - 2(1-\epsilon^2)\lambda\gamma_p, \tag{11.40}$$

$$D_2(\lambda) = [(1+\epsilon)\gamma_p^2 + (1-\epsilon)^2\lambda^2] \tanh(\gamma_p h) - (2+\epsilon)(1-\epsilon)\lambda\gamma_p. \tag{11.41}$$

The complex power due to the pole at $\lambda = k$ is computed by substituting (11.38) in (11.37) as

$$P_{jk} = \omega\epsilon_0\epsilon^* \left[\frac{N(k)}{\partial D_j/\partial k}\right]^*, \tag{11.42}$$

For surface waves of $|k| > |k_p|$, γ_p is almost real and negative for a low-loss plasma. For $k_p \gg k_0$ and for layer thicknesses h comparable to the free space wave length, $|\gamma_p h| \gg 1$, $\tanh(\gamma_p h) \to -1$, and $\cosh^2(\gamma_p h) \gg 1$, which will be used to simplify the above expression.

For a plasma slab in free space ($j = 1$) the surface wave guided along the air-to-plasma interface has a wave number (11.27), and both $N(k)$ and $\partial D_1/\partial k$ of (11.42) are equal to zero. P_{jk} of (11.37) is now recomputed after dividing out a common factor of $N(\lambda)$ and $D_1(\lambda)$ in (11.38). This results in

$$P_{1k} = \omega\epsilon_0 \left[\frac{(1+\epsilon)^2}{1+\epsilon}\right]^* \tag{11.43}$$

for k of (11.27).

For the slot excited plasma slab ($j = 2$), the surface wave guided along the metal-to-plasma interface has a wave number as given by (11.32). Substituting this in (11.39) and (11.41), equation (11.42) yields

$$P_{2k} = \omega\epsilon_0\left[\frac{(1-\epsilon)^2}{2-\epsilon}\right]^*. \tag{11.44}$$

The surface wave guided along the air-to-plasma boundary has a wave number given by (11.33), which makes both $N(k)$ of (11.39) and $D_2(k)$ of (11.41) equal to zero. Equation (11.38) is not singular for this value of k and this surface wave is not excited for thick plasma layers when $\tanh(\gamma_p h) \to -1$. The same result is also obtained by examining (11.42) for the same value of k.

Such simplifications do not apply for surface waves of $|k| < |k_p|$, and the complete expression (11.42) must be used for numerical computations.

The power of the guided waves is computed numerically after substituting the appropriate complex values of k in (11.39), and in (11.42).

The calculations are made first for the plasma slab bounded by free space with $h/\lambda = 0.1$ and $c/u = 100$. The power carried by the individual guided waves with $k_r < k_{pr}$ is shown in Fig. 11.5. The power carried in an individual wave is a maximum near its cut-off point and decreases monotonically with increasing ϵ_r. However, there are more waves and hence there is a larger total power for ϵ_r large.

The sum of all the contributions P_{1k} of $k_r < k_{pr}$ has been added (Fig. 11.6) for several values of h/λ. The curves representing ΣP_{1k} are discontinuous at those values of ϵ_r where an additional mode starts propagating and decrease until the cut-off point of the next mode. These discontinuities are particularly pronounced for thin layers of plasma, but are decreased in magnitude for electrically thick plasma ($\epsilon_r h$ larger), when the number of modes is increased and the power per mode is less. The discontinuities are not shown for $h = 0.3\lambda$ and $\epsilon_r > 0.2$, and for $h = \lambda$, where the plotted curve indicates a smoothed average of ΣP_{1k}. Calculations have been also made for $c/u = 1000$ and $h/\lambda = 0.1$. The corresponding curve of ΣP_{1k} is identical to the one shown in Fig. 11.6 for $c/u = 100$ and $h/\lambda = 1$. The dashed curve represents the power carried by surface waves guided along the air-to-plasma boundary. It can be seen to be several orders of magnitude larger than the sum of the power in the guided waves for $k_r < k_{pr}$, in particular for $\epsilon_r \ll 1$.

The effects of plasma losses have been investigated for $\epsilon_r = 0.3$ and $c/u = 100$. For $h/\lambda = 0.1$, ΣP_{1k} decreases less than 0.5 percent when $\tan\delta$ is changed from 0 to 0.03, and decreases an additional 5 percent when $\tan\delta$ is further increased to 0.1. For $h/\lambda = 0.3$, ΣP_{1k} decreases 2 and 15 percent from the value for a lossless plasma and, for $h/\lambda = 1$, decreases 5 and 17 percent when $\tan\delta = 0.03$ and 0.1, respectively. However, P_{1k} is changed

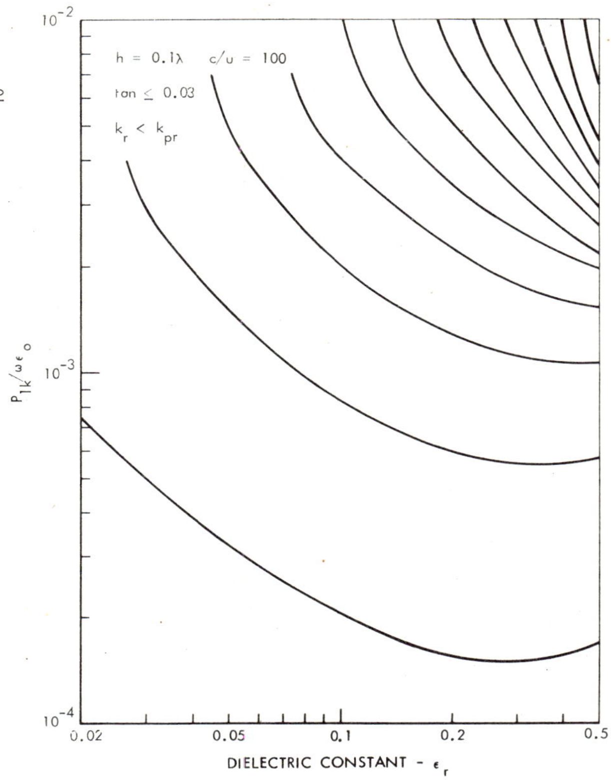

FIG. 11.5. Power carried by the individual surface waves with $k_r < k_{pr}$.

negligibly for $k_r > k_{pr}$ in the same range of tan δ. This variation of the powers ΣP_{1k} or P_{1k} is in line with the heuristical reasoning that suggests that the waves depending upon resonances in the transverse dimension of the plasma slab (or standing waves) should be affected more by plasma losses than waves guided along the plasma boundaries. Increasing losses and an increased thickness of the plasma tend to decrease ΣP_{1k}. This net decrease however, is rather small and ΣP_{1k} is nearly the same as for a lossless plasma if $\tan \delta < 0.01$ and $h < \lambda$.

Similar calculations have been carried out for waves in a plasma bounded by a conducting plane. When $k_r < k_{pr}$, the power carried in the guided waves ΣP_{2k} is almost the same as ΣP_{1k} in Fig. 11.6 for a plasma slab surrounded by free space, particularly for lower values of ϵ_r, but for larger values of $\epsilon_r \Sigma P_{2k}$ is somewhat higher. For plasma layers of a thickness h, which is large relative to the plasma wavelength λ_p, the wave along the air-plasma boundary ($z = h$) is not excited and only the power P_{2k} of the

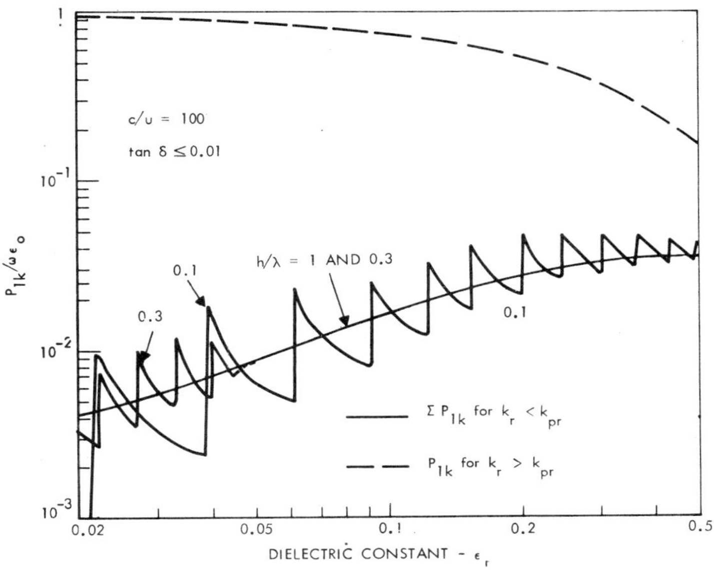

FIG. 11.6. Surface wave power for a plasma slab in free space.

wave guided along the $z = 0$ boundary must be considered. P_{2k} is less than the corresponding power P_{1k} of Fig. 11.6. Nevertheless, P_{2k} is much larger than ΣP_{2k} for small values of ϵ_r.

The power ΣP_{2k} carried by the guided waves exhibits a variation with $\tan\delta$ similar to ΣP_{1k}. For $\epsilon_r = 0.03$, $c/u = 100$, and $\tan\delta = 0.03$, ΣP_{2k} decreases from its value for a lossless plasma by 0.5, 3, and 7 percent for $h/\lambda = 0.1$, 0.3, and 1, respectively. For $\tan\delta = 0.1$, the corresponding figures are 5, 17, and 30 percent for the same three values of h/λ.

11.2. Admittance of a Waveguide Backed Slot

11.2.1. *Field Expressions for a Two-layer Plasma*

The rectangular slot will be examined in the geometry of a two-layer compressible plasma shown in Fig. 11.7. The plasma layers will be further specialized to represent an antenna radiating into a compressible plasma sheath, and an antenna insulated by an ion sheath from a compressible plasma half-space, as shown in Fig. 11.8. The field equations of (11.1) to (11.10) apply also to this problem. It is convenient to separate the vector components of **H** into the TE and TM parts that can be derived from two scalar functions Ψ and Φ, each of which satisfies a wave equation similar to (11.8). H_x, H_y, and H_z are related to Ψ and Φ by (3.13) to (3.15).

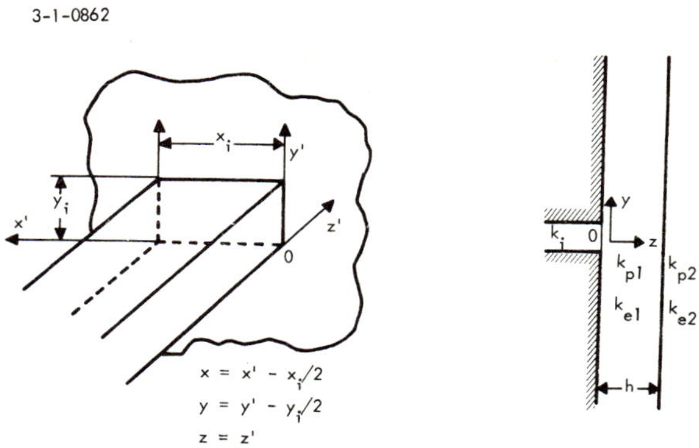

FIG. 11.7. Geometry for admittance calculation.

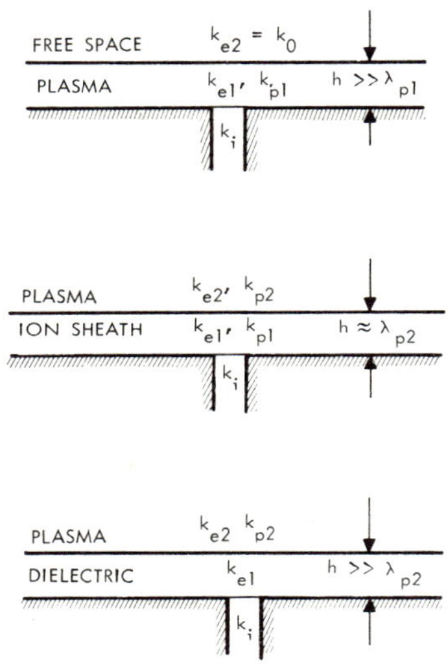

FIG. 11.8. Specialized geometries.

The solutions Ψ, Φ, and p for the two regions of Fig. 11.7 can be expressed in terms of the integrals

$$\Psi_n = \left(\frac{1}{2\pi}\right)^2 \int_{-\infty}^{\infty}\int_{-\infty}^{\infty} A_n(u, v)\mathrm{e}^{iux}\mathrm{e}^{ivy}(\mathrm{e}^{\gamma_{en}z} + R_{an}\mathrm{e}^{-\gamma_{en}z})\,\mathrm{d}u\,\mathrm{d}v, \quad (11.45)$$

$$\Phi_n = \left(\frac{1}{2\pi}\right)^2 \int_{-\infty}^{\infty}\int_{-\infty}^{\infty} B_n(u, v)\mathrm{e}^{iux}\mathrm{e}^{ivy}(\mathrm{e}^{\gamma_{en}z} + R_{bn}\mathrm{e}^{-\gamma_{en}z})\,\mathrm{d}u\,\mathrm{d}v, \quad (11.46)$$

$$p_n = \left(\frac{1}{2\pi}\right)^2 \int_{-\infty}^{\infty}\int_{-\infty}^{\infty} C_n(u, v) e^{iux} e^{ivy} (e^{\gamma_{pn} z} + R_{cn} e^{-\gamma_{pn} z})\, du\, dv, \quad (11.47)$$

where the subscript $n = 1$ or 2 designates the region, where the reflection coefficients $R_{a2} = R_{b2} = R_{c2} = 0$, and where

$$\gamma_{en} = i(k_{en}^2 - u^2 - v^2)^{1/2} \quad \text{and} \quad \gamma_{pn} = i(k_{pn}^2 - u^2 - v^2)^{1/2}$$

are defined in the second quadrant of the complex plane including the positive imaginary and the negative real axes. After computing the components of the **H** vector from (11.45) and (11.47) with the aid of (3.13) to (3.15), the components of **E** and **V** are obtained from (11.5) and (11.9) by making use of (11.47).

For cold plasma (Section 8.1), a stationary expression for the slot admittance was derived in terms of the electric aperture antenna fields, which were related to the coefficients A_1 and B_1 of the functions Ψ_1 and Φ_1 for the region just outside of the aperture. In the present problem of a compressible plasma the fields can be specified with the same coefficients A_1 and B_1 after relating them with the remaining constants R_{a1}, R_{b1}, A_2, B_2, C_1, R_{c1}, and C_2 of the field expressions (11.45) to (11.47) from a total of 7 boundary conditions (E_x, E_y, H_x, H_y, V_z, p continuous at $z = h$, and $V_z = 0$ at $z = 0$). The calculations result in

$$R_{a1} = e^{2\gamma_{e1} h} \frac{\gamma_{e1} - \gamma_{e2}}{\gamma_{e1} + \gamma_{e2}}, \quad (11.48)$$

$$R_{b1} = e^{2\gamma_{e1} h} \frac{\epsilon_2 \gamma_{e1} - \epsilon_1 (\gamma_{e2} - U - W e^{-\gamma_{e1} h})}{\epsilon_2 \gamma_{e1} + \epsilon_1 (\gamma_{e2} - U - W e^{\gamma_{e1} h})}. \quad (11.49)$$

The expressions of the aperture fields remain in the same form as for the incompressible plasma, except that a term $(1 - R_{b1})$ in the field or admittance expressions of the incompressible plasma [Equation (8.11)] is replaced by $(1 - R_{b1} + \Delta)$.

For a homogeneous plasma sheath the gas of region 2 is not ionized, $\epsilon_2 = 1$ and $e_2 = 0$, and it follows that

$$U = -\frac{\epsilon_2 (1 - \epsilon_1)(u^2 + v^2) Q}{\gamma_{p1} [1 + Q \tanh(\gamma_{p1} h)] \epsilon_1} \approx \frac{\epsilon_2}{\epsilon_1} \frac{u^2 + v^2}{\gamma_{p1}} (1 - \epsilon_1), \quad (11.50)$$

$$W = -U / \cosh \gamma_{p1} h, \quad (11.51)$$

$$\begin{aligned} \Delta = {} & (1 - \epsilon_1) \frac{u^2 + v^2}{\gamma_{e1} \gamma_{p1}} (1 + R_{b1}) \tanh(\gamma_{p1} h) \\ & - \frac{1}{\cosh(\gamma_{p1} h)} \left[\left(1 - \frac{\gamma_{e2}}{\gamma_{e1}} \frac{\epsilon_1}{\epsilon_2}\right) e^{\gamma_{e1} h} - R_{b1} \left(1 + \frac{\gamma_{e2}}{\gamma_{e1}} \frac{\epsilon_1}{\epsilon_2}\right) e^{-\gamma_{e1} h} \right] \\ & \approx -(1 - \epsilon_1) \frac{u^2 + v^2}{\gamma_{e1} \gamma_{p1}} (1 + R_{b1}) \end{aligned} \quad (11.52)$$

where

$$Q = -\frac{1-\epsilon_1}{1-\epsilon_2}\left(\frac{N_2 e_2}{N_1 e_1}\right)^2 \frac{\epsilon_2 \gamma_{p1}}{\epsilon_1 \gamma_{p2}} \approx -\frac{m_2 N_2 \epsilon_2 \gamma_{p1}}{m_1 N_1 \epsilon_1 \gamma_{p2}}$$

The approximate expressions are valid for $|\gamma_{p1} h| \gg 1$. U, W, and Δ become negligible for small values of (u^2+v^2).

When an ion layer deplete of electrons is assumed as region 1, $\epsilon_1 \to 1$ and $h \gg \lambda_{p1}$. It follows that

$$U \approx \frac{u^2+v^2}{\gamma_{p2}} \frac{1-\epsilon_2}{1+Q\tanh(\gamma_{p1} h)}, \tag{11.53}$$

$$W \approx \frac{u^2+v^2}{\gamma_{p2}} \frac{(1-\epsilon_2)Q}{\cosh(\gamma_{p1} h)}, \tag{11.54}$$

and Δ approaches zero as $\epsilon_1 \to 1$. Also U and W become negligible in (11.49) when (u^2+v^2) is of the order of $|\gamma_{e2}|$ or $|\gamma_{e1}|$ or less.

For dielectric layer of $h \gg \lambda_{p1} = 2\pi/k_p$ as medium 1, $W = \Delta = 0$, and U can be computed from (11.53) by letting $Q \to 0$.

11.2.2. *Slot Admittance*

The admittance of the rectangular waveguide that radiates into the plasma of Fig. 11.7 or 11.8 is similar to (8.13) to (8.17) except that terms $(1-R_b)$ of (8.15) and (8.16) are replaced by $(1-R_{b1}+\Delta)$, and R_a and R_b are replaced by R_{a1} and R_{b1} respectively. The reflection coefficient of the TE modes R_{a1} of (11.48) is the same as for an incompressible plasma. The compressibility of the plasma introduces the terms U and W in the expression (11.49) for the reflection coefficients R_{b1} of the TM modes. The term Δ of (11.52) is due solely to compressibility.

The slot admittance depends on the assumed electric fields for the waveguide aperture. Several trial functions for E_y and also for E_x have been examined in Section 8.1. The calculations are simplest and are satisfactorily accurate by assuming the principal waveguide mode as the electric aperture field for incompressible plasma layers. However, the pressure gradients of (11.5), or the plasma wave, will also contribute to the aperture fields. The pressure variations are in the direction of the impressed electric field according to (11.3) and the resulting field variations can be expected to have a wavelength of the order of λ_p. The aperture fields are therefore assumed to be of the form

$$E_y(x, y) = \cos(\pi x/x_i)\{A+B(\cos ky + C\sin|k|y)\}, \tag{11.55}$$

where k is of the order of $k_p = 2\pi/\lambda_p$ following Section 11.2.3, and A and B are arbitrary complex constants. The constant C is determined by requiring that $\partial E_y/\partial y = 0$ at the waveguide walls $y = \pm y_i/2$. It follows that $C = \tan(ky_i/2)$. Substituting (11.55), it follows that the slot admittance is given by (7.17) and (7.18) where

$$\gamma_{MN} = \gamma_{MN}^i + \gamma_{MN}^p, \tag{11.56}$$

$$\gamma_{AA}^i = \gamma_{AB}^i = 0, \tag{11.57}$$

$$\gamma_{BB}^i = \sum_{m=2}^{\infty} \frac{2x_i}{y_i} \frac{\beta_{xil}^2 - k_i^2}{\gamma_i} \left[\frac{k \tan(ky_i/2)}{k^2 - \beta_{yi}^2}\right]^2, \tag{11.58}$$

$$\gamma_{NM}^p = \frac{1}{8\pi^2} \int_{-\infty}^{\infty} \int_{-\infty}^{\infty} F(R) \left[\frac{4\pi}{x_i} \frac{\cos(ux_i/2)}{(\pi/x_i)^2 - u^2}\right]^2 g_N(v) g_M(v)\, du\, dv, \tag{11.59}$$

$$g_A(v) = \sin(vy_i/2)/v, \tag{11.60}$$

$$g_B(v) = \frac{1}{v^2 - k^2}\left[v \frac{\sin(vy_i/2)}{\cos(ky_i/2)} - k \tan(ky_i/2)\right], \tag{11.61}$$

$$F_A = x_i y_i/2, \tag{11.62}$$

$$F_B = (x_i y_i/2) \frac{\tan(ky_i/2)}{(ky_i/2)}. \tag{11.63}$$

In (11.58), $n = 1$ and m is restricted to even integers. By noting that $k \gg k_i$, the summation (11.58) can be evaluated analytically for $|\sin(ky_i/2)| \to 1$ and for $\sin(ky_i/2) \to 0$ with $y_i \neq 0$ when the summation is reduced to a single term. In both cases

$$\gamma_{BB}^i = \frac{x_i y_i}{8} \frac{k_i^2 - \beta_{xil}^2}{k} [\cos(ky_i/2)]^{-2} \tag{11.64}$$

following Galejs [1966a]. The summation has been evaluated numerically for other values of $(ky_i/2)$, and was found to differ negligibly from (11.64).

The integrand of γ_{AA}^p in (11.59) does not differ from γ_{AA}^p for an incompressible plasma at values of $(u^2 + v^2) \ll k_{pj}^2$. It approaches zero when $(u^2 + v^2)$ becomes comparable to k_{pj}^2, except for those values of $(u^2 + v^2)$ where $F(R)$ has a pole $(1 - R_{b1} + \Delta = 0)$. The integrand of γ_{AA}^p is much smaller than the integrands of γ_{AB}^p and γ_{BB}^p for values of $(u^2 + v^2)$ near the above poles. The integrals γ_{AA}^p are evaluated numerically as indicated in earlier work [Galejs, 1965d] for $(u^2 + v^2) \ll k_{pj}^2$. For larger values of $(u^2 + v^2)$, the integrals are evaluated in the same way as for γ_{AB}^p and γ_{BB}^p.

Because of $g_B(v)$, the integrands of γ_{BB}^p and γ_{AB}^p are small when $(u^2 + v^2)$ is of the order of k_{e1}^2 and k_{e2}^2, but they become more appreciable as $v^2 \to k^2$ (or k_p^2), if u is of the order of k_{e1} or k_{e2} at the same time. Under these conditions $R_{a1} \to 0$ and (11.59) is changed to

$$\gamma_{NM}^{p} \approx \frac{1}{2\pi^2} \int_{v_i}^{\infty} \frac{dv}{v} \int_{0}^{u_1} du \cdot \left[-u^2 + k_{e1}^2 \frac{1+R_{b1}}{1-R_{b1}+\Delta}\right] \left[\frac{4\pi}{x_i} \frac{\cos(ux_i/2)}{(\pi/x_i)^2 - u^2}\right]^2 g_N(v) g_M(v), \tag{11.65}$$

where NM may have the values of BB or AB, and where $v_1 \gg u_1$. Equation (11.65) is also used for the second part of the γ_{AA}^{p} integrals. Because $v \gg u$, R_{b1} and Δ are functions of v only. After changing the variable of integration from u to $w = (ux_i - \pi)/2$ and expanding the algebraic part of the resultant function of w in partial fractions, the w integrals are evaluated to give

$$\gamma_{NM}^{p} \approx \int_{v_1}^{\infty} \frac{dv}{v} \left[-\frac{\pi}{xi} + \frac{k_{e1}^2 x_i}{\pi} \frac{1+R_{b1}}{1-R_{b1}+\Delta}\right] g_N(v) g_M(v). \tag{11.66}$$

Under the same assumption of $v \geq v_1 \gg u_1$, R_{b1} of (11.51) becomes

$$R_{b1} = e^{-2vh} \frac{\epsilon_2 - \epsilon_1[1 + U/v + We^{vh}/v]}{\epsilon_2 + \epsilon_1[1 + U/v + We^{-vh}/v]}. \tag{11.67}$$

11.2.3. *Surface Waves*

The phase constant k is assumed to be the same as for surface waves supported by the plasma layers, and $k = v$ is determined from the poles of the integrands where

$$1 - R_{b1} + \Delta = 0. \tag{11.68}$$

Substituting (11.67) in (11.68) and neglecting We^{-vh}/v relative to U/v gives

$$(1+\Delta) \frac{\epsilon_2 + \epsilon_1(1 + U/v)}{\epsilon_2 - \epsilon_1(1 + U/v + We^{vh}/v)} = \exp(-2vh). \tag{11.69}$$

The right-hand side of (11.69) will approach zero for a homogeneous plasma sheath of thickness comparable to λ_e, because $k = v$ is of the order of k_p. Using (11.50) and (11.52), the conditions $1 + \Delta = 0$ and $\epsilon_2 + \epsilon_1(1 + U/v) = 0$ result in

$$k = \frac{k_{p1}}{(\epsilon_1)^{1/2}(2 - \epsilon_1)^{1/2}}, \tag{11.70}$$

and

$$k = k_{p1} \frac{1 + \epsilon_1}{2(\epsilon_1)^{1/2}}. \tag{11.71}$$

The phase constants k of (11.70) and (11.71) do not depend explicitly on the thickness of the plasma layer h and apply strictly to a half space of plasma or $h \to \infty$. Equation (11.70) can be identified as the phase constant of a surface wave guided along the conducting plane (11.32), while (11.71) represents the phase constant of a surface wave guided along the air-to-plasma interface (11.27). Similar phase constants k of surface waves are also observed for propagation along cylindrical wires of radius $a \gg \lambda_p$ [equation (10) of Galejs 1965c with $b = a$] and for an insulated wire of insulation radius b if $b - a > \lambda_p$ [equation (11) of Galejs 1965c]. The surface wave guided along the conducting plane will have the dominant effect on the field distribution in the waveguide aperture, and k of (11.70) is used in the trial functions (11.55) for a plasma sheath.

For an ion layer, $\Delta = 0$ and (11.69) is rearranged into

$$\frac{U}{v} + \frac{W}{v} \frac{e^{vh}}{1 + e^{2vh}} = -[1 + \epsilon_2 \tanh(vh)]. \tag{11.72}$$

W is small relative to U. The term proportional to W becomes negligible even for large values of vh, and is therefore neglected. Neglecting also Q of (11.53), (11.74) is rewritten as

$$\epsilon_2 \tanh(vh) = \frac{1 - \epsilon_2}{[1 - (k_{p2}/v)^2]^{1/2}}. \tag{11.73}$$

Letting $h = b - a$, (11.73) can be seen to be identical to (10) of Galejs [1965c], which was derived for a cylindrical wire of radius $a \gg \lambda_p$ surrounded by an insulating shell of radius b by requiring the plasma boundary at b to be rigid. For $h = 0$ and ∞, (11.73) assumes the forms (11.70) and (11.71), which were discussed earlier.

Equation (11.73) applies also to a dielectric layer of $h \gg \lambda_{pl}$.

11.2.4. *Numerical Results*

Numerical calculations have been carried out to show the variation of slot admittance with varying parameters of the geometry, and the slot admittance depends on the resonance of the surface waves within the transverse dimension of the aperture.

The compressibility of the plasma has the largest effect on slot admittance when the slot width y_i is equal to an odd number of one-half wave lengths of the surface waves ($y_i \simeq (2n+1)\pi/k$), which will be termed odd resonances. The slot conductance, shown in Fig. 11.9 is significantly increased for low values of $\tan \delta_1$ if $\epsilon_1 \ll \epsilon_0$ or $\omega \to \omega_p$ and if c/u_1 is 300 or less. The increase of slot conductance is negligible for $\epsilon_p/\epsilon_0 > 0.2$, and may be ignored for $c/u_1 > 1000$. The compressibility of the plasma affects the slot

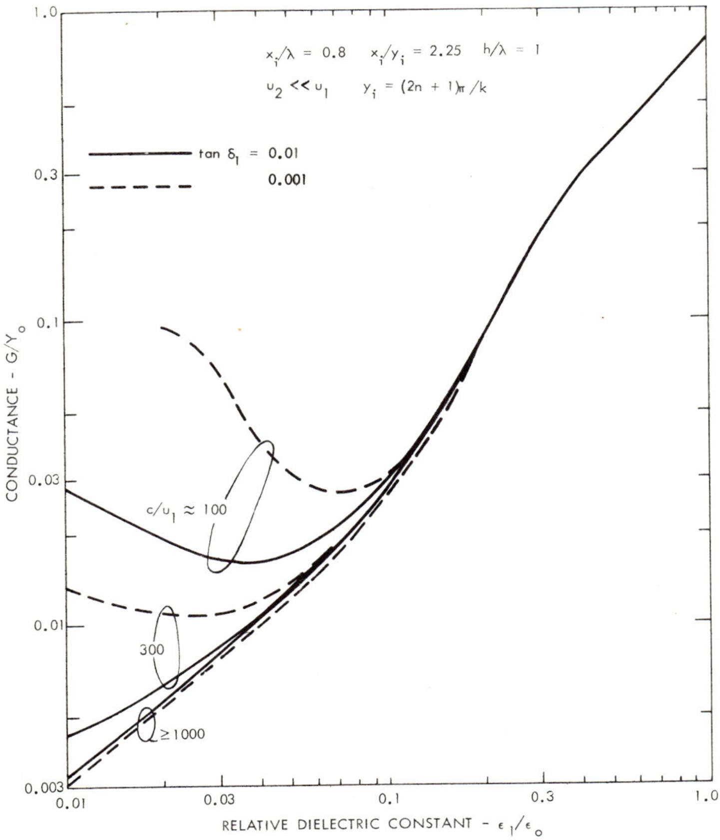

FIG. 11.9. Conductance of a slot radiating into a plasma layer. Effects of dielectric constant.

susceptance only to a minor extent as can be seen from Fig. 11.10. However, Balmain [1965] observed significant changes of the antenna reactance for various temperatures of the plasma, and the present results will be examined more closely.

Since the complexity of the present formulation makes an analytical examination of the admittance behaviour difficult, it was found expedient to compute the complex power flow from a parallel plate waveguide into a halfspace of compressible isotropic plasma. This complex power flow is expressed as an integral [Galejs, 1966a]. The finite plasma temperature gives rise to a surface wave pole that is in the integrand and that corresponds to the surface waves (11.70). For a lossless plasma, the residue evaluation of this pole makes only a real contribution to the complex power flow or to the slot admittance. It is also shown that for slots that are wide relative to

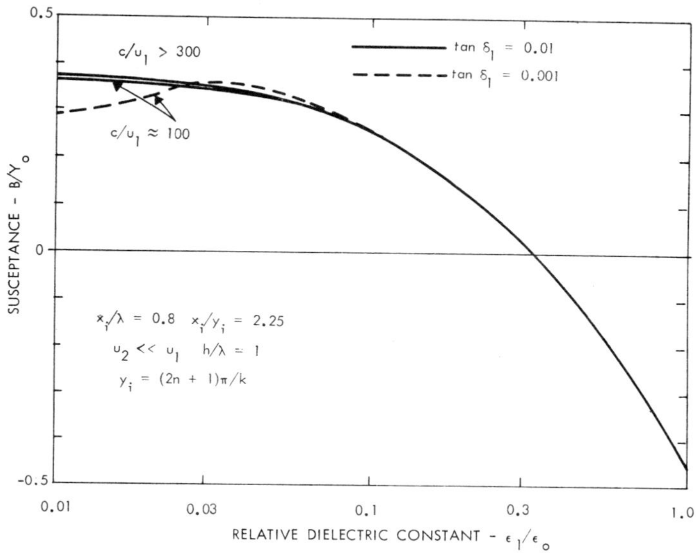

FIG. 11.10. Susceptance of a slot radiating into a plasma layer.

the acoustical wavelength, significant compressibility effects are observed only when the trial function of the aperture fields (11.55) considers the perturbing effects of surface waves. However, for narrow slots ($ky_i \ll 1$), such compressibility effects occur even when neglecting the surface waves in the trial function of aperture fields. This can be seen also from the numerical results of other investigations [Wait and Spies, 1966].

Slight changes of (c/u_1) cause a change in k, which causes corresponding changes of the slot width y_i in relation to the wavelength of the surface waves $(2\pi/k)$. The slot conductance of Fig. 11.11a is rapidly changing near the resonant slot width, and approaches the conductance for cold plasma $(c/u_1 = \infty)$ when the resonance condition is not satisfied. Increasing dielectric losses tend to widen this resonance peak. The slot conductance is slightly increased also for slot width y_i equal to an integer number of wavelengths of the surface waves $(y = 2n\pi/k)$, which is denoted as an even resonance. The conductance of Fig. 11.11b or c depends less critically on the slot width. Also, the shape of the resonance curve is not affected by dielectric losses to any significant degree.

The ratio of the amplitude of the principal wave guide mode A to the amplitude of the surface waves B and BC in the trial functions of the aperture fields (11.55) can be computed for various slot widths. The ratio $|A/BC|$ is of the order of unity for odd resonances of $y_i = (2n+1)\pi/k$, when the slot conductance deviates most from its value for a cold plasma. Plasma losses tend to widen this resonance peak similar to the situation observed in the

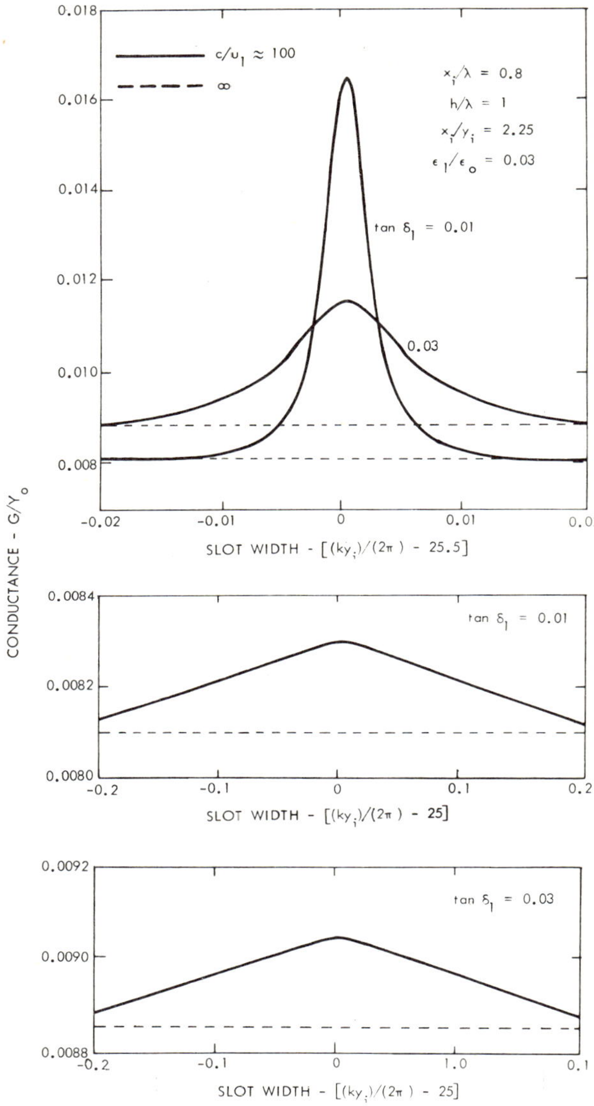

FIG. 11.11. Conductance of a slot radiating into a plasma layer effects of slot width.

conductance curve of Fig. 11.11a. The ratio $|A/B|$ is approximately 40 for even resonances of $y_i = 2n\pi/k$, and is not affected by dielectric losses.

The effects of plasma losses are further illustrated in Fig. 11.12. Increasing losses tend to increase the slot conductance of a cold plasma ($c/u_1 = \infty$). They do not affect the net increase due to compressibility for even resonances

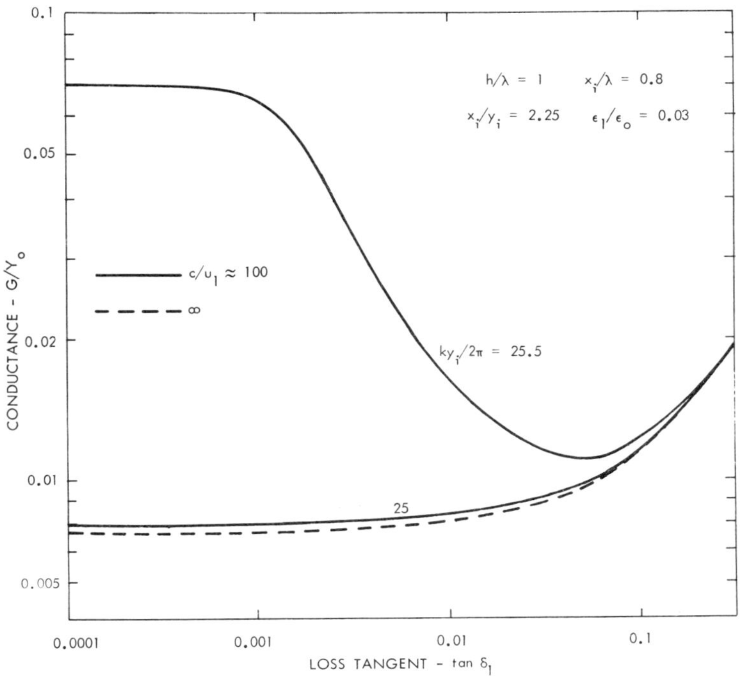

FIG. 11.12. Conductance of a slot radiating into a plasma layer. Effects of plasma losses.

of $y_i = 2n\pi/k$ but gradually decrease the compressibility effects for odd resonances of $y_i = (2n+1)\pi/k$ as can be seen from Fig. 11.12.

The slot conductance can be also computed for varying thicknesses of the plasma layer. For thin layers, the conductance approaches that of free space and the finite temperature of the plasma has no effects. The compressibility effects become significant for $h > \lambda/2$.

Numerical calculations have also been made for an ion sheath that separates the waveguide aperture from a plasma half space. The effects of various sheath thicknesses are illustrated in Fig. 11.13. The geometry with an ion sheath of $h \to 0$ is equivalent to a plasma layer of $h \to \infty$, which is in direct contact with the waveguide aperture. This can be seen by examining (11.59) and (11.65) in these two limiting cases. An increasing thickness of the ion sheath decreases the compressibility effects. Consequently there are only minor changes of the slot conductance for $h \geqslant 0.01\lambda$ with $\epsilon_2/\epsilon_0 = 0.03$. Other calculations show that the compressibility effects can be neglected if $h \geqslant 0.01\lambda$ and/or $\epsilon_2/\epsilon_0 \geqslant 0.1$.

Based on Fig. 11.13, very thin dielectric layers may be sufficient for suppressing the effects of the finite plasma temperature.

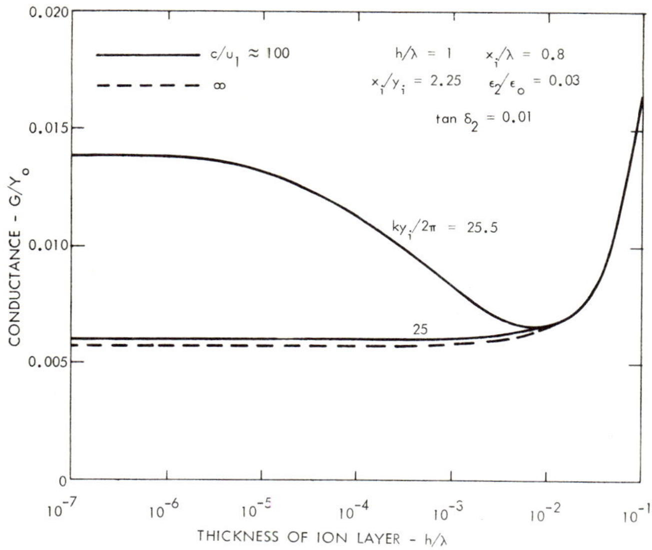

FIG. 11.13. Slot conductance for radiation into an ion sheath and a plasma halfspace.

It is common to assume a rigid boundary with the normal particle velocity $V_z = 0$ at the interfaces between plasma and ion sheath, or at the interface between plasma and air. In the present investigation, this boundary condition has been replaced by the conditions of continuous normal particle velocity V_z and continuous scalar pressure p. This has led to an algebraically more involved formulation. However, the slot admittance computed using the appropriate layer parameters is numerically the same as in the limit of a rigid boundary.

The compressibility of the plasma was shown to affect the slot conductance more significantly than the slot susceptance, but noticeable changes were noted for small values of the relative dielectric constant of the plasma or $\omega \rightarrow \omega_p$. There is no significant radiation for $\omega < \omega_p$. Consequently, this frequency range has not been examined in the present paper.

11.3. Impedance of a Linear Antenna

A flat strip antenna is embedded in a planar dielectric slab of thickness 2Δ, which is covered on both sides by a compressible isotropic plasma of thickness h, as shown in Fig. 11.14. The dielectric layer is intended to approximate the effects of the ion sheath formed around the antenna. The expressions for the fields are derived starting out with linearized field equations (Section 11.1.1). The antenna impedance is expressed in the same way as for a linear antenna in an incompressible plasma or dielectric layers in Section 9.1. The antenna current distribution is assumed to be the

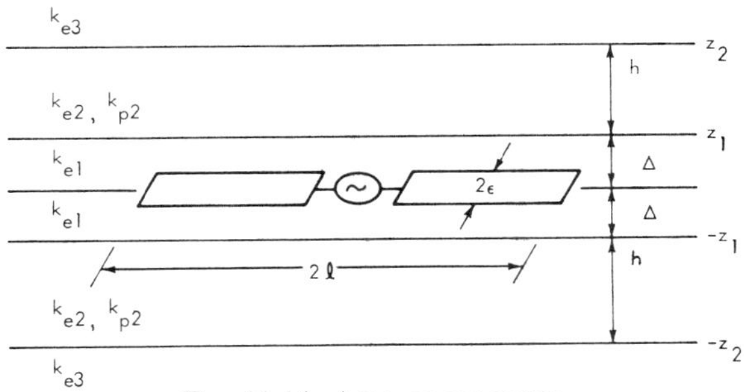

FIG. 11.14. Antenna geometry.

sum of two terms. The first term is a sine wave with the same wave number k_A as in incompressible plasma. The second term is representative of surface waves supported by this plasma geometry. The wave number k_B of the surface wave is determined from the solution of the transcendental equation, which describes the poles of the integrands of the field expressions; one tests the hypothesis that these surface waves modify the current distribution along the antenna.

11.3.1. *Impedance Formulation*

The TE and TM parts of the magnetic field **H** are derived for region n from scalar functions Ψ_n and Φ_n, which are expressed as double Fourier integrals. In the compressible plasma (region $n = 2$), the electric field **E** depends also on the pressure p_n, which is also given by a double integral. [(11.1) to (11.10), and (11.45) to (11.47)]. The fields near the antenna are specified with the amplitudes A_1 and B_1 of Ψ_1 and Φ_1, which are computed after relating them to the amplitudes A_2, B_2, C_2, A_3, and C_3, and to the reflection coefficients R_{a1}, R_{b1}, R_{a2}, and R_{c2} from a total of 10 boundary conditions, (E_x, E_y, H_x, H_y continuous, and $V_z = 0$ at $z = z_1$ and z_2). These equations can be solved for the reflection coefficients R_{aj} and R_{bj}. A lengthy algebraic manipulation shows that the reflection coefficients of the TE modes R_{a1} and R_{a2} are the same as for cold plasma (8.18). The reflection coefficients of the TM modes R_{bn} are given by

$$R_{b1} = e^{2\gamma_{e1}z_1}\frac{P-Q}{P+Q}, \tag{11.74}$$

with

$$P = \left(\frac{k_{e2}}{k_{e1}}\right)^2 (e^{2\gamma_{e2}z_1} + R_{b2}), \tag{11.75}$$

$$Q = \frac{\gamma_{e2}}{\gamma_{e1}}(e^{2\gamma_{e2}z_1} - R_{b2}) - d_1Ct(e^{2\gamma_{e_2}z_1} + R_{b2}) + d_1Cs\, e^{-\gamma_{e2}h}(e^{2\gamma_{e_2}z_2} + R_{b2}), \tag{11.76}$$

$$R_{b2} = e^{2\gamma_{e2}z_2}\frac{-(k_{e2}/k_{e3})^2 + d_3Ct - d_3Cs\, e^{-\gamma_{e2}h} + (\gamma_{e2}/\gamma_{e3})}{(k_{e2}/k_{e3})^2 - d_3Ct + d_3Cs\, e^{\gamma_{e2}h} + (\gamma_{e2}/\gamma_{e3})}, \tag{11.77}$$

where

$$d_j = -(1-\epsilon_2)(u^2+v^2)/(\gamma_{ej}\gamma_{p2}),\; Ct = \coth(\gamma_{p2}h),$$

$$Cs = 1/\sinh(\gamma_{p2}h), \quad k_p^2 = [\omega(\omega+i\nu)-\omega_p^2]/u^2,$$

$$k_e^2 = \omega^2\mu_0\epsilon_0\epsilon, \quad \epsilon = 1-\omega_p^2/[\omega(\omega+i\nu)],$$

and where

$$\gamma_{en} = i(k_{en}^2-u^2-v^2)^{1/2} \quad \text{and} \quad \gamma_{pn} = i(k_{pn}^2-u^2-v^2)^{1/2}$$

are defined in the second quadrant of the complex plane including the positive imaginary and the negative real axes. Also,

$$k_{e1} = k_{e3} = k_0 = \omega(\mu_0\epsilon_0)^{1/2}.$$

The compressibility of the plasma introduces the terms proportional to d_j in the expressions (11.76) and (11.77) for the reflection coefficients R_{bj} of the TM modes. It can be shown that R_{a1} and $R_{b1} \to 0$ as $h \to 0$, if $k_{e1} = k_{e3}$.

The assumption of a rigid boundary of the plasma layers with $V_z = 0$ can be replaced by conditions of continuous scalar pressure p and continuous normal particle velocity V_z across the boundaries of plasma and non-ionized gas (or ion sheath). This leads to increased algebraic complexity of the derivations, but gives essentially the same results as the simpler and less accurate boundary condition of $V_z = 0$ [Galejs, 1966a]. For a nonrigid metallic boundary, the surface waves are attentuated even for a collisionless plasma [Wait, 1966].

The antenna is separated from the plasma layer by an ion sheath, which is approximated by free space. The antenna impedance Z is given by (9.5) and (9.6) where the reflection coefficients R_{b1} of (11.74) should be used for $R_{b0p} = R_{biq}$. The antenna impedance Z depends on the current distribution of $J_x(x, y)$ along the antenna. For antennas in a cold plasma the antenna impedance was determined with a two-term trial function in (9.7). Several free space impedance calculations of Section 2.3 have been made for an assumed sinusoidal current distribution with a wave number k_0. Such calculations are satisfactory only for short antenna of $l < 0.15\lambda$ following Fig. 2.2. However, the wave number of the single term trial function can be selected to be such that the resulting impedance approximates the more accurate impedance values which are based on two-term functions. For free space, $k = 1.2k_0$, $l = 0.25\lambda$ and $\Omega = 10$, the single term impedance $Z_s = 94–i52$, while the two-term calculations shown in Fig. 2.1 result in $Z = 92–i43$. It is therefore possible to select the wave numbers of the single term trial function in such a way that the resulting impedance is a reasonable approximation to the more accurate impedance figures for antenna of length

$l < \lambda/4$. One term of the antenna trial function can be selected to approximate the antenna current distribution of the cold plasma, while the second term can represent the effect of the finite plasma temperature. The antenna current is assumed to be in the form

$$J_x(x, y) = \{A \sin [k_A(l - |x|)] + B \sin [k_B(l - |x|)]\} f(y) \qquad (11.78)$$

with $f(y) = 1/(2\epsilon) = \text{const}$. The phase constant k_A is selected in such a way that the impedance Z computed with $B = 0$ approximates the impedance Z of the corresponding cold plasma geometry, which was derived in Section 9.1 using a more accurate two-term trial function.† The second term of the current distribution (11.78) should approximate the effects of the finite plasma temperature. The phase constant k_B is assumed to be representative of surface waves that are supported by the antenna in the presence of the compressible plasma. It will be discussed in Section 11.3.2. Equations (2.38) to (2.41) apply also to the present problem. F_N and $g_N(u)$ are given by (9.8) and (9.11) if k is replaced by $k_N (N = A \text{ or } B)$. Similarly, k^2 is replaced by $k_N k_M$ in the γ_{NM} expression (9.10).

11.3.2. *Surface Waves*

The phase constant k_B is assumed to be the same as for surface waves supported by the plasma layers, and

$$k_B = (u^2 + v^2)^{1/2}$$

is determined from the poles of the integrand in (9.5), where

$$1 + R_{b1} = 0. \qquad (11.79)$$

Substituting (11.74) to (11.77) in (11.79) results in

$$\frac{\gamma_{e2}}{\gamma_{e1}}\left(\frac{k_{e2}}{k_{e3}}\right)^2 - \frac{\gamma_{e2}}{\gamma_{e3}}\left(\frac{k_{e2}}{k_{e1}}\right)^2 \coth \gamma_{e1} z_1$$
$$- \left(d_1 \frac{\gamma_{e2}}{\gamma_{e3}} + d_3 \frac{\gamma_{e2}}{\gamma_{e1}}\right)\left(\frac{Cs}{\cosh \gamma_{e2} h} - Ct\right)$$
$$+ \tanh \gamma_{e2} h \left\{\left[\left(\frac{k_{e2}}{k_{e3}}\right)^2 - d_3 Ct\right]\left[\left(\frac{k_{e2}}{k_{e1}}\right)^2 \coth \gamma_{e1} z_1 + d_1 Ct\right]\right.$$
$$\left. - \frac{\gamma_2}{\gamma_3}\frac{\gamma_2}{\gamma_1} + d_1 d_3 (Cs)^2\right\} = 0, \qquad (11.80)$$

where the symbols d_n, Cs, and Ct are defined following (11.77).

†In most of the subsequent numerical examples, the antenna is electrically short ($\sqrt{\epsilon_2}\, l/\lambda \approx 0.043$ in Figs. 11.16 and 11.17; $l/\lambda = 0.1$ in Figs. 11.18–20), the current distribution is nearly triangular for a cold plasma, and the impedance is not critically dependent on k_A.

After introducing infinitesimal plasma losses ($\nu \neq 0$) in the limit of $h \to \infty$, it follows that $\tanh \gamma_{e2} h$ and $Ct \to (-1)$, and $Cs \to 0$. Equation (11.82) simplifies now to

$$\left(\frac{k_{e2}}{k_{e1}}\right)^2 \coth \gamma_{e1} z_1 = d_1 + \frac{\gamma_{e2}}{\gamma_{e1}}, \tag{11.81}$$

which can be also obtained from (11.79) and (11.74) by setting $R_{b2} = 0$ in (11.75) and (11.76). Equation (11.81) can be expected to have solutions for $k_B \approx k_{p2} \gg k_{ej}$. Letting $\epsilon_2 = (k_{e2}/k_{e1})^2$, (11.81) can be expressed as

$$\epsilon_2 \coth k_B z_1 = \frac{1-\epsilon_2}{[1-(k_{p2}/k_B)^2]^{1/2}} - 1. \tag{11.82}$$

For thick insulating layers, ($k_B z_1 \gg 1$), (11.82) is changed to (11.27), which represents surface waves guided along the boundary of a semi-infinite plasma. For thin insulating layers, ($k_B z_1 \ll 1$), (11.82) becomes

$$k_B = k_{p2}\left[1 + \frac{1}{2}\left(\frac{1-\epsilon_2}{\epsilon_2} k_{p2} z_1\right)^2 + \dots\right], \tag{11.83}$$

and the wave number of the surface wave k_B approaches the wave number k_{p2} of plasma waves in a homogeneous plasma. In the limit of $\epsilon_2 \to 0$ and $\epsilon_2 \ll k_B z_1$, $k_B \approx k_0 c/(2u)$, where $k_0 = \omega(\mu_0 \epsilon_0)^{1/2}$. A more detailed examination of (11.81) shows that (11.82) represents its only solution and that k_B lies within the limits indicated above if ϵ_2 is in the range $0 < \epsilon_2 < 1$.

For a plasma layer of finite thickness h, (11.80) will be simplified by assuming $k_B \approx k_{p2} \gg k_{ej}$, and h of the order of the free space wavelength $2\pi/k_0$. This gives

$$\left[1 + \epsilon_2 \coth k_B z_1 - \frac{(1-\epsilon_2)k_B}{(k_B^2 - k_{p2}^2)^{1/2}} \coth[(k_B^2 - k_{p2}^2)^{1/2} h]\right.$$
$$\cdot \left\{\left[1 + \epsilon_2 - \frac{(1-\epsilon)k_B}{(k_B^2 - k_{p2}^2)^{1/2}} \coth[(k_B^2 - k_{p2}^2)^{1/2} h]\right\}\right.$$
$$= \frac{(1-\epsilon_2)^2 k_B^2}{k_B^2 - k_{p2}^2} \frac{1}{\sinh^2[(k_B^2 - k_{p2}^2)^{1/2} h]}. \tag{11.84}$$

Equation (11.84) has further solutions with $|k_B| < |k_{p2}|$. However, it has been shown in Section 11.1 that these waves are excited with small amplitudes if the plasma layer has a thickness comparable to the free space wavelength. These waves are not considered at present.

A real root k_B is determined first from the solution of (11.84) for a lossless plasma. For a lossy plasma, the complex root is found by applying Newton's iteration method and by using the former real root for initial estimate. For $h \gg z_1$ the roots of (11.84) differ negligibly from the roots of (11.82).

11.3.3. *Impedance of a Short Antenna in an Unbounded Plasma*

The impedance of a short antenna ($k_0 l \ll 1$) will be computed from (2.41) for an unbounded plasma with an assumed sinusoidal current distribution of $k_A = k_{e2}$. For $z_1 = 0$ and $h \to \infty$, $R_{a2} = R_{b2} = 0$, and the substitution of R_{a1} and R_{b1} in (9.6) shows that

$$F(u, v) = \frac{1}{2(u^2+v^2)}\left(\frac{v^2}{\gamma_{e2}} - \frac{u^2\gamma_{e2}}{k_{e2}^2}\right) + \frac{u^2}{2k_{e2}^2}\frac{1-\epsilon_2}{\gamma_{p2}}. \tag{11.85}$$

The first term of (11.85) does not depend on the finite temperature of the plasma, and substituting it in (9.10) the impedance due to the electromagnetic waves (subscript e) can be computed from

$$Z_e = R_e + iX_e = \frac{2i\omega\mu_0}{(\pi k_A l)^2}\int_0^\infty \mathrm{d}u \frac{(\cos ul - \cos k_A l)^2}{k_A^2 - u^2} \int_0^\infty \mathrm{d}v \left(\frac{\sin \epsilon v}{\epsilon v}\right)^2 \frac{1}{\gamma_{e2}}. \tag{11.86}$$

The antenna resistance R_e is computed by confining the range of integration to values of u and v where $u^2+v^2 < k_A^2$. Over this range of integration the sine and cosine functions can be replaced by their small angle approximations. The integrations are elementary and

$$R_e = 20(\epsilon_2)^{1/2}(k_0 l)^2. \tag{11.87}$$

In the reactance computation the limits of integration are similar to those in (9.22). The v integrations become elementary by noting that $\sin \epsilon v \approx \epsilon v$ over the range of v where

$$(u^2+v^2)^{1/2} \neq v,$$

which applies strictly if $\epsilon \to 0$ and $u \leq u_0$ is finite. The v integrals are evaluated to give

$$X_e = -\frac{2\omega\mu_0}{(\pi k_A l)^2}\int_0^\infty du \frac{(\cos ul - \cos k_A l)^2}{(k_A^2 - u^2)} \cdot\left[\frac{3}{2} - C - \log(\epsilon k_A) - \frac{1}{2}\log\left|1 - \left(\frac{u}{k_A}\right)^2\right|\right]. \tag{11.88}$$

where $C = 0.5772\ldots$ is Euler's constant. The integral that is proportional to the constant term of the square brackets is evaluated directly. The remaining term of the integral, which involves a logarithmic function, is reduced to tabulated integrals after expanding the denominator in partial fractions and changing the variables of integration to $y = (u/k_A) \pm 1$. This results in

$$X_e = \frac{120}{\epsilon_2 k_0 l}\left[\log\frac{2l}{\epsilon} + 0.5 - 2\log 2\right]. \tag{11.89}$$

The contribution of the plasma waves (subscript p) to the antenna impedance is computed using the second term of (11.85). This results in

$$Z_p = R_p + iX_p = \frac{2\omega\mu_0}{\pi^2 l^2}\frac{1-\epsilon^2}{\epsilon_2 k_0^2}\int_0^\infty du\,\frac{(\cos ul - 1)^2}{u^2}\int_0^\infty dv \cdot\left(\frac{\sin\epsilon v}{\epsilon v}\right)^2\frac{1}{(k_{p2}^2 - u^2 - v^2)^{1/2}} \tag{11.90}$$

The computations will be made first for very thin antennas, $(k_{p2}\epsilon) \ll 1$. In the resistance calculations, $u^2 + v^2 < k_{p2}^2$ and sine $\epsilon v \approx \epsilon v$ over this range of integration. It follows that

$$R_p = \frac{120(1-\epsilon_2)}{\epsilon_2 k_0 l}\left\{2Si(k_{p2}l) - Si(2k_{p2}l) - \frac{4\sin^4(k_{p2}l/2)}{k_{p2}l}\right\}. \tag{11.91}$$

For $k_{p2}l \gg 1$, (11.91) simplifies to

$$R_p = \frac{60\pi}{\epsilon_2 k_0 l}(1-\epsilon_2), \tag{11.92}$$

and for $k_{p2}l \ll 1$ (11.91) is approximated by

$$R_p = 10(1-\epsilon_2)(\epsilon_2)^{1/2}\left(\frac{c}{u_2}\right)^3 (k_0 l)^2. \tag{11.93}$$

Also, R_p of (11.93) can be obtained directly from (11.90) by using small argument approximations of the cosine function. In the reactance computations, $u^2 + v^2 > k_{p2}^2$. The v integrals are evaluated using the same approximations as in (11.88), which gives

$$X_p = -\frac{2\omega\mu_0(1-\epsilon_2)}{\epsilon_2(\pi k_0 l)^2}\int_0^\infty du\,\frac{(1-\cos ul)^2}{u^2} \cdot\left[\frac{3}{2} - C - \log(\epsilon k_{p2}) - \frac{1}{2}\log\left|1 - \left(\frac{u}{k_{p2}}\right)^2\right|\right]. \tag{11.94}$$

For $k_{p2}l \gg 1$, the integrals are evaluated to give

$$X_p = -\frac{120(1-\epsilon_2)}{\epsilon_2 k_0 l}\left\{\frac{3}{2} - C - \log(\epsilon k_{p2}) - \frac{1}{(k_{p2}l)^2}\right.$$
$$\left. \cdot\left[2\cos(k_{p2}l) - \frac{1}{2}\cos(2k_{p2}l)\right]\right\}. \qquad (11.95)$$

For antennas that are wide relative to the plasma wavelength, $k_{p2}\epsilon \gg 1$. Only those values of u and v, where $u^2 + v^2 \ll k_{p2}^2$, will contribute significantly to the integrals; u and v can be neglected relative to k_{p2}. The integrations become elementary in (11.90), and

$$Z_p \approx R_p = \frac{60\pi(1-\epsilon_2)}{\epsilon_2 k_0 l \epsilon k_{p2}}. \qquad (11.96)$$

It may be noted that (11.93) was first obtained by Hessel and Shmoys [1962], (11.91) was derived by Seshadri [1965a], and (11.87), (11.92), and (11.93) are given by Wait [1964b]. Balmain [1965] has considered a cylindrical dipole antenna of a finite radius r, and expresses the antenna impedance in terms of Bessel functions. For r small his resistance agrees with (11.92), and the reactance is the same as the leading term of (11.95) if the term (3/2) is replaced by $2 \log 2$. However, there are differences for r large. His resistance is proportional to $\cos^2(k_{p2}r - \pi/4)$ with a peak value $4/\pi$ times larger than R_p of (11.96). However the reactance X_p of the present calculations is negligible relative to R_p for $k_{p2}\epsilon \gg 1$ in (11.96).

11.3.4. *Numerical Results*

The antenna impedance has been computed for very thick plasma layers ($h \to \infty$). The antenna has a finite width ($\Omega = 2 \log 4l/\epsilon = 10$) and the plasma medium has slight losses ($\tan \delta_2 = 0.03$). For small values of ϵ_2, the antenna is thin relative to the acoustic wavelength ($k_{p2}\epsilon \approx 0.43$ for $\epsilon_2 = 0.01$) and the appropriate closed form approximations are indicated by heavy lines in Fig. 11.15. The finite plasma temperature tends to increase the antenna resistance and to decrease the capacitive reactance. Landau damping of the longitudinal plasma oscillations may become significant if k_B becomes comparable to ω_{p2}/u_2. Noting that $k_{p2} \sim k_B$, this damping will be small if $k_{p2} < \omega_{p2}/u_2$. This condition can be rearranged into $\omega < \sqrt{(2)}\omega_{p2}$ or $\epsilon_2 < 0.5$. The damping can be neglected for the smaller values of ϵ_2 shown in Fig. 11.15.

The further numerical calculations are restricted to a constant value of $\epsilon_2 = 0.03$, and the variation of the antenna impedance with the velocity ratio c/u_2 is shown in Fig. 11.16. The closed form approximations of $k_{p2}\epsilon \ll 1$ and $\gg 1$ apply to c/u_2 small and large, respectively, they are

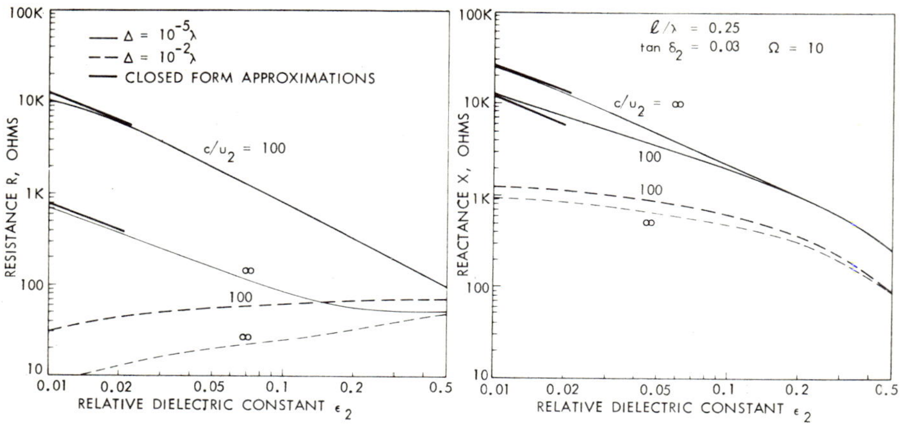

FIG. 11.15. Antenna impedance. Effects of dielectric constant.

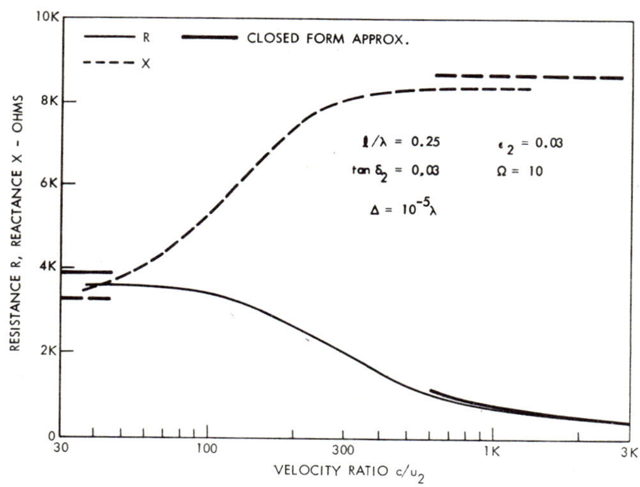

FIG. 11.16. Antenna impedance. Effects of acoustic velocity.

indicated by heavy lines in the figure. The finite width of the antenna can be neglected only if $k_{p2}\epsilon \ll 1$. Approximation of a finite antenna by a current filament will introduce considerable errors if $k_{p2}\epsilon \gg 1$, which corresponds to $c/u_2 > 1000$ in the present numerical examples.

The data shown in Figs. 11.15 and 11.16 refer to an antenna of a constant length $l = 0.5\lambda$, and the impedance $Z = R + iX$ calculated with a two-term trial function differs by less than a few percent from $Z_s = R_s + iX_s$ calculated for the sinusoidal field distribution. Further investigations show that the current ripple with wave number k_B has practically no effect on the antenna impedance if the antenna length $2l$ is an integer multiple of the wavelength of the surface waves, or if Re $k_B l = n\pi$, where n is an integer. The plasma wave components of the current distribution cause a maximum perturbation of the antenna impedance if $\mathrm{Re} k_B l = (n+0.5)\pi$. For a lossless plasma, the wave number k_B is real, and $k_B l = n\pi$ corresponds to a standing wave pattern with a null at the antenna center. Similarly, $k_B l = (n+0.5)\pi$ causes a standing wave pattern with a maximum at the antenna center, it will perturb the antenna current at the feed point, and it will affect the antenna impedance.

The antenna impedance is shown in Fig. 11.17 for $l = 8.5/\mathrm{Re} k_B$ with various amounts of plasma losses. R and R_s are nearly the same for tan δ_2

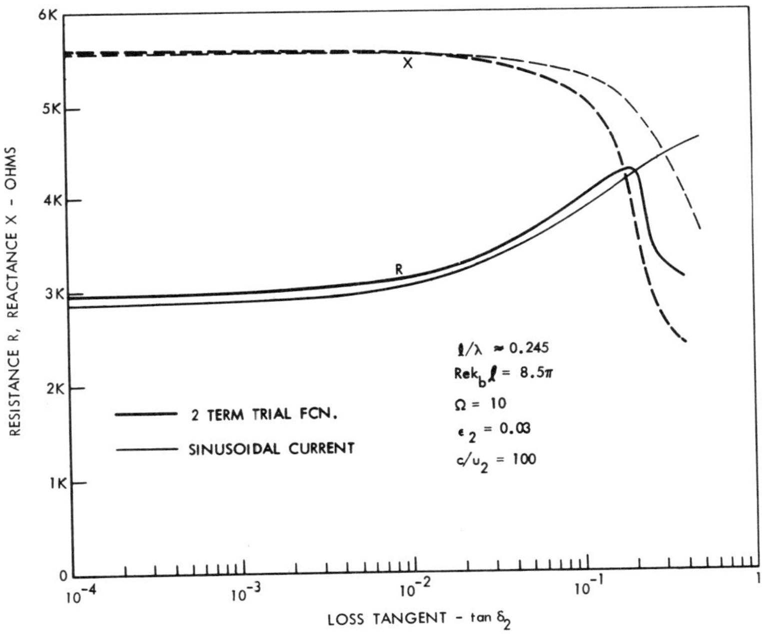

FIG. 11.17. Antenna impedance. Effects of plasma losses.

< 0.1. The amplitude ratio A/B of the two components of the antenna current distribution is computed from (2.38). It is nearly 500 in magnitude for small values of tan δ_2, and is approximately 160 for tan $\delta_2 = 0.3$. The antenna resistance is decreased by increasing thickness of the insulation Δ, as seen in Fig. 11.18. The oscillatory resistance peaks are observed also for a cold plasma for larger values of Δ. The impedance data shown in Fig. 11.18 have been computed for an assumed sinusoidal current distribution ($B = 0$ in (11.5)), and the use of a two-term trial function gives a change of less than 5 percent in the impedance figures.

Antenna impedance for plasma layers of various thicknesses h is shown in Fig. 11.19. The antenna impedance exhibits several high resonance peaks, with the largest peak occurring near $h = 0.028\lambda$, where $k_{p2}h \approx \pi$. These peaks may be attributed to the resonance of the plasma waves in the transverse dimension of the plasma slab, and the surface waves guided along the plasma slab have a considerable effect on the antenna impedance near these

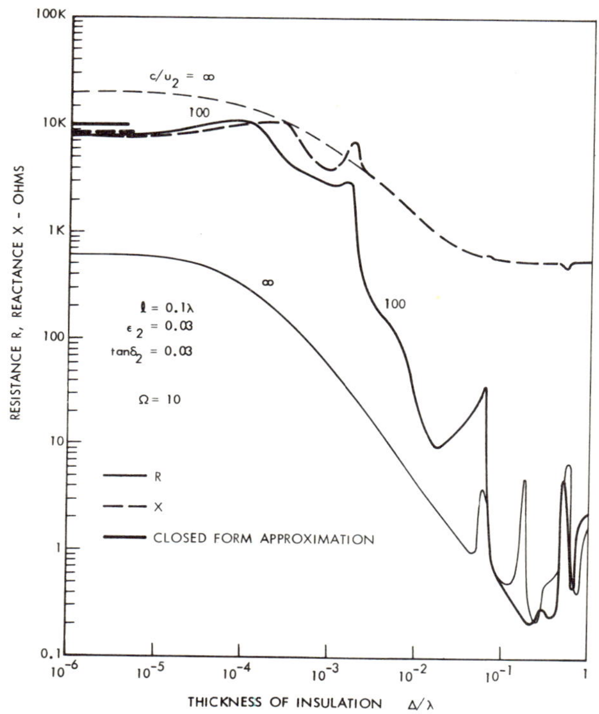

FIG. 11.18. Impedance of an insulated antenna.

resonances. The computed reactance X undergoes 7 sign changes in the h/λ interval between 0.03 and 0.1. However, this could not be shown in Fig. 11.19. The antenna impedance near the resonance peak of $k_{p2}h = \pi$ is shown in Fig. 11.20 for a gradual change of the antenna length. The impedance components R_S and X_S remain nearly constant, whereas the impedance components R and X exhibit significant changes. The two sets of computations give nearly the same impedance figures for $\text{Re}k_B l = n\pi$, but the largest differences in impedance occur near $\text{Re}k_B l = 10.5\pi$, where the amplitude A of the fundamental sine wave is less than the amplitude B of the surface waves.

The present impedance computations tend to support past work, which has been carried out by assuming that the antenna current distribution is the same as for cold plasma [Wait, 1964b; Seshadri, 1965a; Balmain, 1965]. On the other hand, calculations that neglect the presence of the electromagnetic waves and attribute the antenna current variations solely to plasma waves [Cook and Edgar, 1966] are not justified in view of the present results.

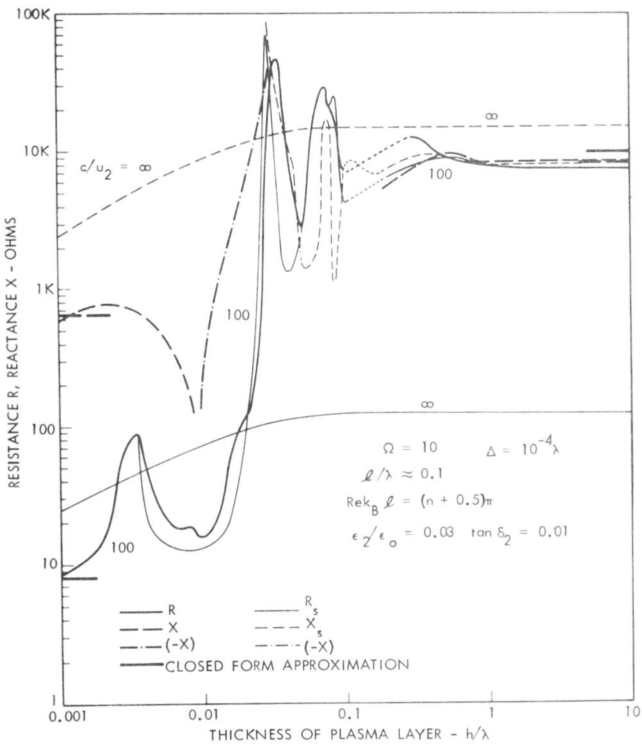

FIG. 11.19. Antenna impedance for plasma layers.

This behavior of a linear antenna is quite different from a wide slot antenna, where the compressibility of the plasma affects the antenna impedance only when considering the surface waves in a direction transverse to the antenna aperture (Section 11.2). The waveguide aperture is wide in terms of the plasma wavelengths, and the presence of a compressible plasma would have negligible effects on the impedance of a linear antenna of comparable dimensions. Furthermore the surface waves in the linear antenna case would perturb only the longitudinal current distribution, which has been shown to have small effects on the antenna impedance.

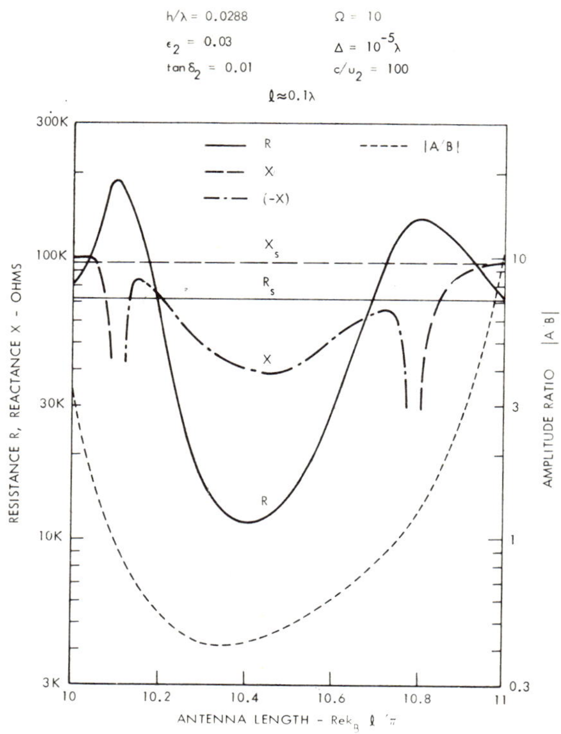

FIG. 11.20. Antenna impedance and current amplitudes for radiation in plasma layer.

The antenna of the present investigations was idealized as a current sheath. For thin insulating layers the surface waves supported in this antenna geometry are different from the waves near a perfectly conducting antenna. However, these differences become small for insulating layers of a thickness of the order of plasma wavelength [Galejs, 1965c]. It is anticipated that the conclusions of the present analysis will apply also to metallic antennas if the ion sheath is of such a thickness.

11.4. Exercises

11.1 The integral (11.35) which represents the power flow in a line source excited compressible plasma slab is evaluated after closing the contour of integration as shown in Fig. 11.4b. Show that the integration along the half circle gives zero.

11.2 Compute $\partial D_j/\partial k$ in (11.42) and derive the surface wave contribution to the power flow in P_{jk} in (11.43) and (11.44).

11.3 Show that the surface waves supported by a plasma slab in free space are symmetrical regardless of the wave number or plasma losses or that the fields near one of the plasma boundaries are of the same magnitude as the fields near the other boundary. Also show that the same amount of power flows from either side into the lossy plasma layer when it is surrounded by free space.

11.4 Derive the reflection coefficients R_{a1} of (11.48) and R_{b1} of (11.49) for the two region plasma geometry of Fig. 11.7. Derive (11.50) to (11.54) in the specialized cases of (a) non-ionized outer region; (b) ion layer deplete of electrons as the inner layer.

11.5 Derive the reflection coefficients R_{b1} of (11.74) for the linear antenna of Fig. 11.14.

11.6 Evaluate the integrals of (11.90) and derive the impedance components (11.91) to (11.96) for a linear antenna radiating into a homogeneous compressible plasma.

11.7 A parallel plate waveguide of height $y_i = 2b$ radiates into a plasma halfspace of $z \geqslant 0$. The isotropic plasma is characterized by a complex electric wave number k_e, and by a plasma wave number k_p, as defined in Section 11.2.1. The fields in the plasma are derived from (11.11) and (11.12). The boundary condition of $V_z = 0$ applies to the conducting plane and the waveguide aperture at $z = 0$. Aperture fields are assumed of the form (11.55) but with no x dependence. The complex power radiated per unit width of the waveguide into the plasma halfspace is given for $V_z = 0$ by the integral

$$P_z = -\int_{-\infty}^{\infty} E_y(y) H_x^*(y)\, dy.$$

The integrand has a simple pole for values of the transform variable given by (11.32). Show that

(a) the residue evaluation of the integral gives a real contribution to P_z for a lossless plasma;

(b) for a constant aperture field ($B = 0$) P_z becomes small if the slot is wide relative to the acoustic wavelength;

(c) the pole contribution to the integral is sizeable if $B \neq 0$ and k of the trial function is the same as k of the surface waves.

11.5. References

BACHYNSKI, M. P. (1966) Sources in unbounded plasmas, *Fifteenth General Assembly of URSI Munich, Germany.*

BALMAIN, K. G. (1965) Impedance of a short dipole in a compressible plasma, *NBS J. Res. (Radio Science)* **69D** (4), 559–66.

BALMAIN, K. G. (1966) Impedance of a radio frequency probe with an absorptive surface, *Radio Science* **1** (1), 1–12.

CARON, P. R. (1965) A study of the waves supported by a warm plasma slab, *NBS J. Res. (Radio Science)* **69D** (4), 729–33.

CHEN, KUN-MU (1964) Interaction of a radiating source with a plasma, *Proc. IEE* **3**, 1668–78.

COHEN, M. H. (1961) Radiation on a plasma. I. Cerenkov effect, *Phys. Rev.* **123**, 711–21.

COHEN, M. H. (1962a) Radiation in a plasma. II. Equivalent sources, *Phys. Rev.* **126**, 389–97.

COHEN, M. H. (1962b) Radiation in a plasma. III. Metal boundaries. *Phys. Rev.* **126**, 398–403.

COOK, K. R. and EDGAR, B. C. (1966) Current distribution and impedance of a cylindrical antenna in an isotropic compressible plasma, *Radio Science*, **1** (1), 13–19.

FEJER, J. A. (1964) Interaction of an antenna with a hot plasma and the theory of resonance probes, *NBS J. Res. (Radio Science)* **68D** (11), 1171–6.

GALEJS, J. (1965a) Admittance of a waveguide radiating into stratified plasma, *IEEE Trans. on Ant. and Prop.* **AP-13** (1), 64–70.

GALEJS, J. (1965b) Self and mutual admittance of waveguides radiating into plasma layers, *NBS J. Res. (Radio Science)* **69D** (2), 179–89.

GALEJS, J. (1965c) Propagation along an insulated cylindrical wire in a compressible plasma, *IEEE Trans. on Ant. and Prop.* **AP-13** (4), 645–6.

GALEJS, J. (1965d) Driving point impedance of linear antennas in the presence of a stratified dielectric, *IEEE Trans. on Ant. and Prop.* **AP-13** (5), 725–37.

GALEJS, J. (1966a) Slot admittance for compressible plasma layers, *Radio Science* **1** (4), 457–74.

GALEJS, J. (1966b) Impedance of an insulated linear antenna between layers of compressible plasma, *Radio Science* **1** (9), 1057–65.

GALEJS, J. (1966c) Guided waves supported by layers of lossy compressible plasma, *Can. J. Phys.* **44** (11), 2875–91.

HESSEL, A., MARCUVITZ, N and SHMOYS, J. (1962) Scattering and guided waves at an interface between air and a compressible plasma, *IRE Trans. on Ant. and Prop.* **AP-10**, 48–54.

HESSEL, A. and SHMOYS, J. (1962) Excitation of plasma waves in a homogeneous isotropic plasma by a dipole, *Proc. Symposium on Electromagnetic and Fluid Dynamics of Gaseous Plasma,* Polytechnic Press, Brooklyn, NY, pp. 173–84.

HOLT, E. H. and HASKEL, R. E. (1965) *Plasma Dynamics*, Macmillan, New York.

MAJUMDAR, S. K. (1960) Electrodynamics of a charged particle moving through a plasma without magnetic field, *Proc. Phys. Soc. (London)* **76**, 657–669.

OSTER, L. (1960) Linearized theory of plasma oscillations, *Rev. Mod. Phys.* **32**, 141–68.

SESHADRI, S. R. (1964) Radiation from an electromagnetic source in a half space of compressible plasma-surface waves, *IEEE Trans. on Ant. and Prop.* **AP-12** (3), 340–8.

SESHADRI, S. R. (1965a) Radiation from electromagnetic sources in plasma, *IEEE Trans. on Ant. and Prop.* **AP-13** (1), 79–88.

SESHADRI, S. R. (1965b) Infinite cylindrical antenna immersed in a warm plasma, *IEEE Trans. on Ant. and Prop.* **AP-13** (5), 789–99.

SESHADRI, S. R. (1965c) Propagation coefficient for the current distribution along a cylindrical antenna immersed in a warm plasma, *Proc. IEE (London)* **112** (5), pp. 877–82.

SESHADRI, S. R. (1965d) Radiation in a warm plasma from an electric dipole with a cylindrical column of insulation, *IEEE Trans. on Ant. and Prop.* **AP-13** (4), 613–29,

STEWART, G. E. and CARON, P. R. (1965) Radiation from a line source in a ground plane covered by a warm plasma slab, *IEEE Trans. on Ant. and Prop.* **AP-13** (4), 600–13.

TIDMAN, D. A. and BOYD, J. M. (1962) Radiation by plasma oscillations incident on a density discontinuity, *Physics of Fluids* **5** (2), 213–18.

WAIT, J. R. (1964a) Theory of a slotted-sphere antenna immersed in a compressible plasma, Part I and II, *NSB J. Res. (Radio Science)* **68D**, 1127–43.

WAIT J. R. (1964b) On radiation of electromagnetic and electroacoustic waves in plasma, *Appl. Sci. Res.*, Sect. B, **11**, 423–32.

WAIT, J. R. (1965a) On radiation of electromagnetic and electroacoustic waves in plasma, II, *Appl. Sc. Res.*, Sect. B, **12**, 130–8.

WAIT, J. R. (1965b) Scattering of electromagnetic and electroacoustic waves by a cylindrical object in a compressible plasma, *Radio Science* **69D** (2), 247–56.

WAIT, J. R. (1966) On the theory of wave propagation in a bounded compressible plasma, *Can. J. Phys.* **44**, 293–302.

WAIT, J. R. and SPIES, K. P. (1966) Theory of a slotted-sphere antenna immersed in a compressible plasma. III, *Radio Science*, **1**, (1), 21–26.

WAIT, J. R. (1967) Antennas in plasma, *Antenna Theory,* ed. by R. E. Collin and F. J. Zucker, McGraw-Hill, New York, NY.

WENGER, N. C. (1966) Radiation from a magnetic line source located on a conducting plane covered by a warm plasma layer, *Radio Science* **1** (5), 577–88.

WHALE, H. A. (1962) The impedance of an electrically short antenna in the ionosphere, *Proc. of the Intl. Conference on the Ionsphere*, London, Chapman & Hall, Ltd., London, pp. 472–7.

WHALE, H. A. (1963) The excitation of electroacoustic waves by antennas in the ionosphere, *J. Geophys. Res.* **68**, 415–22.

WHALE, H. A. (1964) Ion sheath effects near antennas radiating within the ionosphere, *J. Geophys. Res.* **69**, 447–55.

WUNSCH, A. D. (1967) Current distribution on a dipole antenna in a warm plasma *Electronics Letters* **3** (7), 320–321.

CHAPTER 12

LINEAR ANTENNAS IN A MAGNETOIONIC MEDIUM

RADIATION from point sources of electric and magnetic currents or dipoles in a magnetoionic medium is one of the topics which has been studied very extensively. Abraham [1953] has obtained an integral representation for the fields of point sources of electric and magnetic current, and he points out the utility of refractive index surfaces in the discussion of the far fields of the sources. Bunkin [1957] has investigated the radiation fields of an electric dipole in a magnetoionic medium, and Chow [1962] has extended this work to a gyro-electric-magnetic medium. Kuehl [1962] examined the radiation of the electric dipoles and points out that the radiation fields consist of several waves and they may become very large in certain directions. For high frequencies he has observed a similarity in radiation patterns with an isotropic medium, and for low frequencies there is an evidence of a guiding nature of the magnetostatic field. Clemmow [1963a] shows that for an infinite static magnetic field the fields radiated from an arbitrary oriented electric dipole can be related by a simple scaling procedure to the corresponding vacuum fields. Clemmow [1963b] describes the radiation of elementary sources in a magnetoionic medium as an angular spectrum of plane waves. He derives simple expressions for the elementary electric dipoles oriented in a direction normal to the infinite static magnetic field. Arbel and Felsen [1963] treat the radiation from arbitrary sources distributed in a planar stratified anisotropic medium. Their general solutions are specialized to the radiation of an electric dipole in an infinite homogeneous medium oriented with a dipole axis parallel to the magnetic field. They treat in great detail the asymptotic evaluation of the integral expressions of the fields and consider transition phenomena due to points of inflection or of unbounded branches of the refractive index surfaces. They have also obtained simple field expressions for the far fields of the dipoles in the presence of an infinite static magnetic field. Wu [1963] uses the saddle point method and computes the far fields of an electric and magnetic dipole oriented in directions parallel and perpendicular to the static magnetic field. He computes also the radiation patterns of an electric and magnetic dipole in the presence of a grounded plasma slab. Mittra and Deschamps [1963] have examined an elementary electric current source of arbitrary orientation. They obtained a closed form expression for the near fields, and also the integral expressions of the far fields are in a form suitable for numerical calculations. Deschamps and Kesler [1964] have examined the radiation fields from an arbitrary electric or magnetic current density. They show clearly the dependence of the far fields on the free space pattern of the antenna, on the shape of the dispersion surfaces, and on the polarity of the characteristic wave. Wait [1964] reviews the various approaches used for determining the electromagnetic field produced by an electric dipole in an anisotropic medium. He emphasizes that information on the radiation fields may be obtained directly from the shape of the refractive index surfaces. More recently Wait [1966] has examined the radiation from a spherical antenna with a thin anisotropic plasma sheath.

Excitation of plasma slabs by line sources has been also studied. Hodara and Cohn [1962] examine the radiation pattern of the line source in a plasma slab with a parallel static magnetic field using the stationary phase method. Ishimaru [1964] observes trapped surfaces waves which can exist above the interface between the plasma and a perfectly conducting plane. For the plasma slab of finite thickness the trapped surface waves create a leaky wave pole very close to the saddle point of the transformed field components, and the radiation pattern has a very sharp peak. The surface waves and radiation pattern have been examined in a similar geometry also by Seshadri and Pickard [1964] and by Meltz and Shore [1965]. Radiation from a finite slot with a perpendicular static magnetic field is investigated by Hodara [1963]. The signal transmission is enhanced for frequencies where the plasma is opaque in the absence of a static magnetic field, and the radiation pattern is not significantly altered. Radiation patterns of axial slots on plasma clad cylinders have been studied by Yeh [1964], deMarchin and Tyras [1965], and Chen and Cheng [1965].

Integral expressions for radiation resistance of elementary sources have been developed by Kogelnik [1960a, b], Kogelnik and Motz [1963], and Motz [1965]. In this approach the complex power radiated by a current distribution is computed by integrating the product of the source current density and of the electric field, similarly as was indicated in Section 1.5. The electric field is obtained by integrating the product of the source current density and the dyadic Green's function which is expressed by another integral. The source current densities are replaced by their spatial Fourier transforms, and formally the radiated power is computed by evaluating five threefold integrals. For a point source in an unbounded medium 12 of these integrals are trivial and two of the remaining integrals are evaluated in a closed form. The radiation resistance is expressed as a single integral, which must be evaluated over a finite range of integration. Based on these integral expressions Weil and Walsh [1964, 1965] have compiled extensive numerical results on radiation resistance of electric and magnetic dipoles; they observe frequency ranges where the radiation resistance of point sources becomes infinite. Seshadri [1965a] has exploited the cylindrical symmetry of the geometry and has obtained simple closed form expressions of the radiation resistance of a linear current filament in a uniaxial medium. Elementary dipoles, finite current filaments and finite loops have been also investigated [Seshadri, 1965b, 1966a; Seshadri and Tuan, 1965]. It is shown that the radiation resistance of finite current filaments or finite loops can be computed for all frequencies. Seshadri [1966b] shows that an insulating layer keeps the radiation resistance of an electric dipole finite for all frequencies. The impedance of a short current element has been considered using the quasi-static approximation by Bramley [1962], Kaiser [1962], and Balmain [1964]. The impedance of the cylindrical antenna with an arbitrarily directed static magnetic field has been computed by Ament *et al.* [1964] for frequencies above the plasma or cyclotron frequencies. Staras [1964] has computed the impedance of a short antenna with a volumetric current distribution, which is either parallel or perpendicular to the static magnetic field. Numerical results are presented for frequencies that are much lower than the plasma or cyclotron frequencies. Some of this work on radiation resistances has been criticized by Lee and Papas [1965]. They question the definition of Poynting's vector in a magnetoionic medium and maintain that the radiation resistance of an elementary dipole should remain finite. Also, they question the applicability of Parseval's theorem (or the interchange of the order of integrations), which is customarily used in impedance calculations. These arguments have been further discussed by Tai [1966], Staras [1966], Walsh and Weil [1966], and Lee and Papas [1966]. Based on certain symmetries of dyadic functions Staras [1966] shows that the method of Lee and Papas [1965] should give results identical to those of the conventional application of the Poynting's vector. These symmetries appear to be valid if the point of observation coincides with that of the source [$\mathbf{r} = \mathbf{r}'$ in equation (10) of Staras, 1966]. Also, it is generally recognized that only regions of wave propagation contribute to the radiation resistance in a lossless medium [Seshadri, 1965b].

From the circuit analogy, where resistance is defined as a ratio between the in-phase part of the voltage and the impressed current, Seshadri [1966c] has defined the radiation resistance of an electric point source as the ratio between the real part of the electric field and the dipole moment. These calculations do not require the use of the Parseval's theorem and give the same results as conventional calculations [Seshadri, 1965b]. In recent investigation Waineo and Mittra [1966] distinguish between input resistance and radiation resistance in a slightly lossy uniaxial medium. The input resistance is inversely proportional to the antenna length for low frequencies, and it is the same as the radiation resistance computed for a lossless medium [Seshadri, 1965a]. However, the radiated power is proportional to the squared antenna length at large distances, and the long antenna is the most effective radiator for large distances from the source. A very short antenna exhibits a large resistive component of its input impedance, which is analogous to the high input resistance of short antennas in isotropic dissipative media, but the far fields of the short antennas are very weak. A very large real component of the input resistance has been observed for frequencies ω below the plasma frequency ω_p by Stone *et al.* [1966]. In this frequency range there is a qualitative agreement between the experiment and the quasistatic theory of Balmain [1964]. The measured resistance exhibits a peak near the upper hybrid resonance and the reactance changes from large inductive to large capacitive values with increasing frequency. A number of further references on antenna properties in plasma media can be found in the comprehensive review paper by Bachynski [1966].

All of the previous work has been carried out for strictly assumed antenna current distributions. Galejs [1966a, b] has reported a variational impedance computation and he attempts to estimate the current distributions on the finite antenna. These calculations have been carried out for antennas parallel and perpendicular to the static magnetic field. The more general impedance expressions are compared in specialized cases with earlier results on linear current filaments [Seshadri, 1965a, b], with quasi stationary approximations of Balmain [1964], and also with impedance calculations of Ament *et al.* [1964] at high frequencies and with the work of Staras [1964] at low frequencies. The derived propagation coefficients of the antenna current distributions compare with earlier estimates of Ament *et al.* [1965] and with recent numerical results for an infinitely long cylindrical antenna [Lo and Lee, 1966]. These two papers [Galejs, 1966a, b] will be discussed in detail in this chapter and comparisons with pertinent past work will be also given. A brief mention is made of antennas in dielectric layers near a conducting ground plane [Galejs, 1967a]. Also discussed is the distance dependence of the power flow in a lossy uniaxial medium [Galejs, 1967b].

12.1. Cylindrical Antenna with an Axial Static Magnetic Field

In the cylindrical antenna geometry the antenna current is assumed to flow along the surface of a cylindrical shell of length $2l$ and radius a, which is embedded in the center of an infinitely long dielectric region of radius b within a magnetoionic medium and with the static magnetic field in a direction parallel to the antenna axis. The dielectric cylinder is assumed to be either an actual dielectric or to give some indication of the effects of an ion sheath around the antenna. Such a region, depleted of electrons, has been shown to affect the antenna characteristics, particularly for frequencies below the plasma frequency [Seshadri, 1966b]. The field expressions in the anisotropic medium are derived using one dimensional Fourier transforms.

Considerations are restricted to a slightly lossy medium, and the radiation condition requires only that the fields be exponentially attenuated at large distances from the sources. The fields are expressed in terms of Bessel functions of complex argument, and the antenna impedance is formulated after satisfying the boundary conditions at the surface of the current filament and at the interface between the dielectric and the plasma. The antenna impedance is computed using a two-term trial function for the antenna current. For most of the calculations, the trial functions are assumed to be two sine functions. For antennas of finite radius, the impedance is made available in numerical form only. However, closed form expressions are derived for vanishingly thin antennas in the low frequency limit, for short antennas in a uniaxial medium, and for half-wave antennas in free space.

12.1.1. *Field Expressions in the Cold Electron Plasma Medium*

Assuming a suppressed $\exp(-i\omega t)$ harmonic time dependence of the fields for a source free medium, the electric and magnetic vectors **E** and **H** satisfy the Maxwell's equations

$$\nabla \times \mathbf{E} = i\omega\mu_0 \mathbf{H}, \tag{12.1}$$

$$\nabla \times \mathbf{H} = -i\omega\epsilon_0 \epsilon \cdot \mathbf{E}, \tag{12.2}$$

where μ_0 and ϵ_0 are the permeability and permittivity of free space respectively. For a z-directed magnetic field, the dyadic permittivity is of the form traditionally used in the classical magnetoionic theory [Ratcliffe, 1959; Ginzburg, 1964].

$$\epsilon = \begin{bmatrix} \epsilon_1 & i\epsilon_2 & 0 \\ -i\epsilon_2 & \epsilon_1 & 0 \\ 0 & 0 & \epsilon_3 \end{bmatrix}; \tag{12.3}$$

where

$$\epsilon_1 = 1 + \frac{\omega_p^2(\omega + i\nu)/\omega}{\omega_c^2 - (\omega + i\nu)^2}, \tag{12.4}$$

$$\epsilon_2 = -\frac{\omega_p^2\omega_c/\omega}{\omega_c^2 - (\omega + i\nu)^2}, \tag{12.5}$$

$$\epsilon_3 = 1 - \frac{\omega_p^2}{\omega(\omega + i\nu)}; \tag{12.6}$$

where ν is the effective collision frequency; where the plasma frequency ω_p and the cyclotron frequency ω_c of electrons are defined in terms of the electron density N, charge e, mass m and the applied static magnetic induction B_0 as $\omega_p^2 = e^2N/(m\epsilon_0)$ and $\omega_c = |e|B_0/m$. Because of the symmetry of the excitation, the field components will be independent of ϕ, and are related to their Fourier transforms over the z coordinate by

$$F_{ij}(\rho, z) = \frac{1}{2\pi} \int_{-\infty}^{\infty} \bar{F}_{ij}(\rho, w) e^{iwz} \, dw, \tag{12.7}$$

where $F = E$ or H, $i = \rho$, ϕ, or z, and where subscripts j denote particular modes of propagation. Substitution of (12.7) into (12.1) and (12.2) gives the following relations between the transformed field components

$$i\omega\mu_0 \bar{H}_\rho = -iw\bar{E}_\phi, \tag{12.8}$$

$$i\omega\mu_0 \bar{H}_\phi = -\frac{\partial}{\partial\rho}(\bar{E}_z + iw\bar{E}_\rho), \tag{12.9}$$

$$i\omega\mu_0 \bar{H}_z = \frac{1}{\rho}\frac{\partial}{\partial\rho}(\rho\bar{E}_\phi) \tag{12.10}$$

$$-iw\bar{H}_\phi = -i\omega\epsilon_0(\epsilon_1\bar{E}_\rho + i\epsilon_2\bar{E}_\phi), \tag{12.11}$$

$$iw\bar{H}_\rho - \frac{\partial}{\partial\rho}\bar{H}_z = -i\omega\epsilon_0(-i\epsilon_2\bar{E}_\rho + \epsilon_1\bar{E}_\phi), \tag{12.12}$$

$$\frac{1}{\rho}\frac{\partial}{\partial\rho}(\rho\bar{H}_\phi) = -i\omega\epsilon_0\epsilon_3\bar{E}_z, \tag{12.13}$$

where the arguments (ρ, w) and the subscripts j have been omitted. The transverse field components of (12.8), (12.9), (12.11), and (12.12) can be expressed in terms of $\bar{E}_z$ and $\bar{H}_z$ as

$$\bar{E}_\rho = \frac{1}{D}\left[-(\epsilon_i k_0^2 - w^2)\, iw \frac{\partial}{\partial\rho}\bar{E}_z + \omega\mu_0\epsilon_2 k_0^2 \frac{\partial}{\partial\rho}\bar{H}_z\right], \tag{12.14}$$

$$\bar{E}_\phi = \frac{1}{D}\left[\epsilon_2 k_0^2 w \frac{\partial}{\partial\rho}\bar{E}_z + i\omega\mu_0(\epsilon_1 k_0^2 - w^2)\frac{\partial}{\partial\rho}\bar{H}_z\right], \tag{12.15}$$

$$\bar{H}_\rho = \frac{w}{\omega\mu_0 D}\left[-\epsilon_2 k_0^2 w \frac{\partial}{\partial\rho}\bar{E}_z - i\omega\mu_0(\epsilon_1 k_0^2 - w^2)\frac{\partial}{\partial\rho}\bar{H}_z\right], \tag{12.16}$$

$$\bar{H}_\phi = \frac{1}{i\omega\mu_0 D}\left\{[-D + w^2(\epsilon_1 k_0^2 - w^2)]\frac{\partial}{\partial\rho}\bar{E}_z + i\omega\mu_0\epsilon_2 k_0^2 \frac{\partial}{\partial\rho}\bar{H}_z\right\}, \tag{12.17}$$

where

$$D = \epsilon_2^2 k_0^4 - (\epsilon_1 k_0^2 - w^2)^2, \tag{12.18}$$

and $k_0 = \omega(\mu_0\epsilon_0)^{1/2}$. Substitution of (12.14) to (12.17) in (12.10) and (12.13) gives two coupled wave equations:

$$[-D + w^2(\epsilon_1 k_0^2 - w^2)]\mathscr{D}\bar{E}_z + \epsilon_2 k_0^2 w i\omega\mu_0 \mathscr{D}\bar{H}_z = k_0^2\epsilon_3 D\bar{E}_z, \tag{12.19}$$

$$\epsilon_2 k_0^2 w \mathscr{D}\bar{E}_z + i\omega\mu_0(\epsilon_1 k_0^2 - w^2)\mathscr{D}\bar{H}_z = i\omega\mu_0 D\bar{H}_z, \tag{12.20}$$

where the operator $\mathscr{D}$ is defined as

$$\mathscr{D} = \frac{1}{\rho}\frac{\partial}{\partial \rho}\left(\rho \frac{\partial}{\partial \rho}\right). \tag{12.21}$$

Substituting $\mathscr{D}\bar{H}_z$ of (12.20) into (12.19), $\bar{H}_z$ is related to $\bar{E}_z$ by

$$i\omega\mu_0\epsilon_2 w\bar{H}_z = \epsilon_1\mathscr{D}\bar{E}_z + \epsilon_3(\epsilon_1 k_0^2 - w^2)\bar{E}_z. \tag{12.22}$$

This equation for $\bar{H}_z$ can be differentiated and substituted in (12.19), which gives the following fourth order differential equation

$$a\mathscr{D}^2\bar{E}_z + b\mathscr{D}\bar{E}_z + c\bar{E}_z = 0, \tag{12.23}$$

where

$$a = \epsilon_1 k_0^2, \tag{12.24}$$

$$b = k_0^2[k_0^2(\epsilon_1^2 - \epsilon_2^2 + \epsilon_1\epsilon_3) - w^2(\epsilon_1 + \epsilon_3)], \tag{12.25}$$

$$c = -\epsilon_3 k_0^2 D. \tag{12.26}$$

After carrying out the differentiations indicated by the operator (12.21), the resulting equation can be exhibited as a sum of a fourth order differential equation and a second order differential equation (equations (4.26) and (2.162.1a) respectively of Kamke [1948]). Both differential equations have the Hankel functions of order zero $H_0^{(m)}(\beta\rho)$ as common solutions ($m = 1$ or 2), if β^2 satisfies the quadratic equation

$$a\beta^4 - b\beta^2 + c = 0. \tag{12.27}$$

The arguments of the Hankel functions for both differential equations become identical and the solutions of (12.23) have the form

$$\bar{E}_z = B(w)H_0^{(m)}(\beta\rho). \tag{12.28}$$

This result can be verified by direct substitution of (12.28) in (12.23) when noting that $\mathscr{D}H_0^{(m)}(\beta\rho) = -\beta^2 H_0^{(m)}(\beta\rho)$. The solutions of (12.27) are given by

$$\beta^2 = \frac{b \pm (b^2 - 4ac)^{1/2}}{2a}. \tag{12.29}$$

For $w = 0$, the above solutions β correspond to purely radial wave propagation in a direction transverse to the z axis. The two solutions of (12.29) are now simplified to

$$\beta^2 = \begin{cases} \epsilon_3 k_0^2 \\ (\epsilon_1^2 - \epsilon_2^2)k_0^2/\epsilon_1 \end{cases}. \tag{12.30}$$

Following a convention used in the past by Arbel and Felsen [1963], the solution with a plus sign in (12.29), which gives $\beta^2 = \epsilon_3 k_0^2$ for $w = 0$, is denoted as an ordinary wave (subscript $j = +$) and the other solution is denoted as the extraordinary wave ($j = -$). The $\beta = 0$ solutions of (12.28) represent plane waves propagating in the z direction with wave numbers

$$w_{\pm} = k_0(\epsilon_1 \pm \epsilon_2)^{1/2}. \tag{12.31}$$

In the presence of strong magnetic fields, $\epsilon_2 \to 0$ and the anisotropic medium becomes uniaxial with $\epsilon_3 \neq \epsilon_1$. The two solutions of (12.29) become

$$\beta_+^2 = \epsilon_3(k_0^2 - w^2/\epsilon_1), \tag{12.32}$$

$$\beta_-^2 = \epsilon_1 k_0^2 - w^2. \tag{12.33}$$

Equations (12.32) and (12.33) apply also to isotropic media when $\epsilon_2 = 0$ and $\epsilon_3 = \epsilon_1$. The ordinary wave (12.32) corresponds now to TM field components that satisfy the wave equation (12.19). The extraordinary wave (12.33) has TE field components and satisfies the wave equation (12.20). Obviously there is no coupling between TE and TM modes when $\epsilon_2 = 0$. For a lossy plasma, β^2 will generally be complex, and the sign of β can be determined from the requirement that field components should decay for large values of ρ. $\text{Im}(\beta)$ should be positive if Hankel functions of the first kind ($m = 1$) are used for the solutions (12.27).

In an anisotropic medium with finite magnetic field, $\epsilon_2 \neq 0$ and the field components $\bar{E}_{zj}$ and $\bar{H}_{zj}$ for a given mode j are coupled. The coupling is characterized by the admittance

$$Y_j = \frac{\bar{H}_{zj}}{\bar{E}_{zj}}. \tag{12.34}$$

It is computed after substituting (12.28) in (12.22) as

$$Y_j = -\frac{1}{i\omega\mu_0}\frac{\epsilon_1(\beta_j^2 - \epsilon_3 k_0^2) + \epsilon_3 w^2}{\epsilon_2 w}. \tag{12.35}$$

Substituting (12.29) into (12.35) and letting $\epsilon_2 \to 0$, $Y_+ \sim \epsilon_2^2/\epsilon_2 \to 0$, and $\bar{H}_{z+} = 0$ due to an excitation by $\bar{E}_{z+}$. Similarly, it is seen that $|Y_-| \to \infty$ and $\bar{E}_{z-} = 0$ due to an excitation by $\bar{H}_{z-}$, as can be also anticipated from the wave equations (12.19) and (12.20).

The antenna is approximated by a cylindrical current sheet of surface density J_z of radius a and total length $2l$. The transform of the current density $J_z(\rho, z)$ is denoted as $\bar{J}_z(\rho, w)$. This current sheet is embedded in an infinitely long dielectric cylinder of radius b and of relative dielectric constant ϵ_1.

The dielectric cylinder is surrounded by an electron plasma with static magnetic field impressed in the direction of the cylinder axis (z direction). The axial electric field components are assumed to be proportional to $J_0(\beta_i\rho)$, $H_0^{(m)}(\beta_i\rho)$ with $m = 1$ and 2 and to $H_0^{(1)}(\beta_j\rho)$ in the three regions $0 \leqslant \rho \leqslant a$, $a \leqslant \rho \leqslant b$, and $b \leqslant \rho$ respectively. The axial magnetic field components are proportional in $J_0(\beta_i\rho)$ for $0 \leqslant \rho \leqslant b$ and to $H_0^{(1)}(\beta_j\rho)$ for $b \leqslant \rho$. In the above expressions $\beta_i = \sqrt{(\epsilon_i k_0^2 - w^2)}$ and β_j is given by (12.29), $j = +$ or $-$, and $J_n(x)$ is the Bessel function of the first kind of order n. The field components $\bar{E}_{\phi 1}$, $\bar{E}_{\phi 2}$, $\bar{H}_{\phi 1}$ and $\bar{H}_{\phi 2}$ are obtained also from (12.15) and (12.17) by noting that $\epsilon_2 = 0$ and $\epsilon_1 = \epsilon_3 = \epsilon_i$ in the dielectric region. The boundary conditions $\bar{H}_{\phi 2} - \bar{H}_{\phi 1} = \bar{J}_z$ and continuous $\bar{E}_{z1} = \bar{E}_{z2}$ are applied at $\rho = a$, and the continuity of E_ϕ, H_z, H_ϕ and E_z is postulated at $\rho = b$. The antenna impedance will be specified in Section 3 in terms of the axial electric field $\bar{E}_z$ at the surface of the cylindrical current filament. A straightforward, but lengthy algebraic manipulation shows that E_{z2} at $\rho = a$ is related to $\bar{J}_z$ by

$$\frac{\bar{E}_{z2}}{\bar{J}_2} = F(w) \tag{12.36}$$

where $F(w)$ is a complicated function involving combinations of cylindrical functions [Galejs, 1966a].

12.1.2. *Antenna Impedance*

The driving point impedance of a cylindrical antenna that carries a z-directed current may be computed from (1.35) which gives

$$Z = -\frac{2\pi a \int E_z J_z \mathrm{d}z}{[I(z=0)]^2}, \tag{12.37}$$

where E_z is computed at the surface of the current filament. Z of (12.37) is stationary with respect to small changes of the surface current density J_z, or of the current $I(z)$ about its correct value following Section 1.5. J_z is assumed to be an even function about $z = 0$. Substituting the Fourier integral representation (12.7) for $E_z(z)$ and interchanging the order of integrations, it follows that

$$Z = -\frac{a}{[I(z=0)]^2} \int_{-\infty}^{\infty} \mathrm{d}w\, \bar{E}_z(w) \int_{-\infty}^{\infty} J_z(z) \cos wz\, \mathrm{d}z,$$

$$= -\frac{a}{[I(z=0)]^2} \int_{-\infty}^{\infty} \mathrm{d}w\, F(w) \Big[\int_{-\infty}^{\infty} J_z(z) \cos wz\, \mathrm{d}z \Big]^2, \tag{12.38}$$

where $F(w)$ is defined by (12.36). The antenna current density is assumed to be representable by the trial function

$$J_z(z) = A \sin [k_A(l-|z|)] + B \sin [k_B(l-|z|)]. \tag{12.39}$$

Antenna impedances in free space has been calculated using such trial functions in Section 2.3. In a magnetoionic medium, little is known regarding the proper values of k_A and k_B. The form (12.39) was considered to provide a larger variety of current distributions during the search for optimum values of the wave numbers k_A and k_B than the sine and shifted cosine functions of (9.7). The accuracy limitations of this formulation are discussed following equation (12.43). Substituting (12.39) into (12.38), it follows that

$$Z = \frac{A^2\gamma_{AA} + 2AB\gamma_{AB} + B^2\gamma_{BB}}{(AF_A + BF_B)^2}, \tag{12.40}$$

where

$$F_N = \sin k_N l, \tag{12.41}$$

$$\gamma_{NM} = -\frac{2}{\pi^2 a}\int_0^\infty dw\, F(w) g_N(w) g_M(w), \tag{12.42}$$

$$g_N(w) = \int_0^l \sin [k_N(l-z)] \cos wz\, dz$$

$$= \frac{k_N}{k_N^2 - w^2}(\cos wl - \cos k_N l), \tag{12.43}$$

with $N, M = A$ or B.

The impedance (12.40) is a function of the amplitude ratio of the two trial functions (A/B). Because of the stationary character of (12.37), the optimum ratio (A/B) is determined from the condition $dZ/d(A/B) = 0$. This leads to the expressions (2.38) to (2.41).

The impedance formulation (2.39), which uses the antenna trial functions (12.39), becomes inaccurate for very short antennas when the two terms of the current distribution are linearly related (i.e. $\sin k_B z = (k_B/k_A) \sin k_A z$ within the computer accuracy). Inaccuracies can be also anticipated for longer antennas if F_A and F_B approach zero simultaneously. Sine and shifted cosine functions of (9.7) have been used in these cases ($\omega > 1.5\omega_p$ in Figs. 12.3–12.5; $\omega \ll \omega_p$ in Fig. 12.7).

12.1.3. *Impedance of a Thin Antenna without a Dielectric Layer*

For a thin antenna without an insulating shell, $b \to a$. In the limit of $a \to 0$, it follows that $F(w)$ of (12.36) can be expressed as

$$F(w) = d_+ \frac{H_0^{(1)}(\beta_- a)}{H_1^{(1)}(\beta_- a)} - d_- \frac{H_0^{(1)}(\beta_+ a)}{H_1^{(1)}(\beta_+ a)}, \tag{12.44}$$

where

$$d_j = \frac{i\omega\mu_0 b_{1j}}{b_{1+}b_{2-} - b_{1-}b_{2+}}, \tag{12.45}$$

and b_{1j} and b_{2j} are defined by

$$b_{1j} = \frac{\beta_i \beta_j}{i\omega\mu_0 D} \left[i\omega\mu_0 (\epsilon_1 k_0^2 - w^2) Y_j + \epsilon_2 k_0^2 w \right], \tag{12.46}$$

$$b_{2j} = \frac{\beta_j}{D} \left[k_0^4 (\epsilon_2^2 - \epsilon_1^2) + \epsilon_1 k_0^2 w^2 - i\omega\mu_0 \epsilon_2 k_0^2 w Y_j \right], \tag{12.47}$$

12.1.3.1. *Quasistationary approximations.* In the low frequency limit, $k_0 \to 0$, and k_0 will be assumed to be negligible relative to the integration variable w in all of the previous expressions. The wave numbers β_j^2 simplify to $\beta_+^2 = -w^2\epsilon_3/\epsilon_1$ and $\beta_-^2 = -w^2$, which makes $d_+ = 0$ and

$$d_- = \omega\mu_0 w / (\epsilon_3 \epsilon_1 k_0^4)^{1/2}.$$

The antenna impedance is computed from (2.41), (12.41), and (12.42) with k_A of the trial functions approaching zero. This results in

$$Z_s = -\frac{2}{\pi^2 a l^2} \int_0^\infty \mathrm{d}w\, F(w) \left(\frac{1 - \cos wl}{w^2} \right)^2. \tag{12.48}$$

Substituting the above values of d_+ and d_- in $F(w)$ of (12.44), and applying the small argument approximations of the Hankel functions,

$$Z_s = -\frac{2i}{\omega\epsilon_0\epsilon_1\pi^2 l^2} \int_0^\infty \frac{\mathrm{d}w}{w^2} (1 - \cos wl)^2 \cdot \left[\log w + \log \left(\frac{a}{2} \left[\frac{\epsilon_3}{\epsilon_1} \right]^{1/2} \right) + C \right], \tag{12.49}$$

where $C = 0.57721\ldots$ is Euler's constant. The integral can be evaluated to give

$$Z_s = \frac{i}{\omega\epsilon_0\epsilon_1\pi l} \left[\log \left(\frac{l}{a} \right) - 1 + \frac{1}{2} \log \frac{\epsilon_1}{\epsilon_3} \right]. \tag{12.50}$$

Z_s of (12.50) is the same as the quasistationary approximation (63) of Balmain [1964], or (4.13) of Blair [1964]. It will have a resistive component even for a lossless medium if ϵ_1 and ϵ_3 have different signs.

12.1.3.2. *A short antenna in a uniaxial medium* ($k_0 l \ll 1$). In a uniaxial medium, $\epsilon_1 = 1$ and $\epsilon_2 = 0$, and the substitution of β_j^2 of (12.32) and (12.33) into (12.45) shows that $d_+ = 0$. $F(w)$ of (12.44) is simplified to

$$F(w) = \frac{i\omega\mu_0 (k_0^2 - w^2)}{k_0^2 \beta_+} \frac{H_0^{(1)}(\beta_+ a)}{H_1^{(1)}(\beta_+ a)}, \tag{12.51}$$

Substituting (12.51) in the integral of (2.42), assuming $k_A = k_0$ and $k_0 l \ll 1$, and using the small argument approximations of the Hankel functions, the real part of the impedance Z_s is computed from (2.41) as

$$R_s = \begin{cases} 20\,(k_0 l)^2 & \text{for} \quad \omega > \omega_p \\ 60\pi/(k_0 l) & \text{for} \quad \omega < \omega_p. \end{cases} \tag{12.52}$$

This expression for the resistance (12.52) has been obtained also by Seshadri [1965a]. The imaginary part of Z_s is computed from (12.51), (12.42), and (2.41) as

$$X_s = \frac{i}{\omega\epsilon_0 \pi l}\left[\log\left(\frac{l}{a}\right) - 1 - \frac{1}{2}\log|\epsilon_3|\right], \tag{12.53}$$

which is valid for short antennas at all frequencies.

The quasistatic approximation (12.50) gives the same R_s and X_s values as (12.52) and (12.53) for $\omega < \omega_p$. Mittra [1965] observes that the quasi static-field expressions are independent of the off-diagonal terms ϵ_2 of the permittivity matrix (12.3). The impedance expression of an antenna in a uniaxial medium should be equally valid in a general medium under the same range of approximation.

12.1.3.3. *The half-wave antenna in free space.* In free space $\beta_+^2 = k_0^2 - w^2$. Applying the small argument approximation of $H_1^{(1)}(x)$, (12.51) simplifies to

$$F(w) = -\frac{\pi a \omega \mu_0}{2k_0^2}(k_0^2 - w^2)\, H_0^{(1)}[(k_0^2 - w^2)^{1/2} a]. \tag{12.54}$$

For a half-wave antenna, $l = \lambda/4 = \pi/(2k_0)$, k_A of the trial functions (12 39) is set equal to k_0, and the antenna impedance is computed from (2.41). This results in

$$Z = \gamma_{AA} = \frac{\omega\mu_0}{\pi}\int_0^\infty \frac{dw \cos^2\left(\frac{\pi}{2}\frac{w}{k_0}\right)}{k_0^2 - w^2} H_0^{(1)}[(k_0^2 - w^2)^{1/2} a]. \tag{12.55}$$

In the limit of $a \to 0$, the real part of (12.55) is given by

$$\mathrm{Re}Z = \frac{\omega\mu_0}{\pi}\int_0^k \frac{dw \cos^2\left(\frac{\pi}{2}\frac{w}{k_0}\right)}{k_0^2 - w^2}, \tag{12.56}$$

which is equal to 73.1 ohms following equations (9.20) and (9.21). The imaginary part of (12.55) is computed as

$$\mathrm{Im}Z = \frac{\omega\mu_0}{\pi}\int_0^k \frac{dw \cos^2\left(\frac{\pi}{2}\frac{w}{k_0}\right)}{k_0^2} Y_0[(k_0^2-w^2)^{1/2}a]$$

$$+\frac{2\omega\mu_0}{\pi^2}\int_k^\infty \frac{dw \cos^2\left(\frac{\pi}{2}\frac{w}{k_0}\right)}{w^2-k_0^2} K_0[(w^2-k_0^2)^{1/2}a], \qquad (12.57)$$

where $K_0(x)$ is the modified Bessel function of second kind and of order zero. In the limit of $a \to 0$, small argument approximation of the Bessel functions can be applied to both integrals, and (12.57) simplifies to

$$\mathrm{Im}Z = \frac{\omega\mu_0}{\pi^2}\int_0^\infty \frac{dw \cos^2\left(\frac{\pi}{2}\frac{w}{k_0}\right)}{k_0^2-w^2}$$

$$\cdot\left[\log\left|1-\frac{w}{k_0}\right|+\log\left(1+\frac{w}{k_0}\right)\right]. \qquad (12.58)$$

This is equal to −42.5 ohms following equation (9.23). The numerical values of (12.56) and (12.58) may be recognized as standard results of antenna theory, which have been also listed in Section 2.1.

12.1.4. *Discussion of Numerical Results*

12.1.4.1. *Uniaxial medium.* The magnetoionic plasma degenerates into a uniaxial medium with $\epsilon_1 = 1$ and $\epsilon_2 = 0$ in the presence of strong magnetic fields ($\Omega_c = \omega_c/\omega_p \to \infty$). For such a medium, the antenna impedance is shown in Fig. 12.1, where the left-hand scale of the resistance applies for $\omega < \omega_p$ and the right-hand scale for $\omega > \omega_p$. The calculations are made for antennas of length $\lambda/20$ and $\lambda/10$ (indexes 1 and 2) and for very small ($b-a < 10^{-4}\lambda$) and for larger ($b-a = \lambda/100$) radii of an insulating layer around the antenna (indexes p and i), where λ is the free space wave length.

The antenna radius a and length $2l$ are such that $\Omega = 2\log(2l/a) = 8$. The medium has slight losses with an average collision frequency $\nu = 0.001\omega_p$. The resistance is nearly constant for $\omega > \omega_p$, increases suddenly near $\omega = \omega_p$, and reaches its maximal value for $\omega < \omega_p$. The resistance decreases gradually for lower values of ω or higher values of ω_p. The resistance is proportional to l^2 for $\omega > \omega_p$ and is inversely proportional to l for $\omega < \omega_p$. The reactance is capactive for the frequency range indicated in Fig. 12.1. For a lossless plasma, it becomes logarithmically infinite at $\omega = \omega_p$. The reactance is decreasing for higher values of ω_p. and will become inductive for sufficiently large values of ω_p. The same general impedance behavior is

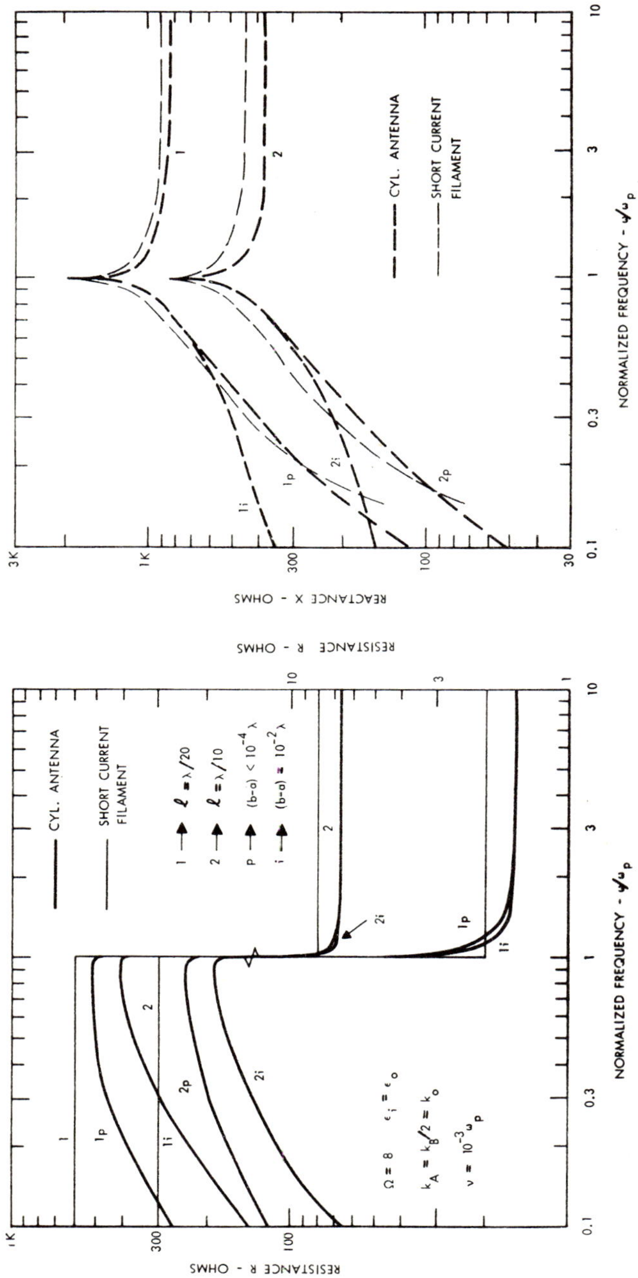

FIG. 12.1. Impedance of a short antenna in a uniaxial medium.

also observed from the approximate impedance expressions (12.52) and (12.53), which are indicated by thin lines in Fig. 12.1. An insulating layer around the antenna of $b-a=\lambda/100$ (index i in Fig. 12.1) has little effect for frequencies $\omega > \omega_p$, but tends to decrease the resistance and make the reactance more capacitive for $\omega < \omega_p$.

12.1.4.2. *Antenna resistance in a magnetoionic medium.* The antenna impedance can be calculated from (2.41) for an assumed sinusoidal current distribution, which becomes triangular for short antennas. The real part of the computed impedance is shown in Fig. 12.2 for an antenna of length

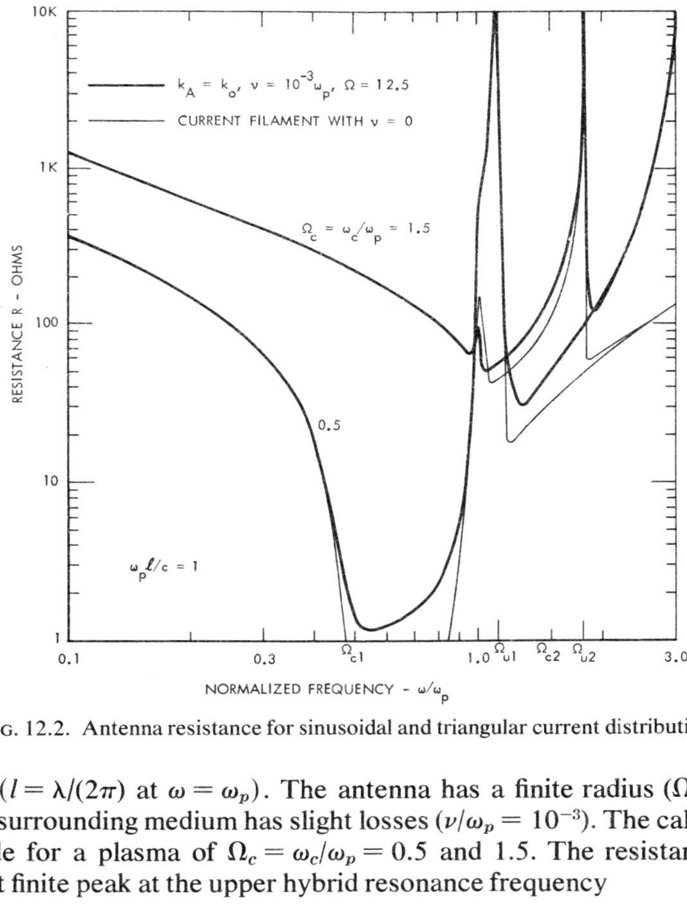

FIG. 12.2. Antenna resistance for sinusoidal and triangular current distributions.

$l=c/\omega_p (l=\lambda/(2\pi)$ at $\omega=\omega_p)$. The antenna has a finite radius ($\Omega=12.5$), and the surrounding medium has slight losses ($\nu/\omega_p=10^{-3}$). The calculations are made for a plasma of $\Omega_c=\omega_c/\omega_p=0.5$ and 1.5. The resistance has a large but finite peak at the upper hybrid resonance frequency

$$\omega_u=(\omega_c^2+\omega_p^2)^{1/2},$$

which is indicated in Fig. 12.2 as $\Omega_u=\omega_u/\omega_p$. The radiation resistance becomes large for small and large values of ω/ω_p. The radiation resistance of a current filament with an assumed triangular current distribution has been calculated recently by Seshadri [1965b] for a lossless medium. These data are indicated by thin lines in Fig. 12.2, and agree with the present calculation, in particular for the lower frequencies. However for a lossless medium and $\omega_c < \omega_p(2)^{1/2}$, the resistance is equal to zero in the frequency range $\omega_c < \omega < \omega_1$, where

$$\omega_1 = 0.5[-\omega_c + (\omega_c^2 + 4\omega_p^2)^{1/2}].$$

Also, the resonance peak observed for $\omega_c = 1.5\omega_p$ at $\omega = \omega_p$ is higher for a lossless medium. The two sets of calculations differ significantly for higher frequencies where the triangular current distribution becomes inaccurate. The assumed sinusoidal current gives a high resistance for $l \approx \lambda/2$ (or near $\omega = 3\omega_p$).

12.1.4.3. *Antenna impedance.* The antenna impedance is computed next using a two-term trial function (12.39) with $k_A = k_B/2 = k_0$. Such trial functions can be seen to be adequate for impedance calculations in free space for antenna lengths $l < 0.35\lambda$ in Fig. 2.2; they will also indicate the frequency ranges where the electrical antenna length becomes large in a magneto-ionic medium and where further resonances may be expected. The calculations of Fig. 12.3 are made for the normalized gyro frequency $\Omega_c = \omega_c/\omega_p = 0.5$. An additional resonance is observed for frequencies ω somewhat less than ω_c if the insulating layer is of negligible thickness $(b-a)$. Increasing losses $(\nu/\omega_p = 10^{-2})$ will decrease this resonance peak. The losses also increase the resistance in the frequency range $\omega_c < \omega < \omega_1$, where the resistance is zero for a lossless plasma. An insulating layer (or ion sheath) of a thickness $b-a = \lambda/100$ shifts the resonance to frequencies $\omega > \omega_c$. Also it makes the reactance more capacitive for $\omega < \omega_u$ and slightly more inductive for $\omega > \omega_u$. The antenna resistance is decreased for $\omega \ll \omega_p$. A similar impedance behavior is also seen in Fig. 12.4 for $\Omega_c = 1.5$. The additional resonance peak is observed for $\omega < \omega_c$ if the insulating layer is thin, and the resonance peak is shifted closer to the frequency ω_c for increased thickness of the insulating layers.

The presence of the resonance peak near ω_c indicates that the antenna becomes electrically long as ω approaches ω_c. Under these conditions ϵ_1 approaches infinity and the resonance can be explained possibly by a propagation coefficient of the current distribution proportional to the $(\epsilon_1)^{1/2}$, following the suggestion of Ament *et al.* [1965]. Further calculations are therefore made using the trial functions $k_A = k_s$, $k_B = 2k_s$, where $k_s = (\text{Re}\,\epsilon_1)^{1/2}k_0$. The impedance curves shown in Fig. 12.5 exclude the frequency range $\omega_c < \omega < \omega_u$ (where $\text{Re}\,\epsilon_1$ is negative) because the present computer program is intended for real coefficients k_A and k_B only in (12.41) and (12.43). The resonance peaks are higher than in Figs. 12.3 and 12.4, and there are

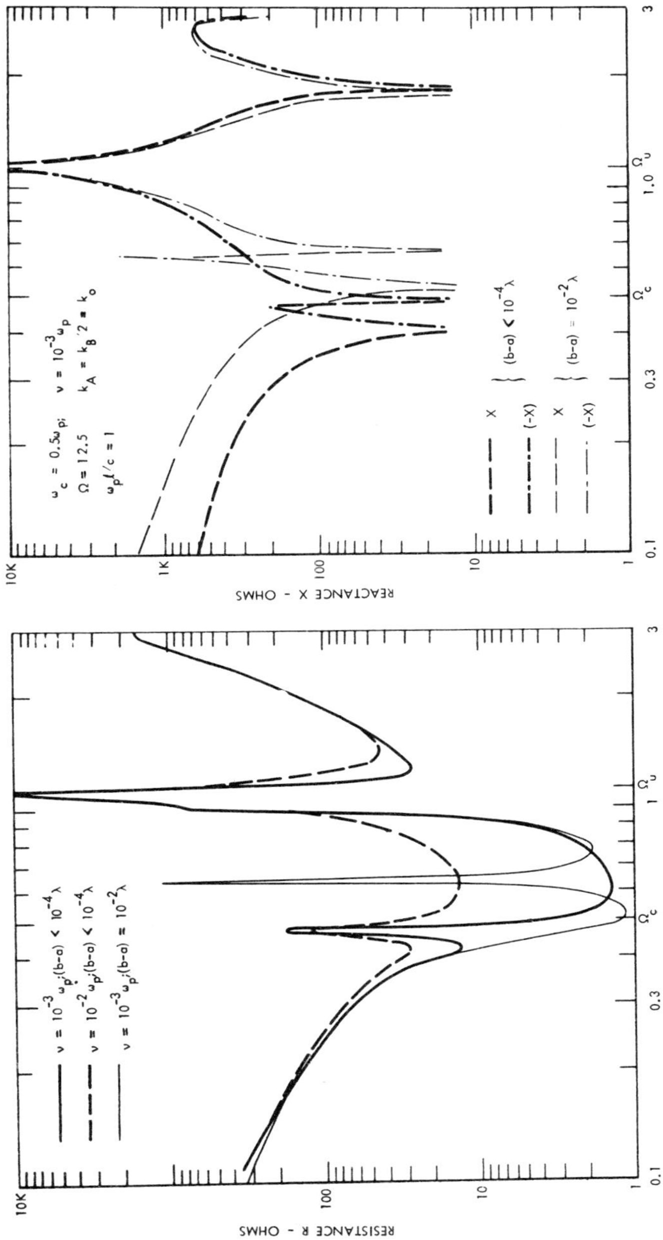

FIG. 12.3. Antenna impedance using the trial functions $k_A = k_B/2 = k_0$; $\Omega_c = \omega_c/\omega_p = 0.5$.

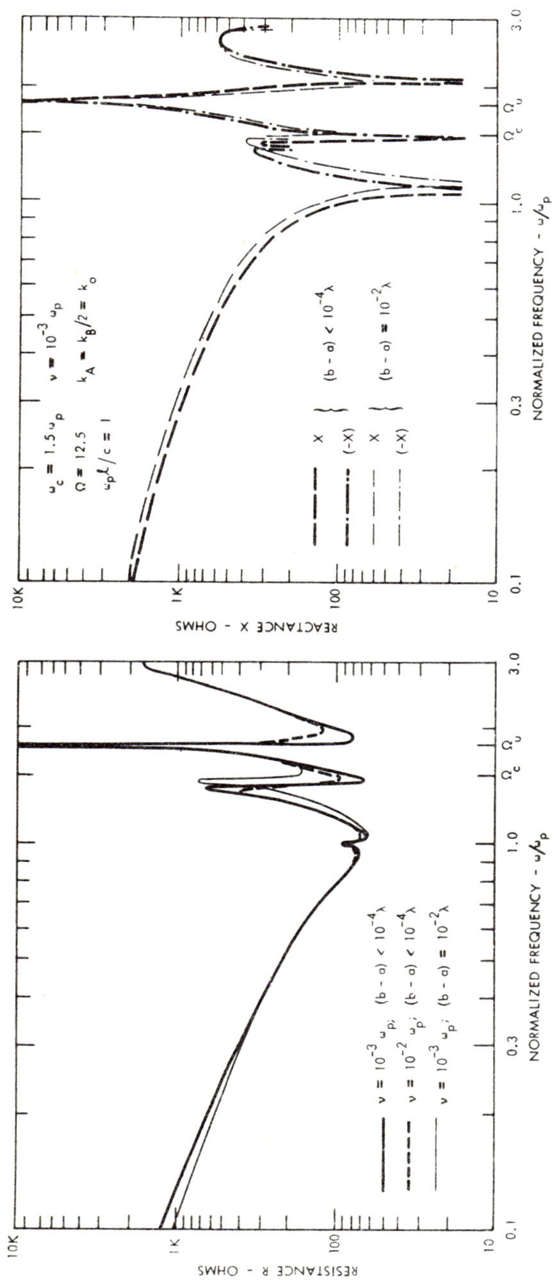

FIG. 12.4. Antenna impedance using the trial functions $k_A = k_B/2 = k_0$; $\Omega_c = \omega_c/\omega_p = 1.5$.

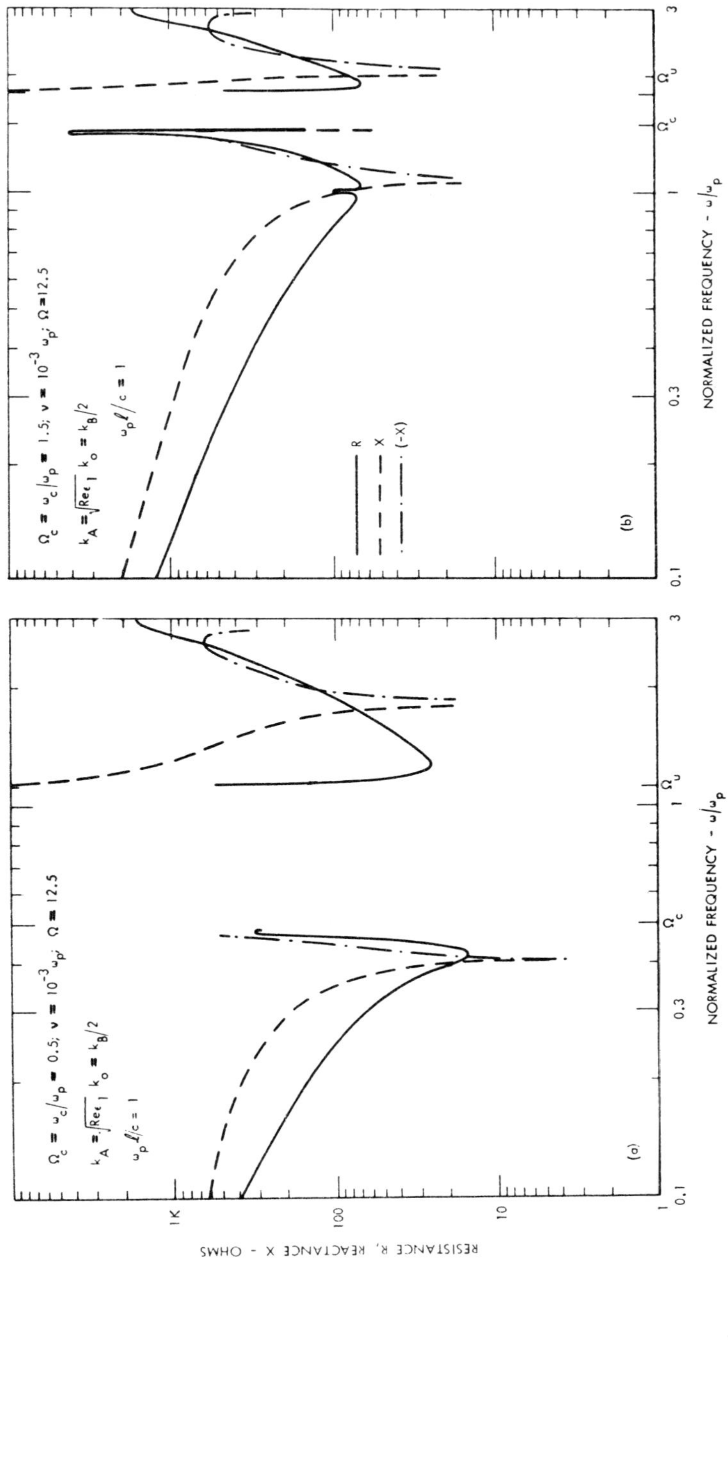

FIG. 12.5. Antenna impedance using the trial functions $k_A = k_B/2 = \sqrt{\text{Re}\epsilon_1}\, k_0$.

also further resonances in the immediate vicinity of ω_c, which could not be shown in the frequency scale used in Fig. 12.5. The impedance computed from two-term trial functions is also nearly the same as the computations based only on $k_A = k_s$ from equation (2.41). The ratio of the trial function amplitudes A/B is computed from (2.38). The magnitude of A/B is of the order 5–10, and this indicates that the predominant term of the antenna current representation (12.39) has a wave number $k_A = k_s$. Further calculations have been made using $k_A = k_s$ and $k_B = k_s/2$. Again the magnitude of A/B is larger than unity, and the impedance is nearly the same as for $k_B = 2k_s$ shown in Fig. 12.5.

The calculations of Figs. 12.6a and b are made for frequencies near the resonance peak $\omega < \omega_c$. For small values of $(b-a)$, the two sets of calculations using $k_A = k_s$ and $k_B = 2k_s$ (dashed curve) or $k_B = 0.5k_s$ (dashed and dotted curve) give essentially similar current distributions. The calculations using free space wave number $k_A = k_0$ and $k_B = 2k_0$, indicated by solid curves, approximate the other current distributions more closely in Fig. 12.6b because ϵ_1 of (12.4) is smaller in magnitude for larger values of ω_c, and k_s differs less from k_0 for ω_c large. The thin line denotes a simple sine wave with a wave number k_s.

For an insulating layer of thickness $(b-a) = \lambda/100$, the current distribution is quite different. It is nearly triangular for $\omega_c/\omega_p = 0.5$, and has a relatively low peak for $\omega_c/\omega_p = 1.5$. The insulating layer tends to make the antenna electrically shorter for frequencies near ω_c. Thus, the simple approximation of the wave number of the current distribution by k_s that may be justified for antennas in direct contact with a surrounding plasma definitely will be in error in the presence of an insulating layer or an ion sheath.

Further calculations are shown in Figs. 12.6c and d for antennas of negligibly thin insulating layers and for frequencies ω in the immediate vicinity of ω_c. The thin line that is a sine wave with a wave number k_s approximates the current distributions computed with $k_A = k_s$ and $k_B = 2k_A$ or $k_B = k_a/2$. Such a propagation coefficient k_s has been recently observed by Lo and Lee [1966] for long antennas. For electrically long antennas in free space, the first current minimum is less than $\lambda/2$ from the end of the antenna. This analogy suggests that the current distributions computed using $k_A = k_s$ and $k_B = k_s/2$ are the more accurate ones in Figs. 12.6c and d. However, the approximation of the propagation coefficient by k_s will not apply to antennas insulated from the surrounding plasma. It may be noted that surface waves do not exist with a lossless plasma for frequencies ω above the lowest of ω_p or ω_c following Seshadri [1965d]. Obviously such surface waves cannot explain the resonances that are shown in Figs. 12.3 and 12.5b, and that occur in the frequency range $\omega_p < \omega < \omega_c$.

The resonances observed with $\omega \to \omega_c$ and the general shape of the current distributions of Fig. 12.6 are further verified by calculations based on more

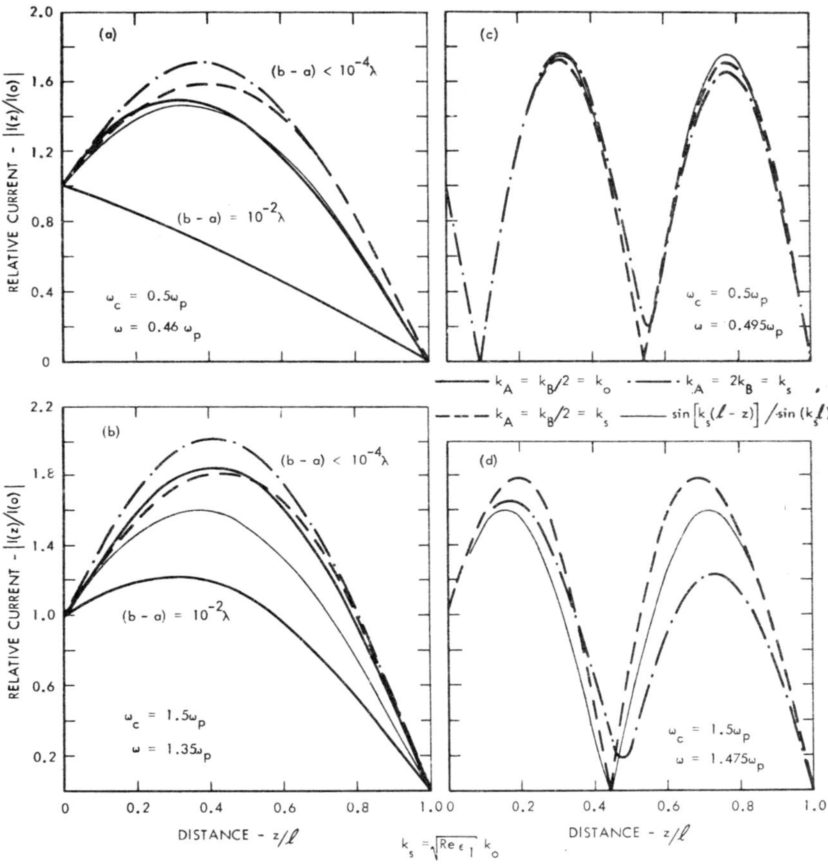

FIG. 12.6. Variational approximations to the current distributions.

general three-term trial functions. In this work, the boundary condition of vanishing tangential electric fields on the antenna surface is satisfied to the same extent as for antennas surrounded by an isotropic medium [Galejs, 1968]. However, the impedance calculations based on two-term trial functions become less accurate for electrically long antennas as $\omega \rightarrow \omega_c$.

12.1.4.4. *Comparisons with other impedance calculations.* The current calculations are compared with the quasistationary approximations and with the computation by Staras [1964] in Fig. 12.7. The quasistationary approximations agree closely with the more accurate calculations for long thin antennas. For short antennas (antenna of a smaller l/a ratio), however, the quasistationary approximation indicates a resonance. The antenna can be shown to remain capacitive and electrically short near this resonance; therefore the accuracy of the quasistationary formula should be questioned.

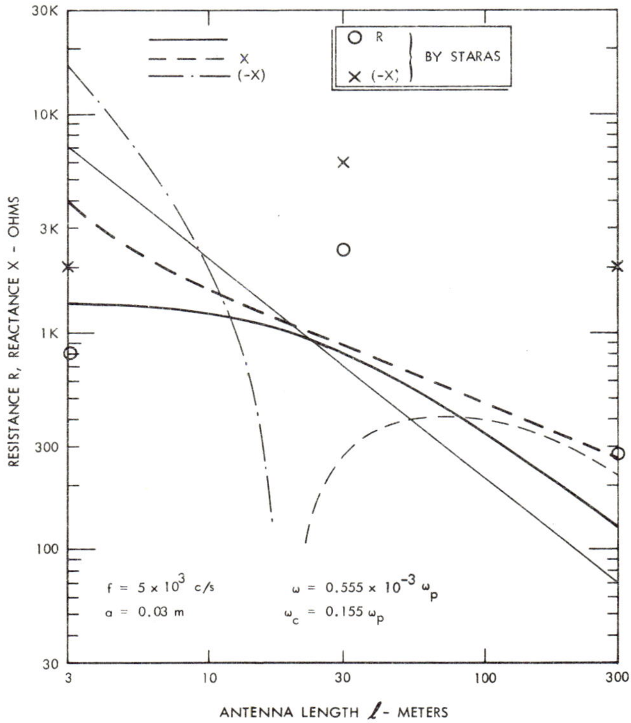

FIG. 12.7. Antenna impedance for $\omega \ll \omega_p$. Comparison with the quasistatic approximation (thin line) and with computations by Staras.

(This apparent resonance is due solely to the decrease of the l/a ratio in equation (12.50), but the quasistationary approximations are known to become inaccurate for antennas of decreasing l/a). The resistance points listed by Staras [1964] show the same trend as the present calculations, but the reactance is of different signs. However, available experimental data indicate only a capacitive reactance for frequencies between 10 to 18 kc/s [Young, 1966].

12.2. Flat Strip Antenna with a Perpendicular Static Magnetic Field

In this Section the impedance of an insulated flat strip antenna is calculated for a cold plasma with a steady transverse magnetic field as depicted in Fig. 12.8. Near the antenna, an isotropic dielectric region is intended to

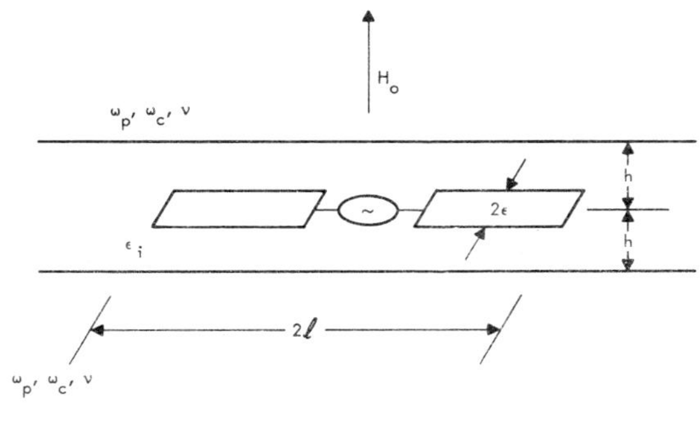

FIG. 12.8. Antenna geometry.

approximate the effects of an ion sheath formed around the antenna immersed in the ionosphere.

The variational impedance expressions developed previously for a linear antenna in an arbitrarily stratified isotropic dielectric (Section 9.1) involve reflection coefficients of the TE and TM modes at the dielectric boundary nearest to the antenna. In the present antenna geometry these reflection coefficients should consider the anisotropy of the surrounding magnetoionic medium. The fields in the anisotropic medium are expressed as double Fourier integrals [Arbel and Felsen, 1963; Hodara, 1963]. They are described as a sum of the field components of the ordinary and extraordinary modes, which are characterized by different wave numbers for propagation in the z direction. The field components of these two modes are separated into coupled TE and TM components, which are matched with the corresponding uncoupled TE and TM modes of the isotropic medium at the plasma boundary. However the relative amplitudes of the ordinary and extraordinary modes depend on the boundary conditions near the antenna. The determination of the reflection coefficients is more involved than for an isotropic dielectric

12.2.1. *Field Expressions in the Plasma Medium*

Assuming a suppressed exp $(-i\omega t)$ harmonic time dependence of the fields for a source free medium, the electric and magnetic vectors **E** and **H** satisfy the Maxwell's equations with the dyadic permittivity defined as in (12.1) to (12.6). The field components F_{ij} in the anisotropic medium are related to their Fourier transforms $\bar{F}_{ij}$ by the integrals

$$F_{ij}(x, y, z) = \left(\frac{1}{2\pi}\right)^2 \int_{-\infty}^{\infty}\int_{-\infty}^{\infty} \bar{F}_{ij}(u, v)\, e^{iux+ivy}\, e^{\gamma_j z} du\, dv, \qquad (12.59)$$

where $F = E$ or H, $i = x$, y or z, and where subscripts j denote a mode characterized by a particular value of γ_j. The relations between γ_j and the transform variables u and v will be determined in the subsequent development. A substitution of (12.59) into (12.1) and (12.2) gives the following relations between the transformed field components:

$$iv\bar{E}_z - \gamma\bar{E}_y = i\omega\mu_0\bar{H}_x, \tag{12.60}$$

$$\gamma\bar{E}_x - iu\bar{E}_x = i\omega\mu_0\bar{H}_y, \tag{12.61}$$

$$iu\bar{E}_y - iv\bar{E}_x = i\omega\mu_0\bar{H}_z, \tag{12.62}$$

$$iv\bar{H}_z - \gamma\bar{H}_y = -i\omega\epsilon_0(\epsilon_1\bar{E}_x + i\epsilon_2\bar{E}_y), \tag{12.63}$$

$$\gamma\bar{H}_x - iu\bar{H}_z = -i\omega\epsilon_0(-i\epsilon_2\bar{E}_x + \epsilon_1\bar{E}_y), \tag{12.64}$$

$$iu\bar{H}_y - iv\bar{H}_x = -i\omega\epsilon_0\epsilon_3\bar{E}_z, \tag{12.65}$$

where the arguments (u, v) and the subscripts j have been omitted. The transverse field components of (12.60) to (12.62) and of (12.65) can be expressed in terms of $\bar{E}_z$ and $\bar{H}_z$ as

$$\bar{E}_x = \frac{1}{\rho^2}\left[-iu\frac{k_z^2}{\gamma}\bar{E}_z - \omega\mu_0 v\bar{H}_z\right], \tag{12.66}$$

$$\bar{E}_y = \frac{1}{\rho^2}\left[-iv\frac{k_z^2}{\gamma}\bar{E}_z + \omega\mu_0 u\bar{H}_z\right], \tag{12.67}$$

$$\bar{H}_x = \frac{1}{\rho^2}[v\omega\epsilon_0\epsilon_3\bar{E}_z + iu\gamma\bar{H}_z], \tag{12.68}$$

$$\bar{H}_y = \frac{1}{\rho^2}[-u\omega\epsilon_0\epsilon_3\bar{E}_z + iv\gamma\bar{H}_z], \tag{12.69}$$

where $\rho^2 = u^2 + v^2$, $k_z^2 = k_0^2\epsilon_3 - \rho^2$ and $k_0 = \omega(\mu_0\epsilon_0)^{1/2}$. Substitution of (12.66) to (12.69) in (12.63) to (12.64) gives two homogeneous equations in the unknowns $\bar{E}_z$ and $\bar{H}_z$. Solving for $\bar{E}_z$ gives

$$[(\rho^2 - \gamma^2 - k_0^2\epsilon_1)(\epsilon_3\gamma^2 + \epsilon_1 k_z^2) + \epsilon_2^2 k_0^2 k_z^2]\bar{E}_z = 0. \tag{12.70}$$

Equation (12.70) of the transformed field components $\bar{E}_z(u, v)$ contains factors ρ^2, γ^2, and their squares. This equation corresponds to a fourth order partial differential equation for the field components $E_z(x, y, z)$. Equation (12.70) is satisfied for values of γ that make the factor contained in the square bracket equal to zero. This factor is rearranged into

$$\left(\frac{\gamma}{k_0}\right)^4 + \left(\frac{\gamma}{k_0}\right)^2\left[2\epsilon_1 - \frac{\epsilon_1+\epsilon_3}{\epsilon_3}\left(\frac{\rho}{k_0}\right)^2\right] + \left[(\epsilon_1^2 - \epsilon_2^2) - \left(\frac{\rho}{k_0}\right)^2\left(\frac{\epsilon_1^2-\epsilon_2^2}{\epsilon_3} + \epsilon_1\right) + \left(\frac{\rho}{k_0}\right)^4\frac{\epsilon_1}{\epsilon_3}\right] = 0, \tag{12.71}$$

the solution of which is

$$\left(\frac{\gamma}{k_0}\right)^2 = \left[-\epsilon_1 + \frac{\epsilon_1+\epsilon_3}{2\epsilon_3}\left(\frac{\rho}{k_0}\right)^2\right]$$
$$\pm\left\{\left[\frac{\epsilon_1-\epsilon_3}{2\epsilon_3}\left(\frac{\rho}{k_0}\right)^2\right]^2 + \epsilon_2^2\left[1-\frac{1}{\epsilon_3}\left(\frac{\rho}{k_0}\right)^2\right]\right\}^{1/2}. \qquad (12.72)$$

Equation (12.71) can be recognized as the Booker quartic, its two solutions representing the ordinary and the extraordinary modes. For $(\gamma/k_0) \rightarrow 0$, (12.71) represents a cylindrical wave that propagates radially, i.e. transverse to the applied magnetic field with no variation in the z direction. Its wave number has the values

$$\rho_{\pm}^2 = \begin{cases} \epsilon_3 k_0^2 \\ (\epsilon_1^2-\epsilon_2^2)\,k_0^2/\epsilon_1 \end{cases} \qquad (12.73)$$

where the subscripts + or − correspond to the + or − signs of (12.72). Following a convention used in the past by Arbel and Felsen [1963], a solution of (12.72), which gives for transverse propagation ($\gamma = 0$) a wave number $\rho = (\epsilon_3)^{1/2} k_0$, is designated as the ordinary mode. The plus sign in the solution of (12.72) represents therefore the ordinary mode (subscript $j = +$), while the minus sign gives the extraordinary mode ($j = -$). For $\rho \rightarrow 0$, equation (12.72) represents a plane wave propagating in the z direction with

$$\gamma_{\pm}^2 = (-\epsilon_1 \pm \epsilon_2)\,k_0^2. \qquad (12.74)$$

In the presence of a strong static magnetic field, $\epsilon_2 \rightarrow 0$, and the anisotropic medium becomes uniaxial with $\epsilon_3 \neq \epsilon_1$. The two solutions of (12.72) become

$$\gamma_+^2 = -\epsilon_1 k_0^2 + \frac{\epsilon_1}{\epsilon_3}\rho^2,$$
$$\gamma_-^2 = -\epsilon_1 k_0^2 + \rho^2. \qquad (12.75)$$

When $\epsilon_2 = 0$ and $\epsilon_3 = \epsilon_1$, (12.75) also applies to an isotropic medium. The ordinary wave ($j = +$) corresponds to TM field components and the extraordinary wave ($j = -$) has TE field components, which are uncoupled by the medium if $\epsilon_2 = 0$.

Equation (12.72) characterizes only γ^2 and there is still an ambiguity of sign in the definition of γ. For complex or positive real values of γ^2, the real part of γ should be less than zero in order to obtain exponentially attenuated fields (12.59) for large values of z. For a lossless plasma, γ may be a pure imaginary. Under this circumstance the sign is determined from the requirement of outward power flow across the plasma boundary at $z = h$. In the present study the plasma is considered slightly lossy and the sign of γ_j is always determined from the requirement $\text{Re}\,\gamma_j < 0$ for $z > 0$.

The amount of coupling between the TE and TM modes in the anisotropic medium is determined by the admittance

$$Y_j = \frac{\bar{H}_{zj}}{\bar{E}_{zj}}. \tag{12.76}$$

With γ_j determined from the solution of (12.72), this admittance is computed by substituting (12.66) to (12.69) in either (12.63) or (12.64). A lengthy algebraic manipulation results in

$$Y_j = \frac{\epsilon_3(\gamma_j^2 + k_0^2\epsilon_1) - \epsilon_1\rho^2}{\omega\mu_0\epsilon_2\gamma_j}. \tag{12.77}$$

Substituting (12.72) in (12.77) and letting $\epsilon_2 \to 0$, it follows that $Y_+ \sim \epsilon_2^2/\epsilon_2 \to 0$ and $\bar{H}_{z+} = 0$ due to an excitation by $\bar{E}_{z+}$. Similarly, it follows that $|Y_-| \to \infty$ and $\bar{E}_{z-} = 0$ due to an excitation by $\bar{H}_{z-}$.

12.2.2. *Reflection Coefficients*

The reflection coefficients of the TE and TM modes in the isotropic region $|z| \leqslant h$ are determined by matching the tangential field components across the plasma boundaries at $z = h$. The fields of the isotropic region of $0 \leqslant z \leqslant h$ are deduced from the scalar functions

$$\Psi = \left(\frac{1}{2\pi}\right)^2 \int_{-\infty}^{\infty}\!\!\int A(u, v)\, e^{iux+ivy}(e^{\gamma_i z} + R_a\, e^{-\gamma_i z})\, du\, dv, \tag{12.78}$$

$$\Phi = \left(\frac{1}{2\pi}\right)^2 \int_{-\infty}^{\infty}\!\!\int B(u, v)\, e^{iux+ivy}(e^{\gamma_i z} + R_b\, e^{-\gamma_i z})\, du\, dv, \tag{12.79}$$

using the formulas (3.10) to (3.15), where,

$$\gamma_i = i(k_i^2 - u^2 - v^2)^{1/2}, \quad k_i^2 = \epsilon_i k_0^2,$$

and $\epsilon_i = \epsilon_{ri}(1 + \tan\delta_i)$. After equating these field components to the field components computed from (12.66) to (12.69), it follows that

$$\gamma_i B(e^{\gamma_i h} - R_b\, e^{-\gamma_i h}) = -\frac{1}{\rho^2}\sum_j \frac{k_z^2}{\gamma_j}\bar{E}_{zj}, \tag{12.80}$$

$$A(e^{\gamma_i h} + R_a\, e^{-\gamma_i h}) = \frac{1}{\rho^2}\sum_j \bar{H}_{zj}, \tag{12.81}$$

$$\gamma_i A(e^{\gamma_i h} - R_a\, e^{-\gamma_i h}) = \frac{1}{\rho^2}\sum_j \gamma_j\bar{H}_{zj}, \tag{12.82}$$

$$\epsilon_i B(e^{\gamma_i h} + R_b e^{-\gamma_i h}) = \frac{\epsilon_3}{\rho^2} \sum_j \bar{E}_{zj}. \tag{12.83}$$

Solving (12.33) and (12.34) for R_a and (12.32) and (12.35) for R_b gives

$$R_a = e^{2\gamma_i h} \frac{\sum_j (\gamma_i - \gamma_j)\bar{H}_{zj}}{\sum_j (\gamma_i + \gamma_j)\bar{H}_{zj}}, \tag{12.84}$$

$$R_b = e^{2\gamma_i h} \frac{\sum_j \left(\frac{k_z^2}{\gamma_j} + \frac{\epsilon_3 \gamma_i}{\epsilon_i}\right)\bar{E}_{zj}}{\sum_j \left(\frac{-k_z^2}{\gamma_j} + \frac{\epsilon_3 \gamma_i}{\epsilon_i}\right)\bar{E}_{zj}}, \tag{12.85}$$

where $j = +$ or $-$. Noting that $\bar{H}_{zj}$ is related to $\bar{E}_{zj}$ by (12.76), equations (12.84) and (12.85) contain one as yet undetermined quantity $\bar{E}_{z-}/\bar{E}_{z+}$, which will be designated as X.

The above calculation of R_a and R_b was carried out for $z > 0$. For $z < 0$, the signs of γ_j in (12.59) and γ_i in (12.78) and (12.79) are reversed. At the boundary $z = -h$, (12.80) to (12.83), and the final expressions for R_a and R_b in (12.84) and (12.85), remain unaltered. However $\bar{E}_{zj}$ and $\bar{H}_{zj}$ refer now to their values at $z = -h$. After matching the E_x, E_y, and H_x field components across the $z = 0$ boundary and applying the boundary condition $H_y(0^-) = H_y(0^+) + J_x$, where J_x is the x-directed current density on the antenna, it can be shown that the signs of E_{zj} and Y_j are reversed at $z = -h$. However $X = \bar{E}_{z-}/\bar{E}_{z+}$ remains the same for both half-spaces of the plasma. The ratio X can be determined by using the R_a and R_b expressions (12.84) and (12.85) in conjunction with the boundary conditions at $z = 0$. It follows that

$$X = \frac{[-\gamma_+ Y_+ \beta/\alpha + \epsilon_3\gamma_i/\epsilon_i] \cosh \gamma_i h + [\gamma_i Y_+ \beta/\alpha + k_z^2/\gamma_+] \sinh \gamma_i h}{[\gamma_- Y_- \beta/\alpha - \epsilon_3\gamma_i/\epsilon_i] \cosh \gamma_i h - [\gamma_i Y_- \beta/\alpha + k_z^2/\gamma_-] \sinh \gamma_i h}, \tag{12.86}$$

where

$$\alpha = ivI/(\gamma_i \rho^2), \tag{12.87}$$

$$\beta = \omega\mu_0 uI/(k_i^2 \rho^2), \tag{12.88}$$

$$I = \int_{-\infty}^{\infty} \int_{-\infty}^{\infty} J_x e^{-iux} e^{-ivy} dx\, dy. \tag{12.89}$$

This completes the formal specification of the reflection coefficients. However, for the subsequent numerical work, it is desirable to derive an explicit expression for combinations of reflection coefficients R_a and R_b as they will be encountered in the impedance calculations. Substituting $X = \bar{E}_{z-}/\bar{E}_{z+}$ in (12.84) and (12.85) and applying (12.76), it follows that

$$\frac{1+R_a}{1-R_a} = \frac{N_a + P_a}{D}; \quad \frac{1-R_b}{1+R_b} = \frac{N_b + P_b}{D}, \tag{12.90}$$

where

$$N_a = \epsilon_3\gamma_i^2(Y_+ - Y_-)\cosh^2\gamma_i h + \left[\epsilon_i\gamma_i k_z^2\left(\frac{Y_+}{\gamma_-} - \frac{Y_-}{\gamma_+}\right) - \epsilon_3\gamma_i(\gamma_+ Y_+ - \gamma_- Y_-)\right]$$

$$\cdot \cosh\gamma_i h \sinh\gamma_i h - k_z^2\epsilon_i\left(\frac{\gamma_+}{\gamma_-}Y_+ - \frac{\gamma_-}{\gamma_+}Y_-\right)\sinh^2\gamma_i h, \qquad (12.91)$$

$$D = -\gamma_i k_z^2\epsilon_i\left(\frac{Y_+}{\gamma_-} - \frac{Y_-}{\gamma_+}\right)\sinh^2\gamma_i h + \left[k_z^2\epsilon_i\left(\frac{\gamma_+}{\gamma_-}Y_+ - \frac{\gamma_-}{\gamma_+}Y_-\right) - \epsilon_3\gamma_i^2(Y_+ - Y_-)\right]$$

$$\cdot \sinh\gamma_i h\cosh\gamma_i h + \epsilon_3\gamma_i(\gamma_+ Y_+ - \gamma_- Y_-)\cosh^2\gamma_i h, \qquad (12.92)$$

$$P_a = (\beta/\alpha)\gamma_i(\gamma_+ - \gamma_-)Y_+ Y_-\epsilon_i, \qquad (12.93)$$

$$P_b = \frac{\alpha}{\beta}\epsilon_3\gamma_i k_z^2\left(\frac{1}{\gamma_-} - \frac{1}{\gamma_+}\right), \qquad (12.94)$$

and where N_b is obtained from N_a by interchanging $\cosh\gamma_i h$ and $\sinh\gamma_i h$ functions. It may be noted that N_a, N_b, and D are even functions of the transform variables u and v, and that P_a and P_b are odd functions because they involve the ratio β/α of (12.87), and (12.88). The expressions for the reflection coefficients (12.84) and (12.85) with X of (12.86), plus the expression (12.90), are valid for both positive and negative values of z.

For $\epsilon_2 \to 0$, $Y_+ \to 0$ and $|Y_-| \to \infty$, but Y_+Y_- remains finite. Noting that P_a and P_b are negligible relative to N_a and N_b, (12.90) can be simplified to

$$\frac{1+R_a}{1-R_a} = \frac{\gamma_i - \gamma_- \tanh\gamma_i h}{\gamma_- - \gamma_i\tanh\gamma_i h}, \qquad (12.95)$$

$$\frac{1+R_b}{1-R_b} = -\frac{\epsilon_i k_z^2 + \epsilon_3\gamma_i\gamma_+\tanh\gamma_i h}{\epsilon_3\gamma_i\gamma_+ + \epsilon_i k_z^2\tanh\gamma_i h}, \qquad (12.96)$$

where γ_+ and γ_- are defined by (12.75). Equations (12.95) and (12.96) can be also derived directly from (12.84) and (12.85) by noting that the reflection coefficients R_a of the TE modes contain only the $j = -$ terms, and that the reflection coefficients R_b of the TM modes have only the $j = +$ terms in the limit of $\epsilon_2 \to 0$.

12.2.3. *Antenna Driving Point Impedance*

The driving point impedance of a flat strip antenna may be computed from (1.35), which is rewritten as

$$Z = -\frac{\iint \mathbf{E}\cdot\mathbf{J}_s\,dx\,dy}{[I(x=0)]^2}, \qquad (12.97)$$

This expression is stationary with respect to small changes of the surface current density $\mathbf{J}_s$ or of the current $I(x, y)$ about its correct value following Section 1.5. The surface current $\mathbf{J}_s$ is assumed to have only an x-component J_x, which in addition is an even function about $x = 0$, $y = 0$. It follows from the field expressions (12.78), (12.79), and (3.10) and from the boundary relations near the antenna that

$$Z = \frac{i\omega\mu_0}{[2\pi I(x=0)]^2} \int_{-\infty}^{\infty}\int_{-\infty}^{\infty} F(u, v)\left[\iint J_x(x, y) \cos ux \cos vy \, dx \, dy\right]^2 du \, dv, \tag{12.98}$$

where

$$F(u, v) = \frac{v^2}{2\rho^2\gamma_i} \frac{1+R_a}{1-R_a} - \frac{u^2}{2\rho^2} \frac{\gamma_i}{k_i^2} \frac{1-R_b}{1+R_b}, \tag{12.99}$$

and $\rho^2 = u^2 + v^2$. Only the even parts of $F(u, v)$ function will contribute to the integral over u and v. The odd parts P_a and P_b of (12.90) can be neglected and $F(u, v)$ can be simplified to

$$F(u, v) = \frac{1}{2\rho^2 D}\left[\frac{v^2}{\gamma_i} N_a - \frac{u^2\gamma_i}{k_i^2} N_b\right], \tag{12.100}$$

where D, N_a and N_b are obtained from (12.91) and (12.92).

Based on the reasoning summarized following (12.39) the antenna current density is represented by the trial function

$$J_x(x, y) = \{A \sin [k_A(l - |x|)] + B \sin [k_B(l - |x|)]\} f(y), \tag{12.101}$$

with $f(y) = 1/(2\epsilon) = \text{const}$. Substituting (12.101) in (12.98) and using the stationary character of the impedance expression for determining the complex amplitudes A and B, it follows that the amplitude ratio (A/B) and the impedance Z are given by (2.38) to (2.41) with

$$F_N = \sin k_N l, \tag{12.102}$$

$$\gamma_{NM} = \frac{4i\omega\mu_0}{\pi^2} \int_0^{\infty} dv \left(\frac{\sin \epsilon v}{\epsilon v}\right)^2 \int_0^{\infty} du \, F(u, v) g_N(u) g_M(u), \tag{12.103}$$

$$g_N(u) = \frac{k_N}{k_N^2 - u^2} (\cos ul - \cos k_N l), \tag{12.104}$$

where N, $M = A$ or B. The accuracy of this impedance formulation is briefly discussed in the last paragraph of Section 12.1.2.

12.2.4. *Impedance of a Thin Antenna Without a Dielectric Layer*

12.2.4.1. *Quasistationary Approximations.* In the low frequency limit, $k_0 \to 0$ and k_0^2 will be assumed to be negligible relative to the integration variables $u^2+v^2=\rho^2$ in all of the previous expressions. The propagation constants γ_j of (12.93) simplify to $\gamma_+=-(\epsilon_1/\epsilon_3)^{1/2}\rho$, $\gamma_-=-\rho$, which makes $Y_+=0$. Substituting this in the N_a, N_b, and D expressions of (12.91) and (12.92), and letting $h=0$ and $\epsilon_i=1$, it follows that $F(u,v)$ of (12.100) becomes

$$F(u,v)=-\frac{v^2}{2\rho^3}+\frac{1}{(\epsilon_1\epsilon_3)^{1/2}}\frac{u^2}{2\rho k_0^2}. \tag{12.105}$$

The antenna impedance is computed from (2.41), (12.102), and (12.103) with k_A of the trial functions approaching zero, which implies a triangular current distribution along the antenna. This results in

$$Z_s=R_s+iX_s=\frac{4i\omega\mu_0}{\pi^2 l^2}\int_0^\infty du\left[\frac{\cos ul-\cos k_A l}{k_A^2-u^2}\right]^2\int_0^\infty dv\left(\frac{\sin \epsilon v}{\epsilon v}\right)^2 F(u,v), \tag{12.106}$$

where $k_A\to 0$. The resistive part of Z_s is computed by substituting the second term of (12.105) as $F(u,v)$ in (12.106), which gives

$$R_s=\frac{2i\omega\mu_0}{(\pi k_0 l)^2(\epsilon_1\epsilon_3)^{1/2}}\int_0^\infty du\frac{(\cos ul-1)^2}{u^2}\int_0^\infty dv\left(\frac{\sin\epsilon v}{\epsilon v}\right)^2\frac{1}{(u^2+v^2)^{1/2}}. \tag{12.107}$$

The v integrations become elementary by noting that $(\sin \epsilon v/\epsilon v)\approx 1$ over the range of v where

$$(u^2+v^2)^{1/2}\neq v,$$

which applies strictly if $\epsilon\to 0$ and $u\leqslant u_0$ is finite. The subsequent u integrals are evaluated using tabulated definite integrals. Thus

$$R_s\approx\frac{2i\omega\mu_0}{(\pi k_0 l)^2(\epsilon_1\epsilon_3)^{1/2}}\int_0^\infty du\frac{(\cos ul-1)^2}{u^2}\left[\frac{3}{2}-C-\log(\epsilon u)\right]$$

$$=\frac{(\mu_0/\epsilon_0)^{1/2}}{\pi k_0 l(|\epsilon_1\epsilon_3|)^{1/2}}\left[\log\frac{2l}{\epsilon}+\frac{1}{2}-2\log 2\right], \tag{12.108}$$

where $C=0.57721\ldots$ is Euler's constant. The reactive part of Z_s is computed by substituting the first term of (12.105) as $F(u,v)$ in (12.106) which gives

$$X_s=-\frac{2\omega\mu_0}{\pi^2 l^2}\int_0^\infty du\frac{(\cos ul-1)^2}{u^4}\int_0^\infty dv\left(\frac{\sin\epsilon v}{\epsilon v}\right)^2\frac{v^2}{(v^2+u^2)^{3/2}}. \tag{12.109}$$

The v integral is evaluated using the same approximations as in (12.107) and the reactance becomes

$$X_s \approx -\frac{2\omega\mu_0}{\pi^2 l^2}\int_0^\infty du\,\frac{(\cos ul-1)^2}{u^4}\left[\frac{1}{2}-C-\log(\epsilon u)\right]. \tag{12.110}$$

The integral, which is proportional to the constant part of the square brackets, is evaluated directly. The remaining term of the integral is reduced to tabulated integrals with u^3 and u^2 in the denominator by repeating the integration by parts twice. This gives

$$X_s = -\frac{k_0 l}{3\pi}\left(\frac{\mu_0}{\epsilon_0}\right)^{1/2}\left[\log\frac{2l}{\epsilon}+\log 2-\frac{4}{3}\right]. \tag{12.111}$$

The antenna resistance (12.108) is inversely proportional to the length l and compares closely with the resistance of the quasi-static approximation (64) of Balmain [1964], and with the leading term of the impedance shown by Ament *et al.* [1965]. A flat strip antenna of width $w = 2\epsilon$ corresponds to a cylindrical antenna of radius a if $a = \epsilon/2$. For an antenna of $\Omega = 2\log(4l/\epsilon) = 8$, the calculations of Balmain [1964] give a 25 percent higher resistance while the results of Ament *et al.* [1965] show a 5 percent lower resistance than (12.108). These differences will become even smaller for larger values of Ω.

The antenna reactance X_s is negative (i.e. inductive) in (12.111), which can be also seen from (12.109). X_s is proportional to the antenna length l. The reactance (12.111) was computed from the first term of (12.105), which represents the TE modes. (Alternately, (12.111) can be obtained by integrating the first term of (12.112) for a uniaxial medium.) TE modes exhibit $E_z = 0$ by definition; they do not depend on ϵ_3, and in the low frequency limit, they are the same as for free space around the antenna. The TE fields represent an equivalent magnetic dipole and should exhibit an inductive reactance. The integration of the second term of (12.105) or (12.112), which represent the TM modes, gives a pure resistance in the limit of a lossless plasma at low frequencies.

12.2.4.2. *A short antenna in a uniaxial medium.* In a uniaxial medium (with $\epsilon_1 = 1$ and $\epsilon_2 = 0$), γ_+^2 and γ_-^2 are given by (12.75). Substituting (12.95) and (12.96) in (12.100) and applying (12.75), it is seen that

$$F(u, v) = \frac{1}{2\rho^2}\left[\frac{v^2}{i(k_0^2-\rho^2)^{1/2}}-\frac{iu^2(\epsilon_3 k_0^2-\rho^2)^{1/2}}{k_0^2(\epsilon_3)^{1/2}}\right], \tag{12.112}$$

where $\rho^2 = u^2+v^2$. In the low frequency limit $k_0 \to 0$, (12.112) reduces to $F(u, v)$ of the quasi-stationary approximation (12.105). The quasi-stationary impedance approximations (12.108) and (12.111) also apply to a uniaxial medium at low frequencies. Separate approximations applicable to a uniaxial medium will be derived only for frequencies $\omega > \omega_p$.

The antenna impedance is computed by substituting (12.112) in (12.106) and by assuming $k_A = (\epsilon_3)^{1/2}k_0$ for purposes of simplifying the integrations. For $\omega > \omega_p$, the permittivity element $\epsilon_3 > 0$ and the current distribution is triangular for sufficiently short antennas. Both terms of $F(u, v)$ contribute now to the resistance and to the reactance of the antenna.

The first term of $F(u, v)$ contributes to the antenna resistance only for $\rho \leqslant k_0$ and the second term for $\rho \leqslant (\epsilon_3)^{1/2}k_0$. Over this range of ρ, $\cos ul - \cos k_A l \approx (k_A^2 - u^2)(l^2/2)$, and $\sin \epsilon v = \epsilon v$. After introducing polar coordinates

$$\rho = (u^2 + v^2)^{1/2}$$

and $\phi = \tan^{-1}(v/u)$, the double integrals are changed into products of two elementary integrals, and

$$R_s = (15 + 5\epsilon_3)(k_0 l)^2. \tag{12.113}$$

This expression for the antenna resistance has been given by Seshadri [1966a] also.

The first term of $F(u, v)$ contributes to the antenna reactance for $\rho \geqslant k_0$ and the second term for $\rho \geqslant (\epsilon_3)^{1/2}k_0$. The contribution of the second term over the range of ρ given by $k_0 \geqslant \rho \geqslant (\epsilon_3)^{1/2}k_0$ can be shown to be proportional to $(k_0 l)^2$, and will be neglected. Both terms of $F(u, v)$ will be considered only for $\rho \geqslant k_0$. $F(u, v)$ of (12.112) is now simplified by assuming that $1 + \epsilon_3 = 2(\epsilon_3)^{1/2}$, which will be most accurate for $\epsilon_3 \approx 1$. This gives

$$X_s = \frac{2\omega\mu_0}{(\epsilon_3)^{1/2}(\pi k_0 l)^2}\left[\int_0^{k_0} du \int_{(k_0^2-u^2)^{1/2}}^{\infty} dv + \int_{k_0}^{\infty} du \int_0^{\infty} dv\right] \cdot \frac{(\cos ul - \cos k_A l)^2}{(u^2 - k_A^2)(u^2 + v^2 - k_0^2)^{1/2}}\left(\frac{\sin \epsilon v}{\epsilon v}\right)^2. \tag{12.114}$$

The v integrals are evaluated using the same approximations as in (12.107), which gives

$$X_s = \frac{2\omega\mu_0}{(\epsilon_3)^{1/2}(\pi k_0 l)^2}\int_0^{\infty} du \frac{(\cos ul - \cos k_A l)^2}{(u^2 - k_A^2)}\left[\frac{3}{2} - C - \log(\epsilon |k_0^2 - u^2|^{1/2})\right]. \tag{12.115}$$

The integral, which is proportional to the constant term of the square brackets, is evaluated directly. The remaining term of the integral, which involves a logarithmic function, is reduced to tabulated integrals after expanding the denominator in partial fractions and changing the variables of integration to $y = (u/k_0) \pm 1$. This results in

$$X_s = \frac{(\mu_0/\epsilon_0)^{1/2}}{\pi k_0 l (\epsilon_3)^{1/2}} \left[\log \frac{2l}{\epsilon} + 0.5 - 2 \log 2 \right]. \qquad (12.116)$$

In the high frequency limit, $\epsilon_3 \to 1$ and (12.116) gives nearly the same capacitive reactance as a standard formula for antennas in free space, where the two terms (0.5–2 log 2) are replaced by (–1) in (2.48).

For a slightly lossy plasma, ϵ_3 is complex and (12.116) exhibits a small real part

$$\Delta R_s = \frac{\nu}{2\omega} \frac{\omega_p^2}{\omega^2 - \omega_p^2} |X_s|, \qquad (12.117)$$

which should be added to R_s of (12.113).

12.2.5. *Discussion of Numerical Results*

12.2.5.1. *Uniaxial medium.* The magnetoionic plasma degenerates into a uniaxial medium with $\epsilon_1 = 1$ and $\epsilon_2 = 0$ in the presence of strong magnetic fields ($\omega_c/\omega_p \to \infty$). For such a medium, the antenna impedance is shown in Fig. 12.9 where the left-hand scale of the resistance applies for $\omega < \omega_p$ and the right-hand scale for $\omega > \omega_p$. The calculations are made for antennas of length $\lambda/20$ and $\lambda/10$ (indices 1 and 2) and for very small ($h = 10^{-6}\lambda$) and for larger ($h = \lambda/100$) thicknesses of the insulating layer around the antenna (indices p and i). The antenna width 2ϵ and length $2l$ are such that $\Omega = 2 \log (4l/\epsilon) = 8$. The medium has slight losses with an average collision frequency $\nu = 0.001\omega_p$. The calculated resistance is nearly constant for $\omega > \omega_p$, increases suddenly, and reaches its maximum value for $\omega = \omega_p$. The resistance is decreased for lower values of ω or higher values of ω_p. The resistance is proportional to l for $\omega > \omega_p$ and is inversely proportional to l for $\omega < \omega_p$. For very thin insulating layers ($h = 10^{-6}\lambda$), the antenna reactance is inductive for $\omega < \omega_p$, but is capacitive for $\omega > \omega_p$. The reactance reaches its maximum value at $\omega = \omega_p$, and is numerically equal to the resistance. The same general impedance behavior is also observed from the approximate impedance expressions (12.108), (12.111), (12.113), and (12.116), which are indicated by heavy lines in Fig. 12.9. The closed form approximations tend to agree more closely with reactance data computed for a sinusoidal current distribution (2.41), which are indicated by thin lines in Fig. 12.9. This indicates that the current distribution along the antenna may differ from the usually assumed sinusoidal or triangular distributions. These differences appear to be more pronounced for frequencies $\omega < \omega_p$. An insulating layer around the antenna of $h = \lambda/100$ (index i in Fig. 12.9)

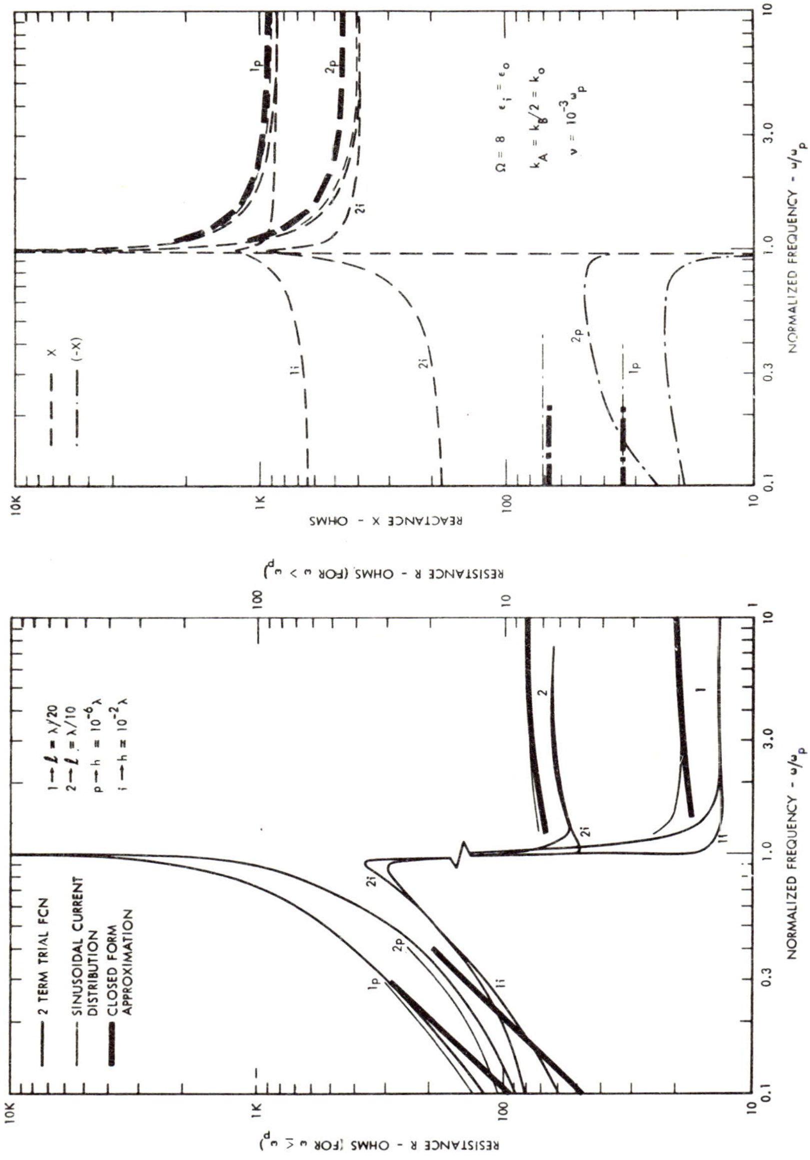

FIG. 12.9. Impedance of a short antenna in a uniaxial medium.

has little effect for frequencies $\omega > \omega_p$; it decreases the magnitude of the impedance at $\omega = \omega_p$. The reactance of the antenna, however, becomes capacitive, and the magnitude of the resistance is decreased for frequencies $\omega < \omega_p$.

12.2.5.2. *Antenna impedance for a finite magnetic field.* The impedances are computed for an antenna length $l = c/\omega_p$ ($l = \lambda/2\pi$ at $\omega = \omega_p$). The antenna has a finite radius ($\Omega = 8$) and the surrounding medium has slight losses ($\nu/\omega_p = 10^{-3}$). The calculations are made for a normalized gyrofrequency of the plasma of $\Omega_c = \omega_c/\omega_p = 0.5$ and 1.5. They are shown in Figs. 12.10 and 12.11 respectively.

For $\omega = 0.5\omega_c$ and for antennas with very thin insulating layers ($h = 10^{-6}\lambda$), the resistance decreases with increasing values of ω/ω_p, as indicated in Fig. 12.10. The resistance has low values for $\omega > \omega_c$ and exhibits finite peaks at $\omega = \omega_p$ and $\omega = \omega_u$, where the upper hybrid resonance frequency $\omega_u = (\omega_c^2 + \omega_p^2)^{1/2}$. The larger peak occurs at $\omega = \omega_p$. The reactance is inductive (negative) for $\omega < \omega_p$. It becomes capactive in the vicinity of ω_u and inductive again at higher frequencies where the antenna is electrically long. This impedance behavior is quite similar to the impedance shown in Fig. 12.3 for an antenna aligned with static magnetic field, except that the low frequency reactance is of opposite signs and more pronounced impedance peaks occur at $\omega = \omega_p$ in the present computations. The resonance peak observed for $\omega < \omega_c$ with a parallel magnetic field is completely absent in the present calculations for a transverse magnetic field. This suggests that the antenna remains electrically short with a transverse magnetic field, although ϵ_1 becomes very large in the vicinity of $\omega = \omega_c$. For larger thicknesses of the insulating layer ($h = \lambda/100$), the impedance exhibits a resonance peak near $\omega \approx 0.7\omega_p$, which was also observed for a parallel magnetic field in Fig. 12.3. The resonance peaks that occurred at $\omega = \omega_p$ and ω_u for thin insulating layers are lowered in frequency. The reactance is now capacitive for $\omega < \omega_c$ and becomes somewhat more inductive for $\omega > \omega_u$. The numerical calculation showed that the reactance changes sign near the resonance peak at $\omega \approx 0.7\omega_p$, but this could not be shown in the frequency scale used in Fig. 12.10. The quasi-stationary approximations (12.108) and (12.111) are also indicated in Fig. 12.10. The numerically computed resistance agrees closely with the closed form expression, and the closed form reactance agrees better with the impedance calculations based on a sinusoidal field distribution (2.41), which are also indicated in the reactance plots in Fig. 12.10.

For $\omega_c = 1.5\omega_p$ and thin insulating layers, the impedance curves shown in Fig. 12.11 exhibit similar peaks at $\omega = \omega_p$ and near $\omega = \omega_u$. A further resonance peak occurs, however, near ω_c, which is not present when computing the impedance for an assumed sinusoidal current distribution from (2.41). This resonance peak is similar to the one observed for a parallel

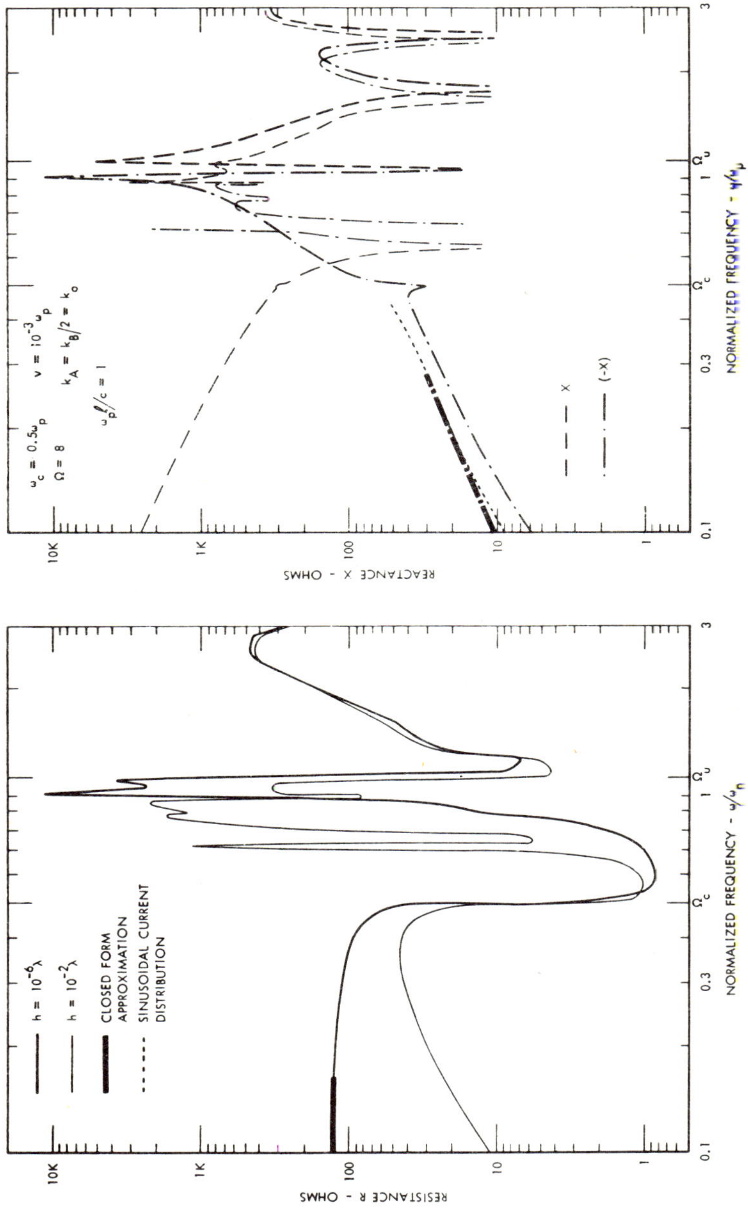

FIG. 12.10. Antenna impedance for $\Omega_c = \omega_c/\omega_p = 0.5$.

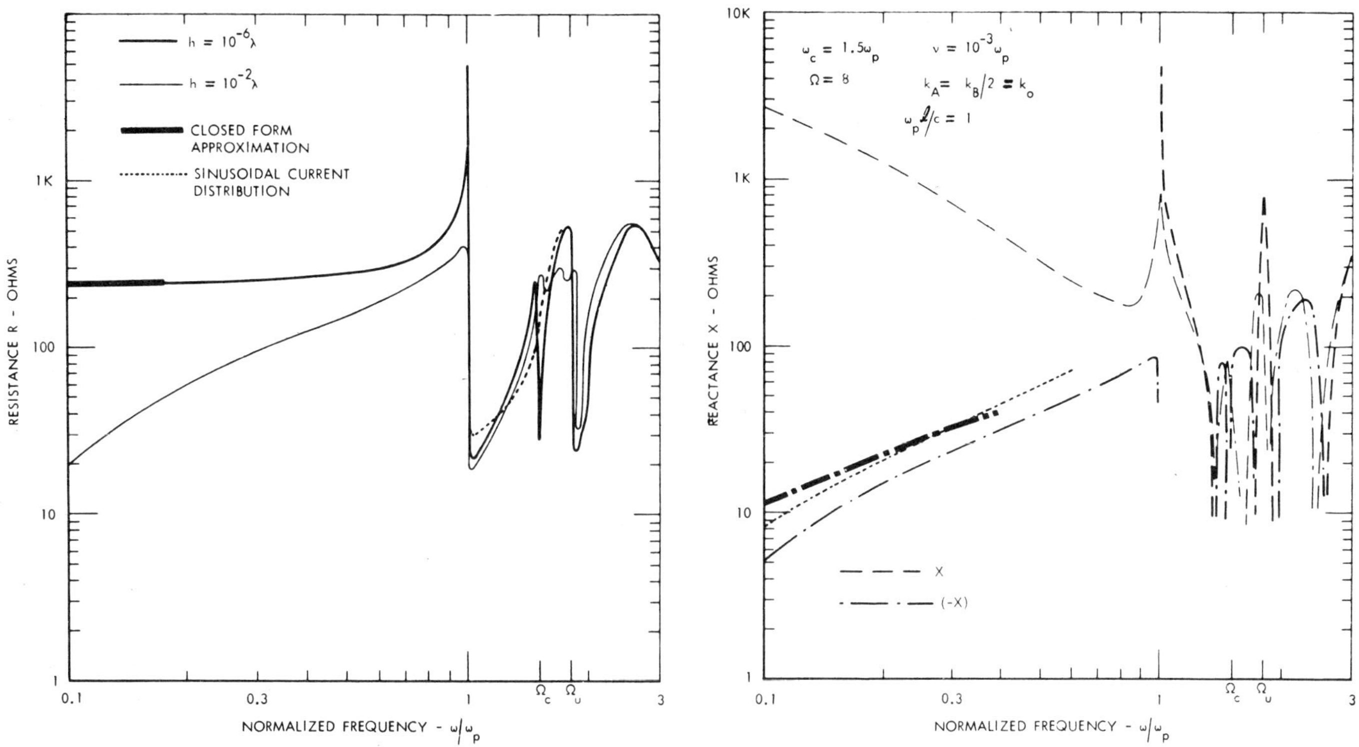

FIG. 12.11. Antenna impedance for $\Omega_c = \omega_c/\omega_p = 1.5$.

magnetic field near $\omega = \omega_c$, where the antenna becomes electrically long.

12.2.5.3. *Current distributions for antenna in a plasma with a finite magnetic field.* Variational approximations to the current distributions have been computed from (2.38), (12.101), and (12.103) for the antenna parameters of Figs. 12.10 and 12.11, and are shown in Figs. 12.12 and 12.13

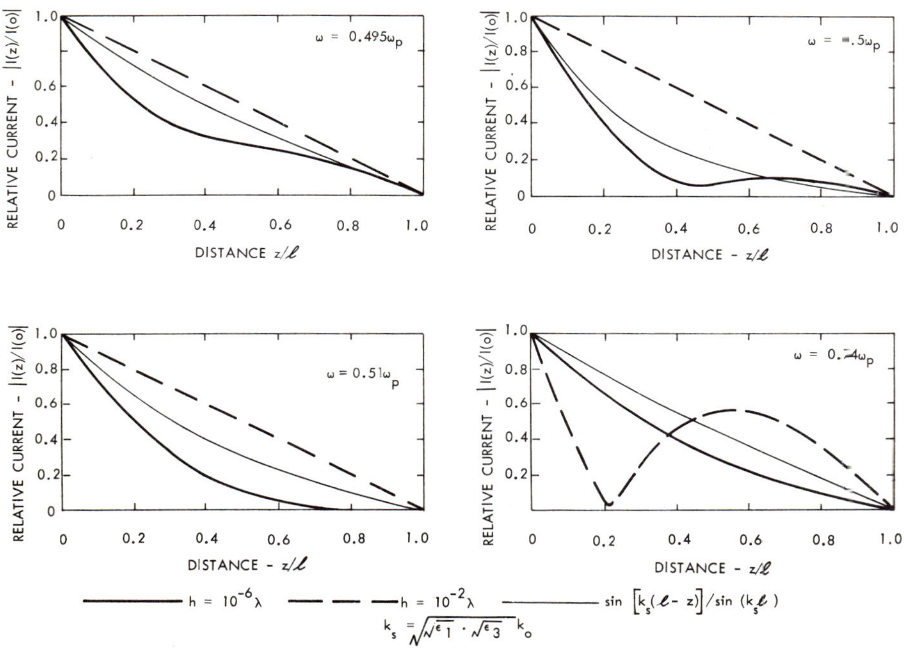

FIG. 12.12. Variational approximations to the current distributions antenna of Fig. 4 with $\omega_c = 0.5\omega_p$.

respectively. For $\omega_c = 0.5\omega_p$ and $\omega \approx \omega_c$, the current is decaying rapidly along the antenna for thin insulating layers in Fig. 12.12. For $\omega_c = 1.5\omega_p$ and $\omega < \omega_c$, the current distributions are oscillatory (as shown in Fig. 12.13), which indicates that the electrical antenna length is increased. For $\omega = \omega_c$, the current distribution is attenuated and approaches a triangular shape for somewhat larger values of ω.

Ament *et al.* [1965] have suggested that the wave number of the current distribution along a thin antenna oriented in a direction perpendicular to the static magnetic field is given by $k_s = [\epsilon_1\epsilon_3]^{0.25}k_0$, which is multivalued. The permittivity element ϵ_1 of (12.4) becomes large as ω approaches ω_c, and its real part is negative in the frequency range $\omega_c < \omega < \omega_u$. The element ϵ_3 of (12.6) is negative for $\omega < \omega_p$. For $\omega \approx \omega_c$, the product $(\epsilon_1\epsilon_3)$ is positive for $\omega > \omega_c$ if $\omega_c < \omega_p$, and for $\omega < \omega_c$ if ω_c is larger than ω_p. Large real values

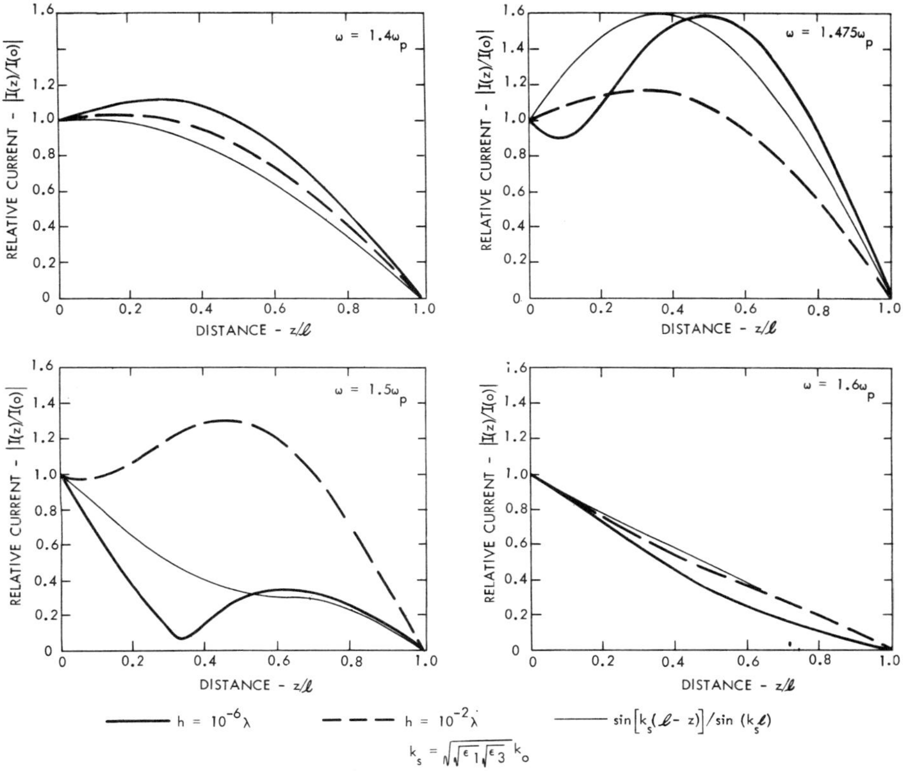

FIG. 12.13. Variational approximations to the current distributions antenna of Fig. 5 with $\omega_c = 1.5\omega_p$.

of k_s may be anticipated in both cases for frequencies near ω_c, but the computations show an oscillatory current distribution only for $\omega_c > \omega_p$ in Fig. 12.13. The wave number k_s is therefore defined as

$$k_s = [(\epsilon_1)^{1/2} (\epsilon_3)^{1/2}] k_0. \tag{12.118}$$

The imaginary parts of ϵ_1 and ϵ_3 are always positive, and all the square roots of (12.118) can be defined in the first quadrant of the complex plane including the real and imaginary axis. Using this convention a large positive value of k_s can be observed only for $\omega_c > \omega_p$, when both ϵ_1 and ϵ_3 are positive. This explains qualitatively the difference between the current distributions shown in Figs. 12.12 and 12.13 in the vicinity of $\omega = \omega_c$. The accuracy of (12.118) is further examined by computing a sine wave with a complex wave number k_s and plotting it along with the variational approximations to the current distributions. These curves, indicated by thin lines in

Figs. 12.12 and 12.13 follow the general shape of the variationally computed current distributions for small values of h (solid curves), but k_s seems to provide a consistently low estimate for the wave number. The variationally computed current distributions are either more attenuated or are more oscillatory than the simple sine waves with the wave number k_s.

The previous discussion refers only to antennas with a negligibly thin insulating layer. For insulating layers of larger thickness $(h = \lambda/100)$, the current distribution becomes less attenuated for $\omega \approx \omega_p$ in Fig. 12.12. However, near the resonance frequency of $\omega \approx 0.7\omega_p$ of Fig. 12.10, the current distribution is oscillatory for large values of h, whereas it is nearly triangular in the absence of insulation. For $\omega_c = 1.5\omega_p$, the insulating layer tends to make the antenna electrically short for $\omega < \omega_c$, but oscillatory current distributions may occur at larger values at ω, where the current distribution is attenuated in the absence of insulation. The wave number k_s of (12.118) cannot be used for estimating current distribution along an insulated antenna.

12.2.5.4. *Comparisons with other impedance calculations.* In the low frequency limit, closed form approximations that have been computed in Section 12.2.4.1 have also been compared with the corresponding quasi-stationary approximations of Balmain [1964] and with the approximate impedance expression of Ament *et al.* [1965]. The expressions for antenna resistance were in a reasonable agreement. Balmain [1964] deduces an antenna reactance that is capacitive and is inversely proportional to the antenna length. However the closed form expression (12.111) and numerical calculations shown in Figs. 12.9, 12.10, and 12.11 indicate an inductive reactance proportional to the antenna length.

For an antenna oriented in a direction transverse to the static magnetic field, Staras [1964] has computed only the antenna resistance. His numerical examples are restricted to low frequencies. His results can be compared with the resistance computed from (12.108), as can be seen from the data shown from Table 12.1.

The calculations of Seshadri [1966a] are made for a current strip in a uniaxially anisotropic plasma, where the static magnetic field is in a direction perpendicular to the antenna axis, but parallel to the surface of the strip.

TABLE 12.1. LOW FREQUENCY ANTENNA RESISTANCE

$f = 5 \times 10^3$ c/s $\quad \omega = 0.555 \times 10^{-3}\,\omega_p$

$\omega_c = 0.155\,\omega_p$ $\quad \epsilon = 0.06\,m$

l	3	30	300
R (Staras)	94	14	3
R [from equation (12–108)]	122	19.8	2.77

For high frequencies, Seshadri obtains also the antenna resistance (12.113), but his results are different for frequencies $\omega < \omega_p$ even in the limit of ω approaching zero.

The differences between the present calculation and the results of Seshadri [1966a] and Balmain [1964] indicate that the impedance is quite sensitive to changes in the antenna shape for $\omega < \omega_p$.

This will be illustrated by computing the antenna impedance in a geometry where the static magnetic field is parallel to the antenna surface but perpendicular to the antenna axis. As shown in the Appendix the formulation of this Section can be readily adapted to such an antenna. The calculations are made for short antennas in a uniaxial medium and only a simple sinusoidal current distribution is considered. In the low frequency limit a closed form approximation is derived for the input resistance of the antenna which agrees with numerical results of Seshadri [1966a]. For low frequencies the antenna reactance is capacitive and it is inversely proportional to the antenna length, as has been indicated in the quasi-static calculations of Balmain [1964] for a cylindrical current element.

12.2.6. *Insulated Antennas Above a Ground Plane*

The antenna geometry of Fig. 12.9 can be modified by inserting a conducting ground plane at one of the boundaries of the dielectric layer. The strip antenna is now embedded in a dielectric layer overlain by a half space of a cold homogeneous plasma. The impedance of such an antenna [Galejs, 1967a] can be computed by suitable modification of the development reported in Sections 12.2.2 and 12.2.3.

The numerical calculations are made for dielectric layers of $\epsilon_{ir} = 2$ and $\tan \delta_i = 0.03$ and 0.01. The antenna length $l = 0.2\lambda$, where $\lambda =$ free space wave length, the plasma has slight losses ($\nu/\omega_p = 0.001$), and the antenna impedance is computed for a wide range of plasma frequencies ω_p in Fig. 12.14.

In the absence of a static magnetic field ($\omega_c = 0$) the antenna resistance increases with an increasing ratio (ω/ω_p), reaches a peak and drops suddenly near $\omega/\omega_p = 0.6$, it has a nearly constant low value until $\omega \approx \omega_p$ and the resistance increases for larger values of ω/ω_p. The reactance is negative (inductive) for most values of ω/ω_p and it is positive only near the resistance peak. A line source excited geometry of a free space region which is bounded by a perfectly conducting plane from below and by a homogeneous lossless cold plasma from above has been considered by Seshadri [1965c]. For a source near the plasma boundary (P_0 of his Fig. 4c) the power carried by the waves guided along the plasma free space boundary increases with (ω/ω_p) until the cutoff frequencies of the guided waves. For an isotropic plasma the cutoff frequency is at $\omega/\omega_p = 1/\sqrt{2}$. For a plasma bounded by a dielectric layer this corresponds to $\epsilon_3/\epsilon_{ir} = -1$, where ϵ_3 is defined by (12.6).

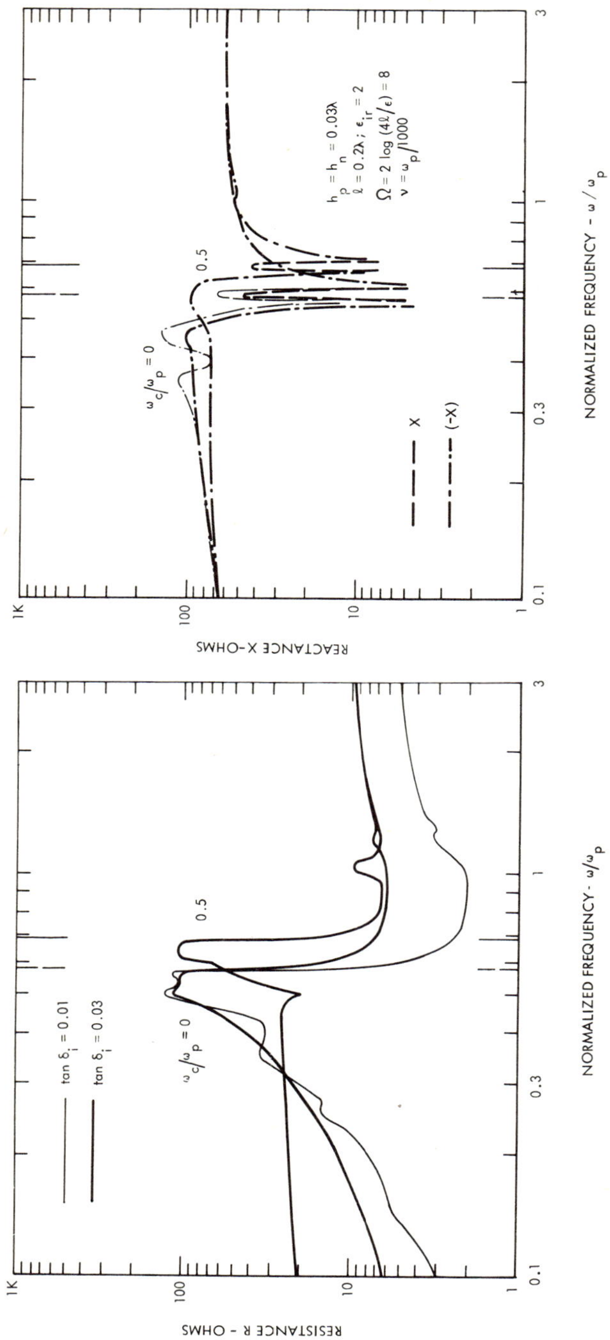

FIG. 12.14. Impedance of a strip antenna centered in a grounded dielectric layer.

This gives

$$\omega/\omega_p = 1/\sqrt{(1-\epsilon_3)} = 1/\sqrt{(1+\epsilon_{ir})} \tag{12.119}$$

This value of ω/ω_p is indicated in Fig. 12.14 by a dashed vertical line. The maximum values of the resistance occur for smaller value of ω/ω_p. Decreasing the dielectric losses ($\tan \delta_i = 0.01$) decreases the resistance for lower values of ω/ω_p, increases slightly the maximum resistance values, and decreases also the resistance in the range of ω/ω_p above the cutoff frequency of the guided waves.

In the presence of a static magnetic field ($\omega_c/\omega_p = 0.5$) the resistance is nearly constant for low frequencies, it increases rapidly for $\omega/\omega_p > \omega_c/\omega_p$, reaches a peak and decays at higher frequencies. There is also a secondary resistance peak for $\omega > \omega_p$. From analogy with an isotropic plasma the major resistance peak may be associated with surface waves guided along the plasma-dielectric boundary. The surface waves guided along the interface between free space and a lossless plasma have been investigated by Seshadri and Sivaprasad [1966]. They indicate that surface waves exist only for $\omega_c < \omega_p$ in the frequency range of $\omega_c < \omega < \sqrt{(\omega_c^2+\omega_p^2)}/\sqrt{2}$. The low frequency cutoff point depends only on the plasma characteristics and should be independent of the presence of dielectric layer. The high frequency cutoff point is determined by (37) of Seshadri and Sivaprasad. For a dielectric layer this cutoff condition can be modified to

$$\frac{\epsilon_1}{\epsilon_i} + \sqrt{\left(\frac{\epsilon_1}{\epsilon_3}\right)} = 0 \tag{12.120}$$

This cutoff point is indicated by solid vertical lines in Fig. 12.14. The major resistance peak of Fig. 12.14 corresponds therefore to the frequency range of propagating surface waves.

A linear antenna immersed in a cold plasma exhibits a large resistance peak in the frequency range $\omega_p < \omega < \sqrt{(\omega_p^2+\omega_c^2)}$, which is lower in the presence of an insulating layer (Fig. 12.10). In the present discussion the insulating layer is of a larger thickness and this resistance peak is much lower.

The high resistance peaks observed in Fig. 12.14 occur for frequencies just below the high frequency cutoff of the surface waves which are supported by the dielectric to plasma interface for both the isotropic and the magnetized plasma. This impedance characteristic may be useful for diagnostic studies of re-entry plasma.

12.3. Radiation from a Short Antenna in a Lossy Uniaxial Medium

The previous discussion of this chapter was restricted to the input impedances of antennas in a magnetoionic medium and it was shown that the input impedance of an antenna in a slightly lossy medium does not differ appreciably from the impedance for a lossless plasma. However significant differences may be expected between the radiated fields in the two cases. Waineo and Mittra [1966] have obtained the field expressions in a uniaxial medium from free space fields of the antenna using a scaling procedure similar to the one described by Clemmow [1963a]; the power flow was computed by numerically integrating the Poynting's vector over spherical surfaces.

In this section the analog problem is investigated by considering the transform field components of Section 12.1 and by integrating the Poynting's vector over a cylindrical surface of radius ρ. In this latter approach simple closed form expressions are derived for the equivalent radiation resistance of the antenna, and the distance dependence of the radiation resistance can be examined quite readily.

12.3.1. *Formulation of the Effective Resistance*

The TE (E_ϕ, H_ρ, H_z) and TM (E_ρ, E_z, H_ϕ) field components are not coupled in a uniaxial medium, and the former field components are not excited in a ϕ-symmetrical geometry with z-directed current elements following Section 12.1. The axial electric field is related to the source current density by (12.36) and (12.51). It follows that

$$\bar{E}_z(\rho, w) = \frac{i\omega\mu_0(k_0^2 - w^2)}{k_0^2\beta}\frac{H_0^{(1)}(\beta\rho)}{H_1^{(1)}(\beta\rho)}\bar{J}_z(a, w) \tag{12.121}$$

where $H_n^{(m)}(x)$ is the Hankel function of order n and kind m, w is a transform variable defined in (12.7), $\beta = \sqrt{[\epsilon_3(k_0^2 - w^2)]}$, $k_0 = \omega\sqrt{(\mu_0\epsilon_0)}$, the permittivity element ϵ_3 is defined as $\epsilon_3 = 1 - \omega_p^2/[\omega(\omega + i\nu)]$, and ν is the effective collision frequency. The wave number β is defined with $\mathrm{Im}\beta > 0$, and the sign of the square root should be selected accordingly. The field component $\bar{H}_\phi(\rho, w)$ is computed from (12.17) as

$$\bar{H}_\phi(\rho, w) = -\frac{k_0^2}{i\omega\mu_0(k_0^2 - w^2)}\frac{\partial}{\partial\rho}\bar{E}_z(\rho, w) \tag{12.122}$$

The power radiated across a cylindrical surface of radius ρ is related to the RMS values of the field components or their transforms as

$$P(\rho) = -2\pi\rho\,\mathrm{Re}\int_{-\infty}^{\infty} \mathrm{d}z\, E_z(\rho, z)H_\phi^*(\rho, z) = -\rho\,\mathrm{Re}\int_{-\infty}^{\infty} \mathrm{d}w\, \bar{E}_z(\rho, w)\bar{H}_\phi^*(\rho, w) \tag{12.123}$$

The sinusoidal source current distribution is assumed of the form

$$J_z(a, z) = (I_0/2\pi a)\sin[k_A(l - |z|)]/\sin k_A l, \tag{12.124}$$

where I_0 is the RMS value of the source current at $z = 0$. After computing the transform of (12.124) and applying equations (12.121) to (12.123) a ρ-dependent effective resistance may be defined as

$$R(\rho) = \frac{P(\rho)}{I_0^2} = \frac{2\omega\mu_0\rho}{(\pi a \sin k_A l)^2} \mathrm{Im} \int_0^\infty \frac{(k_0^2 - w^2)}{k_0^2 \beta} g_A^2(w) \frac{H_0^{(1)}(\beta\rho)}{H_1^{(1)}(\beta a)} \left[\frac{H_1^{(1)}(\beta\rho)}{H_1^{(1)}(\beta a)}\right]^* \mathrm{d}w \tag{12.125}$$

where $g_A(w)$ is defined by (12.43) as

$$g_A(w) = \frac{k_A}{k_A^2 - w^2} (\cos wl - \cos k_A l) \tag{12.126}$$

For current elements of sufficiently small radius a the Hankel functions of arguments (βa) can be replaced by the small argument approximations. For large values of ρ it can be anticipated that Hankel functions of arguments $(\beta\rho)$ can be replaced by their asymptotic expansions. These approximations are substituted in (12.125), but it will be pointed out in Section 12.3.3 that the large argument approximations are inadequate if integration near $\omega \approx k_0$ (or $\beta \to 0$) contributes significantly to the integral. It follows that

$$R(\rho) = \frac{\omega\mu_0}{\pi k_0^2 \sin^2 k_A l} \mathrm{Re} \int_0^\infty \frac{\beta^*}{|\beta|} (k_0^2 - w^2) g_A^2(w)\, \mathrm{e}^{-2\rho \mathrm{Im}\beta}\, \mathrm{d}w. \tag{12.127}$$

12.3.2. *Radiation in the Limit of Low Losses*

The evaluation of (12.127) will be considered separately for high and low frequencies when $(\omega^2 + \nu^2)$ is larger or smaller than ω_p^2 respectively.

In a low loss plasma $(\nu \ll \omega)$, $\mathrm{Re}\epsilon_3 > 0$ and $\mathrm{Im}\epsilon_3 \ll \mathrm{Re}\epsilon_3$ for $(\omega^2 + \nu^2) > \omega_p^2$. The wave number β is essentially real $(\mathrm{Re}\beta \gg \mathrm{Im}\beta > 0)$ and the waves are propagating for $w < k_0$. The waves are attenuated for $w > k_0$, when β is almost imaginary $(\mathrm{Im}\beta \gg |\mathrm{Re}\beta|)$. Only values $w < k_0$ will contribute significantly to the integral (12.127) and the upper limit of the integral may be replaced by k_0. If $[2\rho k_0 \mathrm{Im}\sqrt{(\epsilon_3)}] \ll 1$, the exponential of (12.127) can be replaced by unity over this range of integration. The two cosine functions of (12.126) are replaced by the two first terms of a power series expansion. An elementary integration shows that

$$R(\rho) = 20(k_0 l)^2. \tag{12.128}$$

For a lossless medium $[\mathrm{Im}\sqrt{(\epsilon_3)} = 0]\,\rho$ can be selected arbitrarily, but for a lossy medium the validity of (12.128) is restricted by the inequality $[2\rho k_0 \,\mathrm{Im}\sqrt{(\epsilon_3)}] \ll 1$. Equation (12.128) is also identical to the input resistance (12.52).

For $(\omega^2+\nu^2) < \omega_p^2$, $\text{Re}\epsilon_3 < 0$ and $|\text{Re}\epsilon_3| \gg \text{Im}\epsilon_3 > 0$ in a low loss plasma ($\nu \ll \omega$). The wave number β is essentially real ($-\text{Re}\beta \gg \text{Im}\beta > 0$) and the waves are propagating for $w > k_0$. The waves are attenuated for $w < k_0$, when β is almost imaginary ($\text{Im}\beta \gg \text{Re}\beta > 0$). Only values $w > k_0$ will contribute significantly to the integral (12.127) and the lower limit of the integral may be replaced by k_0. The wave number of the current distribution k_A is assumed to be that of free space, $\cos k_A l$ is replaced by 1 in (12.126), and $\sin k_A l$ by $k_A l$ in (12.127). After expanding the denominator of the integrand

$$R(\rho) = \frac{\omega\mu_0}{\pi k_0^2 l^2}\int_{k_0}^{\infty} \frac{dw}{w^2}\left[1+\left(\frac{k_0}{w}\right)^2+\ldots\right][\cos wl-1]^2\, e^{-2\rho\text{Im}\beta}. \qquad (12.129)$$

This integral can be readily evaluated for a lossless medium ($2\rho\text{Im}\beta = 0$). It follows that

$$R(\rho) = \frac{60\pi}{k_0 l}[1+(k_0 l)^2/3+\ldots]. \qquad (12.130)$$

The leading term of (12.130) agrees with the input resistance (12.52). The neglect of the term proportional to $(k_0/w)^2$ in (12.129) is equivalent to neglecting $(k_0 l)^2/3$ in (12.130) which introduces an error of the same magnitude as the earlier approximation of $\cos k_A l$ by 1. The neglect of $(k_0/w)^2$ in (12.129) can be justified for a lossless plasma, and the same approximation will be also applied in the limit of small loss and small distances ρ, when the exponential term becomes significant only for $w \gg k_0$. Under this restriction the exponent is approximated as

$$2\rho\text{Im}\beta \approx [-2\rho\text{Im}\sqrt{(-\epsilon_3)}]\, w = pw. \qquad (12.131)$$

Applying the integral (4.7.77) of Erdelyi [1954] it follows that

$$R(\rho) = \frac{120}{k_0 p}F(x) \qquad (12.132)$$

with

$$F(x) = \frac{1}{4x^2}\log(1+4x^2) - \frac{1}{x^2}\log(1+x^2) + \frac{2}{x}\tan^{-1}x - \frac{1}{x}\tan^{-1}2x \qquad (12.133)$$

where $x = l/p$ and p is defined by (12.131). For $l \gg p$

$$R(\rho) \approx \frac{120}{k_0 p}\left[\frac{\pi}{2x}+\frac{1}{2x^2}\log 2-\frac{3}{2x^2}(1+\log x)\right] \approx \frac{60\pi}{k_0 l} \qquad (12.134)$$

which is in agreement with (12.130). For $l \ll p$

$$R(\rho) \approx \frac{120}{k_0 p}\frac{x^2}{2} = 60\frac{(k_0 l)^2}{(k_0 p)^3}. \qquad (12.135)$$

$F(x)$ of (12.133) is plotted in Fig. 12.15 and the approximation corresponding to (12.134) and (12.135) for (x) large and small respectively are indicated by dashed lines. The effective resistance changes from the $(1/l)$ dependence for large x or small values of the distance parameter p to an l^2 dependence

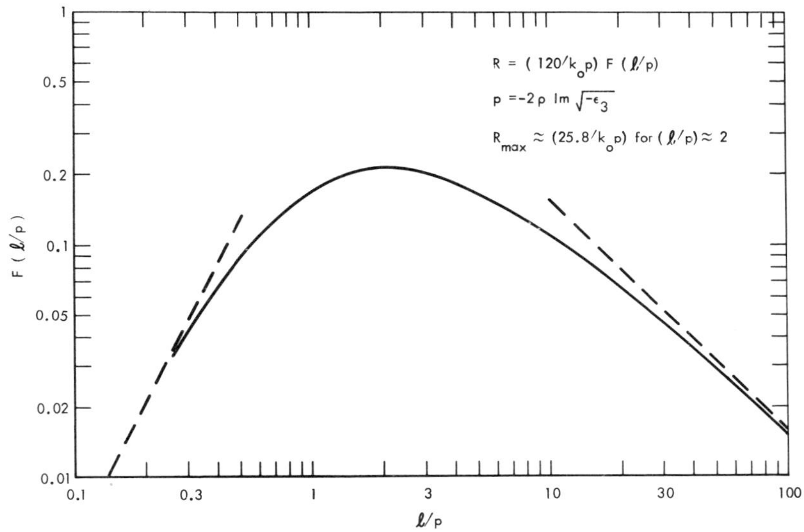

FIG. 12.15. Antenna resistance for cylindrical powerflow in a lossy uniaxial medium.

for small x or large values of p. For a fixed value of the distance parameter p the radiation resistance depends on l and an examination of (12.132) shows that the effective resistance has a broad maximum for $x = l/p \approx 2$. This maximum value of R is obtained as

$$R_{\max} \approx \frac{25.8}{k_0 p}. \tag{12.136}$$

The distance parameter $p = -2\rho \operatorname{Im} \sqrt{(-\epsilon_3)}$ is proportional to ρ and to the imaginary part of $\sqrt{(-\epsilon_3)}$, which depends on $\operatorname{Re}\epsilon_3$ as well as on $\operatorname{Im}\epsilon_3$. The radiation resistance is independent of p (and of ϵ_3) only in the limit of $l \gg p$ shown in (12.134). A further numerical example of this distance dependence of the radiation resistance is shown in Fig. 12.16 for several antenna lengths by assuming $\epsilon_p = -1 + i0.03$. The shorter antennas exhibit maximum effective resistance only for distances $\rho < 0.1\lambda$. The longer antennas exhibit the higher resistances for $\rho \approx 0.5\lambda$. Also the values of $R_{\max}$ computed from (12.136) are indicated by dashed lines in Fig. 12.16. The parameters of the antenna are assumed to be the same as in Fig. 4 of

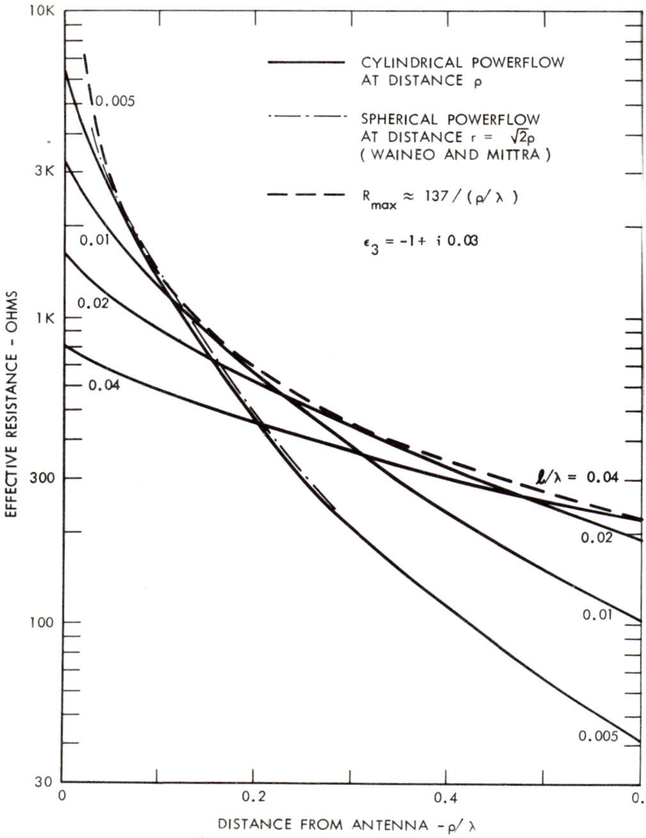

FIG. 12.16. Effective antenna resistance for a lossy uniaxial medium.

Waineo and Mittra [1966], which indicates the distance variations of spherical power flow. The sphere of radius r encompasses a smaller volume than a cylinder of radius ρ if $r = \rho$ and the spherical power flow should be larger than the cylindrical power flow. However, it may be noted that the field component E_z has its maximum values when $\epsilon_3\rho^2 + z^2 = (\epsilon_3 - 1)\rho^2 + r^2$ has a minimum magnitude.† When ϵ_3 has a negative real part maximum fields

†This can be seen after computing the field components $H_\phi(\rho, z)$ and $E_z(\rho, z)$ by substituting (12.121) and (12.122) in the inversion integral (12.7). After replacing the Hankel functions of agrument ρ and a by their respective large and small argument approximations, the inversion integral can be evaluated using the stationary phase method. The resulting field components are the same as obtained by Waineo and Mittra [1966] using a scaling procedure similar to the one described by Clemmow [1963a]. The E_z fields are inversely proportional to $\sqrt{(\zeta_j^2 + \epsilon_3\rho^2)}$ where $\zeta_j = z$ or $(z \pm l)$. These can become large for $\mathrm{Re}\,\epsilon_3 < 0$ if $|\epsilon_3\rho^2| \rightarrow z^2$ or $(z \pm l)^2$, while the H_ϕ components do not exhibit such a singular behavior.

and also the maximum power flow occur when

$$\rho|1-\epsilon_3|^{0.5} \approx r. \tag{12.137}$$

The spherical power flow at a distance r from the source should therefore be approximately the same as the cylindrical power flow at a distance ρ if ρ and r are related by (12.137). The resistance data of Waineo and Mittra [1966] are therefore plotted in Fig. 12.16 for $\rho = r/\sqrt{2}$ and their radiation resistance based on spherical power flow compares very closely with radiation resistance (12.132) which is based on a cylindrical power flow. The computations of Waineo and Mittra [1966] were made by numerically integrating the Poynting's vector, while the present computations imply the use of the Parseval's theorem in equation (12.123). The close agreement between the results of the two computations indicates that the use of the Parseval's theorem is justified also in the present computations.

The integrand of (12.127) should be approximated more accurately for small arguments w when considering increased values of ρ and $[-\text{Im}\sqrt{(-\epsilon_3)}]$; the approximate representation of the denominator of the integrand shown in (12.129) and the approximation of the exponent by (12.131) may not be valid. However, $g_A(w)$ of (12.126) is simplified by approximating the cosine functions by the first two terms of their power series. After changing the variable of integration from w to $t = \sqrt{(w^2 - k_0^2)}$, the effective resistance (12.127) becomes

$$R(\rho) = \frac{\omega\mu_0 l^2}{4\pi k_0^2} \int_0^\infty \frac{\mathrm{d}t\, t^3}{\sqrt{(t^2 + k_0^2)}} \mathrm{e}^{-pt} \tag{12.138}$$

Starting out with the integral (4.2.27) of Erdelyi [1954] and using small argument approximations ($k_0 p \ll 1$) of the functions, the application of (4.1.13) of Erdelyi [1954] shows that (12.138) is identical to (12.135). However (12.138) will assume a different numerical value and (12.135) will not be valid if the condition $k_0 p \ll 1$ is not strictly satisfied.

12.3.3. *Radiation at Large Distances in a Lossy Medium*

The calculations of Section 12.3.2 are based on the asymptotic expansion of Hankel functions which are not valid for small arguments $\beta\rho$. However with increasing values of the parameter p of (12.131) or with an increased distance ρ and losses, only that part of the integration which is carried out over the range of w where β is very small will contribute significantly to the integral (12.127). It is therefore not permissible to employ the asymptotic expansion of Hankel functions if $p \gg 1$.

In an alternate formulation small argument approximations are applied to the Hankel functions of argument $\beta\alpha$ in (12.125) and the two term approximation of the cosine functions is used in $g_A(w)$ of (12.126).† The variable of integration is changed to $t = \sqrt{(k_0^2 - w^2)}$ and $\sqrt{(w^2 - k_0^2)}$, when ω^2 is greater and smaller than $(\omega_p^2 + \nu^2)$ respectively. The effective resistance (12.125) becomes

$$R(\rho) = \frac{\omega\mu_0\rho l^2}{8k_0^2} |\sqrt{(\epsilon_3)}| \operatorname{Im} \int_0^{k_0} \frac{t^4 \mathrm{d}t}{\sqrt{(k_0^2 - t^2)}} H_0^{(1)}(\alpha t) H_1^{(2)}(\alpha^* t) \tag{12.139}$$

with $\alpha = \sqrt{(\epsilon_3)}\rho = \alpha_r + i\alpha_i$, and

$$R(\rho) = -\frac{\omega\mu_0\rho l^2}{8k_0^2} |\sqrt{(-\epsilon_3)}| \operatorname{Im} \int_0^{\infty} \frac{t^4 \mathrm{d}t}{\sqrt{(k_0^2 + t^2)}} H_0^{(2)}(\alpha t) H_1^{(1)}(\alpha^* t) \tag{12.140}$$

with $\alpha = \sqrt{(-\epsilon_3)}\rho = \alpha_r - i\alpha_i$ for these two ranges of frequency. It can be shown that only small values of t contribute significantly to the integrals (12.139) and (12.140), and that the denominators can be approximated by k_0. These simplified integrals are evaluated by a repeated integration by parts. A lengthy calculation shows that

$$R(\rho) = 180 \frac{(k_0 l)^2}{(k_0 p)^4}, \tag{12.141}$$

where

$$p = 2\alpha_i = \begin{cases} 2\rho \operatorname{Im} \sqrt{\epsilon_3} & \text{for} \quad \omega^2 > (\omega_p^2 + \nu^2), \\ -2\rho \operatorname{Im} \sqrt{-\epsilon_3} & \text{for} \quad \omega^2 < (\omega_p^2 + \nu^2). \end{cases} \tag{12.142}$$

In the limit of large distances from the antenna, the radiation resistance is therefore of the same form for high and low frequencies.

The expression (12.141) is strictly valid for $k_0 p \gg 1$. A numerical evaluation of the integrals (12.139) and (12.140) shows that the approximation (12.141) can be used for $k_0 p > 3$ in the example of Fig. 12.17 where $k_0 p = 1 \approx k_0 \rho = 28$ is indicated by a vertical line. The forms (12.128) and (12.135) approximate the effective resistance $R(\rho)$ for $k_0 p < 1$. However for $\operatorname{Re}\epsilon_3 < 0$, the integral expression (12.140) and the approximation (12.135) are valid only if $l \ll p$.

12.3.4. *Summary of Radiation Characteristics*

It has been shown that the integrated power flow near the antenna is inversely proportional to the antenna length for frequencies less than the

†This implies that the spectrum of the current distribution remains nearly constant over the range of integration where the exponentially decaying Hankel functions differ from zero. In the absence of exponentially decaying terms this assumption would imply a point source of $l \to 0$.

plasma frequency, but the radiated power becomes proportional to the antenna length squared as the distance increases. The long antenna exhibits therefore a relatively low input resistance, but it radiates more power at

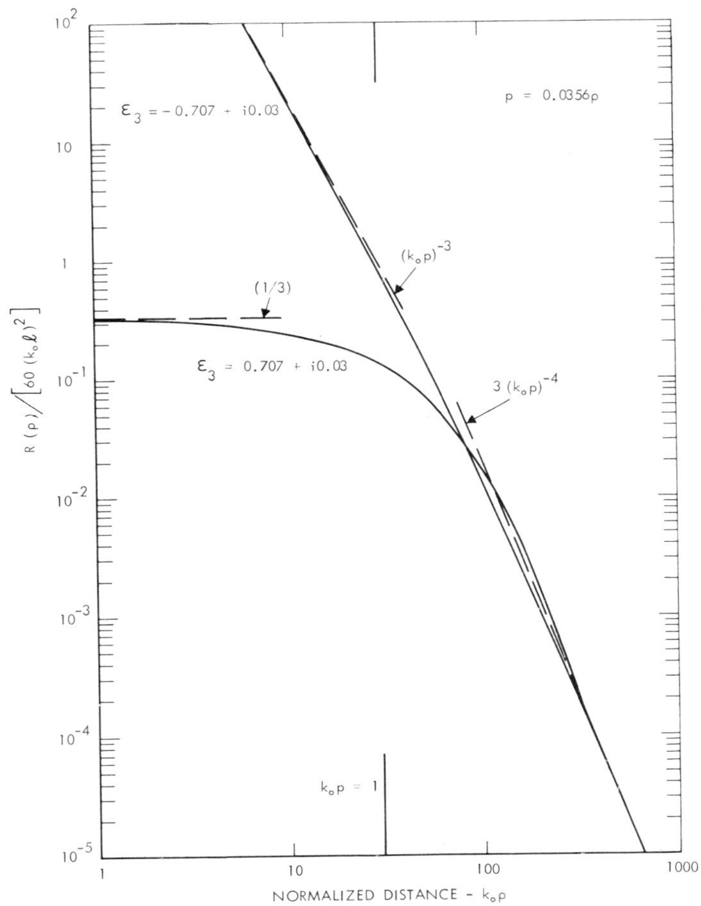

FIG. 12.17. Effective resistance of a short antenna ($l \ll \rho$ for $\mathrm{Re}\,\epsilon_3 < 0$).

large distances from the source and it is the more efficient radiator. This has been also observed by Waineo and Mittra [1966]. At very large distances from the source the radiation resistance remains functionally the same also for frequencies above the plasma frequency, but this result can be established only with a finite amount of medium loss.

Also quasi-static calculations for a lossy plasma [Balmain, 1964] indicate a transitional distance where the extended near-fields of the antenna operating below the plasma frequency ($\omega < \omega_p$) are changed to fields of a static dipole. This transition distance occurs at a range ($l/p < 0.5$) where the effective resistance of the present paper is approximately proportional to the squared distance. Balmain [1964] has also pointed out that beyond this transitional distance the antenna field variation is the same for frequencies ω above and below the plasma frequency ω_p. However, in the present development the effective resistance is the same for frequencies ω above and below ω_p only at very large distances ($k_0 p \gg 1$, where the numerical distance p is defined in (12.142)), which are not directly related to the antenna length l if $l \ll p$.

A radiation resistance which is proportional to the squared dipole moment (or proportional to the antenna length squared) has been obtained also by Lee and Papas [1966]. However, their work has not considered the effects of finite source dimensions or the presence of losses in the uniaxial medium.

The definition of the limits of low losses and large distances can also produce ambiguities. In a lossless medium the effective resistance is proportional to l^{-1} for $\omega < \omega_p$ at arbitrary large distances. However, for an infinite distance the effective resistance remains proportional to l^2 even for negligibly small losses. This apparent ambiguity can be resolved by using finite distances and finite losses typical of those encountered in an actual antenna geometry. The presence of medium losses may be possibly neglected when considering the input resistance of the antenna, but it cannot be neglected when computing the antenna fields or the effective resistance at a finite distance from the source. The input resistance of the antenna in a lossless medium does not characterize the power flow in a slightly lossy medium at a finite distance from the antenna, and it cannot be used even for estimating the relative power flow of antennas of different lengths.

12.4. Appendix. Strip Antenna with a Magnetic Field perpendicular to the Axis, and parallel to its Surface

A flat strip antenna is located in a uniaxial medium in the plane of $y = 0$ with the antenna axis in the x direction and the antenna surface parallel to the z-directed static magnetic field. The antenna length is $2l$ and its width is 2ϵ. The field components F_{ij} in the anisotropic medium are related to Fourier transforms $\bar{F}_{ij}$ by the integrals

$$F_{ij}(x, y, z) = \left(\frac{1}{2\pi}\right)^2 \int_{-\infty}^{\infty} \int_{-\infty}^{\infty} \bar{F}_{ij}(u, w)\, e^{iux+iwz+iv_j y}\, du\, dw \qquad (12.143)$$

where $F = E$ or H, $i = x$, y or z and where the subscripts $j = +$ or $-$ denote a mode characterized by a particular value of v_j. The relations (12.60) to (12.70) apply to the present geometry if iw is substituted for γ and v_j for v.

For a uniaxial medium $\epsilon_2 = 0$ and $\epsilon_1 \to 1$. The roots v_+^2 and v_-^2 are computed from (12.70) as

$$v_+^2 = \epsilon_3 k_0^2 - u^2 - \epsilon_3 w^2/\epsilon_1, \tag{12.144}$$

$$v_-^2 = \epsilon_1 k_0^2 - u^2 - w^2, \tag{12.145}$$

where v_j is defined with $\text{Im}\, v_j > 0$ for lossy media. Application of the boundary conditions to the $y = 0$ plane shows that

$$\begin{aligned} \bar{E}_{z+} &= Iuw/(2\omega\epsilon_0\epsilon_1 v_+), \\ \bar{H}_{z-} &= I/2, \end{aligned} \tag{12.146}$$

where I is the transform of the current density $J(x, z)$. Substituting (12.146) in (12.66)

$$F(u, w) = \frac{\bar{E}_x}{I} = -\frac{1}{2(\epsilon_1 k_0^2 - w^2)} \left[\frac{u^2 w^2}{\omega\epsilon_0\epsilon_1 v_+} + \omega\mu_0 v_-\right]. \tag{12.147}$$

The antenna impedance is defined by (1.35). For a sinusoidal current distribution

$$J_x(x, z) = A \sin [k_A(l - |x|)] f(z) \tag{12.148}$$

with $f(z) = 1/(2\epsilon) =$ const the antenna impedance is expressed as

$$Z = -\frac{4}{(\pi \sin k_A l)^2} \int_0^\infty \int_0^\infty F(u, w)\, g_A^2(u) \left(\frac{\sin \epsilon w}{\epsilon w}\right)^2 du\, dw, \tag{12.149}$$

where $g_A(u)$ is defined by (12.104). It can be separated in two parts $Z = Z_+ + Z_-$ corresponding with the two terms of $F(u, w)$, where the + and − subscripts refer to TM and TE modes respectively. In the limit of k_A approaching zero the current distribution along the antenna becomes triangular and for ϵ_1 approaching 1 the real parts of the integral expressions Z_- and Z_+ can be seen to be identical to the radiation resistances (28) and (36) of Seshadri [1966a]. In the computation of $R_- = \text{Re}Z_-$ the integrations are restricted to $u^2 + w^2 < k_0^2$. The integrals are evaluated after replacing the sine and cosine functions by their small argument approximations. It follows that

$$R_- = 15(k_0 l)^2 \tag{12.150}$$

In the computation of $R_+ = \text{Re}Z_+$ for high frequencies $(\omega_2 > \omega_p^2 + \nu^2)$, $\text{Re}\epsilon_3 > 0$ and the integrations are restricted to $u^2 + \epsilon_3 w^2 < \epsilon_3 k_0^2$. Applying the small argument approximations of the trigonometric functions

$$R_+ = 5\epsilon_3(k_0 l)^2. \tag{12.151}$$

The sum of (12.150) and (12.151) gives the same resistance as (12.113). For low frequencies ($\omega^2 < \omega_p^2 + \nu^2$), $\mathrm{Re}\,\epsilon_3 < 0$. Defining v_+ with $\mathrm{Im}\,v_+ > 0$ it follows that

$$R_+ = \frac{2}{\omega\epsilon_0\epsilon_1\pi^2}\int_0^\infty \frac{\mathrm{d}u\,(\cos ul-1)^2}{(ul)^2}\int_{u/\sqrt{|3|}}^\infty \left(\frac{\sin\epsilon w}{\epsilon w}\right)^2 \frac{\mathrm{d}w}{\sqrt{(|\epsilon_3|w^2-u^2)}}. \quad (12.152)$$

Assuming that $(\sin\epsilon w/\epsilon w) \approx 1$ near the lower limit of w it follows that

$$R_+ \approx \frac{2}{\omega\epsilon_0\epsilon_1\pi^2}\int_0^\infty \frac{\mathrm{d}u\,(\cos ul-1)^2}{(ul)^2}\left[\frac{3}{2}-C-\log(\epsilon u)+\frac{1}{2}\log|\epsilon_3|\right]$$

$$= \frac{120}{k_0 l\sqrt{|\epsilon_3|}}\left[\log\frac{2l}{\epsilon}+\frac{1}{2}-2\log 2+\frac{1}{2}\log|\epsilon_3|\right] \quad (12.153)$$

For $\omega \ll \omega_p$ this closed form approximation can be seen to give the same resistance values as the numerical calculations shown in fig. 2b of Seshadri [1966a]. The resistance (12.153) differs by the last logarithmic term from the resistance (12.108) for an antenna where the static magnetic field is perpendicular to the antenna surface. For $\omega^2 < (\omega_p^2+\nu^2)$ the reactance $X_+ = \mathrm{Im}\,Z_+$ is defined as

$$X_+ = \frac{2}{\omega\epsilon_0\epsilon_1\pi^2}\int_0^\infty \mathrm{d}u\,\frac{(\cos ul-1)^2}{(ul)^2}\int_0^{u/\sqrt{(|\epsilon_3|)}} \left(\frac{\sin\epsilon w}{\epsilon w}\right)^2 \frac{\mathrm{d}w}{\sqrt{(u^2-|\epsilon_3|w^2)}}. \quad (12.154)$$

Over the range of integration shown in (12.154) the variable w is less than at the lower limit of (12.152), where it was permissible to replace $\sin\epsilon w/\epsilon w$ by 1. Applying the same approximation to (12.154) an elementary integration shows that

$$X_+ = \frac{60\pi}{k_0 l\sqrt{|\epsilon_3|}}. \quad (12.155)$$

This reactance is capacitive and is inversely proportional to the antenna length. The capacitive reactance of Balmain [1964] is also inversely proportional to l, but for a cylindrical current element it is inversely proportional to $|\epsilon_3|$.

It is also possible to work out an approximate expression for $X_- = \mathrm{Im}\,Z_-$. This reactance of the TE modes is inductive and proportional to the antenna length. It becomes negligible relative to (12.155) in the limit of low frequencies.

The different impedance characteristics for antennas oriented with their surfaces parallel and perpendicular to the static magnetic field can be explained qualitatively by examining the dielectric tensor near the antenna surface. For a strip antenna the tangential electric surface fields should be equal to zero, and the normal field components will predominate in the vicinity of the antenna surface. For antennas perpendicular to the static magnetic field (Section 2.2) the z- component is perpendicular to the antenna surface and the electric displacement component $\epsilon_3 E_z$ will be dominant near the antenna surface. For antennas parallel to the static magnetic field (Section 12.4) the y-component is perpendicular to the antenna surface and the electric displacement component $\epsilon_1 E_y$ is dominant. In a uniaxial medium $\epsilon_1 = 1$ and ϵ_3 has a large negative real part for low frequencies; for the static magnetic field parallel to the antenna surface the antenna reactance should be closer to the free space conditions that for the antenna oriented with its surface perpendicular to the magnetic field. A capacitive reactance has been observed for the parallel antenna orientation but the reactance is of opposite sign for the perpendicular antenna orientation in Section 12.2. However, for high frequencies ϵ_3 approaches one and the antenna impedances should be comparable in both configurations.

12.5. Exercises

12.1. Derive the admittance Y_j of (12.35), which characterizes the coupling of $\bar{E}_{zj}$ and $\bar{H}_{zj}$ field components.

12.2. Derive the ratio between the axial electric field and the current density defined by $F(w)$ of (12.36) and (12.44).

12.3. Derive the ratio of the axial electric fields of the two modes defined by X of (12.86).

12.4. Evaluate the integrals (12.106) and derive the quasistationary impedance components (12.108) and (12.111).

12.5. Derive the ratio X corresponding to (12.86) and the impedance Z corresponding to (12.98) when the insulating layer of Fig. 12.8 is bounded on one side by a conducting ground plane.

12.6. References

ABRAHAM, L. G. (1953) *Extensions of the Magnetoionic Theory for Radio-Wave Propagation in the Ionosphere including Antenna Radiation and Plane Wave Scattering,* School of Electrical Engineering, Cornell University, Tech. Report No. 13.

AMENT, W. S., KATZIN, J. C., KATZIN, M. and KOO, B. Y. C. (1964) Impedance of a cylindrical dipole having a sinusoidal current distribution in homogeneous anisotropic ionosphere, *NBS J. Res. (Radio Science)* **68**D, (4), 379–405.

AMENT, W. S., KATZIN, M., MCLAUGHLIN, J. R. and ZACHARY, W. W. (1965) *Satellite Antenna Radiation Properties at VLF in the Ionosphere,* Final Report on Contract Nonr-4250 (00) (X), Electromagnetic Research Corporation, College Park, Md.

ARBEL, E. and FELSEN, L. B. (1963) Theory of radiation from sources in anisotropic media, in *Electromagnetic Waves,* edited by E. C. Jordan, Pergamon Press, New York, NY, Part I, p. 391–459.

BACHYNSKI, M. P. (1966) Sources in unbounded plasmas, *Fifteenth General Assembly of URSI, Munich, Germany.*

BALMAIN, K. G. (1964) The impedance of a short dipole antenna in a magnetoplasma, *IEEE Trans. on Ant. and Prop.* **AP-12**, 605–17.

BLAIR, W. E. (1964) *The Driving-Point Impedance of an Electrically Short Cylindrical Antenna in the Ionosphere,* Univ. of N. Mex. Eng. Exp. Sta. Tech. Rep. EE-109.

BRAMLEY, E. N. (1962) The admittance of a short cylindrical dipole in the ionosphere, *Planet. Space Sci.* **9**, 445–54.

BRANDSTATTER, J. J. and PENICO, A. J. (1964) *A Study of the Impedance of a Cylindrical Dipole in an Anisotropic Plasma,* Final Report, Project PMU-3417, Stanford Research Institute, Menlo Park, California.

BUNKIN, F. V. (1957) On radiation in anisotropic medium, *Soviet Physics, JETP* **5**, 277–83 (English translation).

CHEN, H. C. and CHENG, D. K. (1965) Radiation from an axially slotted anisotropic plasma clad cylinder, *Trans. IEEE* **AP-13**, 395–401.

CHOW, Y. (1962) A note on radiation in a gyroelectric-magnetic medium—an extension of Bunkin's calculation, *IRE Trans. on Ant. and Prop.* **AP-10**, 464–9.

CLEMMOW, P. C. (1963a) The theory of electromagnetic waves in a simple anisotropic medium, *Proc. IEE (London)*, **110**, 101–6.

CLEMMOW, P. C. (1963b) On the theory of radiation from a source in a magnetoionic medium, in *Electromagnetic Waves*, edited by E. C. Jordan, Pergamon Press, New York, NY, Part I, p. 461–75.

DEMARCHIN, P. and TYRAS, G. (1965) Radiation from an infinite axial slot on a circular cylinder clad with magnetoplasma, *Radio Science* **69D** (4), 529–38.

DESCHAMPS, G. A. and KESLER, O. (1964) Radiation field of an arbitrary antenna in a magnetoplasma, *IEEE Trans. on Ant. and Prop.* **AP-12**, 783–5.

ERDELYI, A. (1954) *Table of Integral Transforms*, McGraw-Hill, New York, NY, vol. 1.

GALEJS, J. (1965) Driving point impedance of linear antenna in the presence of a stratified dielectric, *IEEE Trans. on Ant. and Prop.* **AP-13** (4), 725–7.

GALEJS, J. (1966a) Impedance of a finite indulated cylindrical antenna in a cold plasma with a longitudinal magnetic field, *IEEE Trans. on Ant. and Prop.* **AP-14** (6), 727--36.

GALEJS, J. (1966b) Impedance of a finite indulate antenna in a cold plasma with a perpendicular magnetic field, *IEEE Trans. on Ant. and Prop.* **AP-14** (6), 737–8.

GALEJS, J. (1967a) Impedance of a strip antenna in a dielectric layer above a ground plane and below a half space of cold plasma, *IEEE Trans. on Ant. and Prop.* **AP-15** (2).

GALEJS, J. (1967b) Power flow from a short antenna in a lossy uniaxial medium, *Radio Science*, **2** (12).

GALEJS, J. (1968) The insulated cylindrical antenna in a cold plasma with a longitudinal magnetic field, *IEEE Trans. on Ant. and Prop.* **AP-16** (3).

GINZBURG, V. L. (1964) *The Propagation of Electromagnetic Waves in Plasma*, Pergamon Press, Oxford.

HODARA, H. (1963) Radiation from a gyro-plasma sheathed aperture, *IEEE Trans. on Ant. and Prop.* **AP-11**, 2–12.

HODARA, H. and COHN, G. I. (1962) Radiation from a gyro-plasma coated magnetic line source, *IRE Trans. on Ant. and Prop.* **AP-10**, 581–593.

ISHIMARU, A. (1964) The effect of a unidirectional surface wave along a perfectly conducting plane on the radiation from a plasma sheath, E-M Aspects of Hypersonic Flight, Second Symposium on the Plasma Sheath—Its Effect Upon Re-entry Communication and Detection, W. Rotman, H. K. Moore, R. Papa, editors, Spartan Books Inc., Baltimore, 147–68.

KAISER, T. R. (1962) The admittance of an electric dipole in a magnetoionic environment, *Planet. Space Sci.* **9**, 639–57.

KAMKE, E. (1948) *Differential Gleichungen*, Chelsea Publishing Co., New York, NY.

KOGELNIK, H. (1960a) On electromagnetic radiation in magnetoionic media, *NBS. J. Res.* **64D** (5), 515–22.

KOGELNIK, H. (1960b) The radiation resistance of an elementary dipole in anisotropic plasma, *Proc. 4th International Conf. on Ionization Phenomenon in Gases*, North Holland Publishing Co.

KOGELNIK, H. and MOTZ, H. (1963) Electromagnetic radiation from sources embedded in an infinite anisotropic medium and the significance of the Poynting vector, in *Electromagnetic Waves*, edited by E. C. Jordan, Pergamon Press, New York, NY, Part I, 477–93.

KUEHL, H. H. (1962) Radiation from an electric dipole in an anisotropic cold plasma, *Phys. of Fluids* **5**, 1095–1103.

LEE, K. S. H. and PAPAS, C. H. (1965) Irreversible power and radiation resistance of antennas in anisotropic ionized gases, *Radio Science* **69D** (10), 1313–20.

LEE, K. S. H. and PAPAS, C. H. (1966) A further explanation of a new theory of antenna radiation with particular reference to uniaxial media, *Radio Science* **1** (9), 1020–3.

LO, Y. T. and LEE, S. W. (1966) Current distribution and input impedance of an infinite cylindrical antenna in anisotropic plasma, *Electronics Letters* **2** (2), 72–73.

MELTZ, G. and SHORE, R. A. (1965) Leaky waves supported by uniaxial plasma layers. *IEEE Trans. on Ant. and Prop.* **AP-13** (1), 94–105.

MITTRA, R. (1965) Radiation resistance of antennas in anisotropic media, *Electronics Letters* **1** (9), 258–9.

MITTRA, R. and DESCHAMPS, G. A. (1963) Field solutions for a dipole in an anisotropic medium, in *Electromagnetic Waves,* edited by E. C. Jordan, Pergamon Press, New York, NY, Part I, 495–512.

MOTZ, H. (1965) On electromagnetic radiation from a magnetic dipole with arbitrary orientation embedded in a lossless magnetoionic medium, *NBS J. of Res.* **69D** (5), 671–9.

RATCLIFFE, J. A. (1959) *The Magnetoionic Theory in Application to the Ionosphere,* Cambridge University Press.

SESHADRI, S. R. (1965a) Radiation resistance of a linear current filament in a simple anisotropic medium, *IEEE Trans. on Ant. and Prop.* **AP-13** (5), 819–20.

SESHADRI, S. R. (1965b) Radiation resistance of elementary electric current sources in a magnetoionic medium, *Proc. IEE* **112** (10), 1856–68.

SESHADRI, S. R. (1965c) Excitation of guided VLF modes below a lossless anisotropic, *Proc. IEE* **112** (2), 242–8.

SESHADRI, S. R. (1965d) Guided waves on a perfectly conducting infinite cylinder in a magnetoionic medium, *Proc. IEE* **112** (8), 1497–1500.

SESHADRI, S. R. (1966a) Radiation from a current strip in a uniaxially anisotropic plasma medium, *Can. J. Phys.* **44**, 207–7.

SESHADRI, S. R. (1966b) Effect of insulation on the radiation resistance of an electric dipole in a simple anisotropic medium, *Proc. IEE* **113** (4), 593–600.

SESHADRI, S. R. (1966c) Singularity in the radiation resistance of the point electric dipole in a magnetoionic medium, *Electronic Letters* **2** (7), 263–5.

SESHADRI, S. R. and PICKARD, W. S. (1964) Surface waves on an anisotropic plasma sheath, *IEEE Trans. on Microwave Theory and Techniques* **MTT-12** (5), 529–41.

SESHADRI, S. R. and SIVAPRASAD, K. U. (1966) Surface waves at a plane interface between vacuum and plasma with the magnetostatic field normal to the interface, *IEEE Trans. on Ant. and Prop.* **AP-14** (3), 297–302.

SESHADRI, S. R. and TUAN, M. S. (1965) Radiation resistance of a circular current filament in a magnetoionic medium, *Proc. IEE* **112** (12), 2192–8.

STARAS, H. (1964) The impedance of an electric dipole in a magnetoionic medium, *IEEE Trans. on Ant. and Prop.* **AP-12**, 695–702.

STARAS, H. (1966) The infinity catastrophe associated with radiation in magnetoionic media, *Radio Science* **1** (9), 1013–20. [Equation (10) is valid if $r = r' = 0$.]

STONE, R. G., ALEXANDER, J. K. and WEBER, R. R. (1966) Measurements of antenna impedance in the ionosphere, Part I, *Planetary and Space Science* **14** (7), 631–9. Part II, *Ibid.* (10), 1007–16.

TAI, C. T. (1966) Review 2029, *Mathematical Reviews* **32** (2), 342–3.

WAINEO, D. K. and MITTRA, R. (1966) *The Radiation Resistance of a Dipole in a Uniaxial Medium,* Antenna Laboratory Report N. 66–14, University of Illinois, Urbana, Illinois.

WAIT, J. R. (1964) Theory of radiation from sources immersed in anisotropic media, *NBS J. of Res.* **68B** (3), 119–36.

WAIT, J. R. (1966) Radiation from a spherical antenna with an anisotropic plasma sheath, *Can. J. Physics* **34** (10), 2303–13.

WALSH, D. and WEIL, H. (1966) Universal power and radiation resistance of antennas in magnetoionic media, *Radio Science* **1** (9), 1025–7.

WEIL, H. and WALSH, D. (1964) Radiation resistance of an electric dipole in a magnetoionic medium, *IEEE Trans. on Ant. and Prop.* **AP-12**, 297–304.

WEIL, H. and WALSH, D. (1965) Radiation resistance of an elementary loop antenna in a magnetoionic medium, *IEEE Trans. on Ant. and Prop.* **G-AP-13**, 21–27.

WU, C. P. (1963) Radiation from dipoles in a magnetoionic medium, *IEEE Trans, on Ant. and Prop.* **AP-11**, 681–9.

YEH, C. (1964) Electromagnetic radiation from an arbitrary slot on a conducting cylinder coated with a uniform cold plasma sheath with an axial static magnetic field, *Can. J. Phys.* **42**, 1369–81.

YOUNG, C. E. (1966) VLF electric and magnetic antenna impedance characteristics in the ionosphere, Paper 3-11, URSI 1966 Fall Meeting, Stanford University, Stanford, California.

AUTHOR INDEX

SUBJECT INDEX

OTHER TITLES IN THE SERIES IN
ELECTROMAGNETIC WAVES

Vol. 1 FOCK—Electromagnetic Diffraction and Propagation Problems
Vol. 2 SMITH an
Vol. 3 WAIT—Electromagnetic Waves in Stratified Media
Vol. 4 BECKMANN and SPIZZICHINO—The Scattering of Electromagnetic Waves from Rough Surfaces
Vol. 5 KERKER—Electromagnetic Scattering
Vol. 6 JORDAN—Electromagnetic Theory and Antennas
Vol. 7 GINZBURG—The Propagation of Electromagnetic Waves in Plasmas
Vol. 8 DU CASTEL—Tropospheric Radiowave Propagation beyond the Horizon
Vol. 9 BANOS—Dipole Radiation in the Presence of a Conducting Half-space
Vol. 10 KELLER and FRISCHKNECHT—Electrical Methods in Geophysical Prospecting
Vol. 11 BROWN—Electromagnetic Wave Theory
Vol. 12 CLEMMOW—The Plane Wave Spectrum Representation of Electromagnetic Fields
Vol. 13 KERNS and BEATTY—Basic Theory of Waveguide Functions and Introductory Microwave Network Analysis
Vol. 14 WATT—VLF Radio Engineering